NATIONAL GEOGRAPHIC LEARNING
Exploring
Science
Focus on the
NEXT GENERATION SCIENCE STANDARDS
NATIONAL GEOGRAPHIC LEARNING
CENGAGE Learning
Content Consultants
Randy L. Bell, Ph.D.
Malcolm B. Butler, Ph.D.
Kathy Cabe Trundle, Ph.D.
Judith S. Lederman, Ph.D.

Physical Science

Energy 2

Waves: Waves and Information 2

Batter Up! 4

Investigate Speed 6

Hit the Ball 8

Investigate Motion 10

Sounds of the Game 12

Investigate Sound 14

The Sun's Light 16

Investigate Light 18

Heat It Up! 20

Investigate Heat 22

It's Electric 24

Electric Circuits 26

Investigate Electric Circuits 28

Spin It! 30

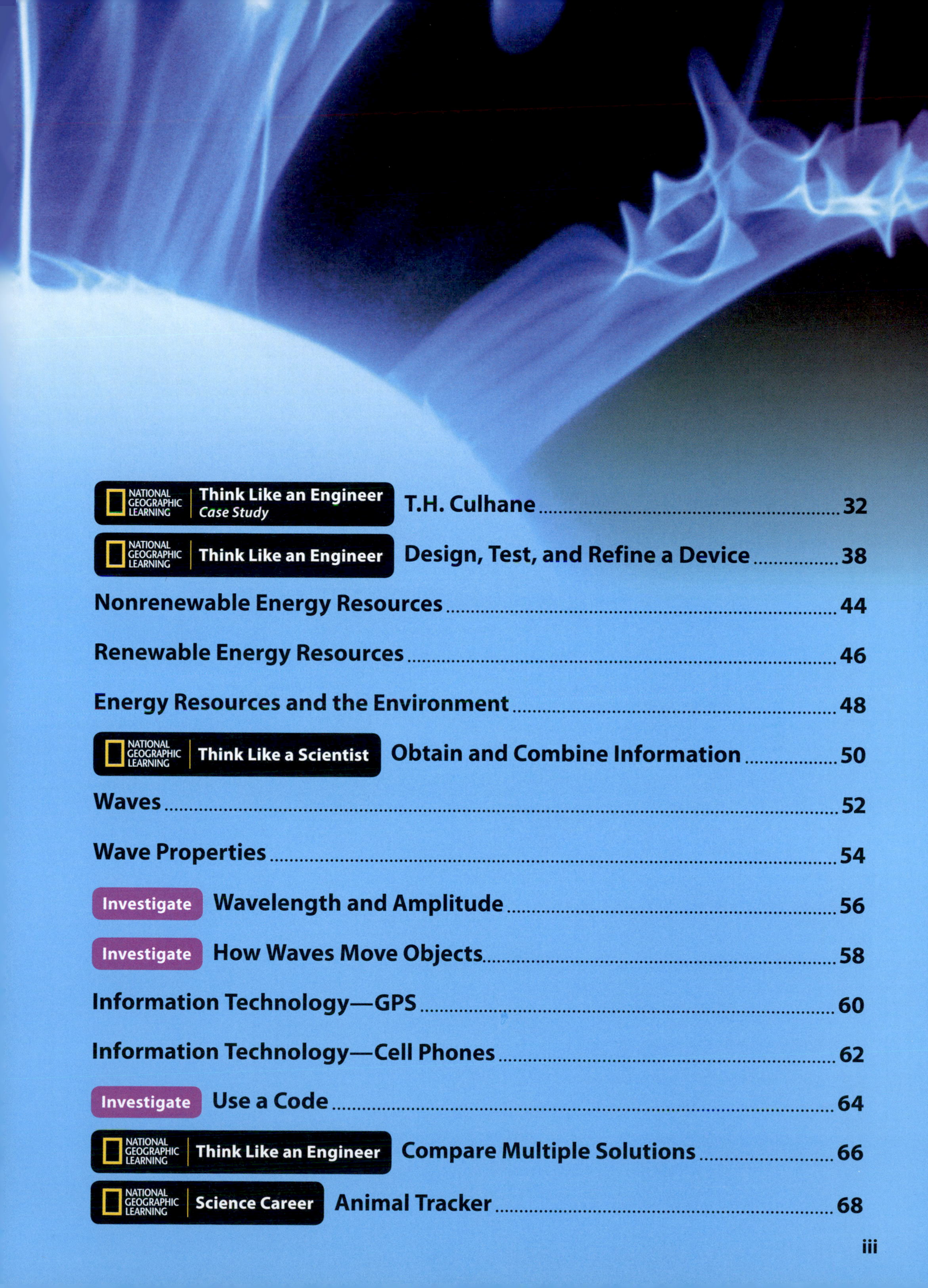

NATIONAL GEOGRAPHIC LEARNING | Think Like an Engineer *Case Study* T.H. Culhane 32

NATIONAL GEOGRAPHIC LEARNING | Think Like an Engineer Design, Test, and Refine a Device 38

Nonrenewable Energy Resources 44

Renewable Energy Resources 46

Energy Resources and the Environment 48

NATIONAL GEOGRAPHIC LEARNING | Think Like a Scientist Obtain and Combine Information 50

Waves 52

Wave Properties 54

Investigate Wavelength and Amplitude 56

Investigate How Waves Move Objects 58

Information Technology—GPS 60

Information Technology—Cell Phones 62

Investigate Use a Code 64

NATIONAL GEOGRAPHIC LEARNING | Think Like an Engineer Compare Multiple Solutions 66

NATIONAL GEOGRAPHIC LEARNING | Science Career Animal Tracker 68

Life Science

Structure, Function, and Information Processing 70

External Structures of a Wild Rose 72

Internal Structures of a Wild Rose 74

NATIONAL GEOGRAPHIC LEARNING | Think Like a Scientist Construct an Argument 76

External Structures of an Elephant 78

Internal Organs of an Elephant 80

Bones and Muscles of an Elephant 82

NATIONAL GEOGRAPHIC LEARNING **Think Like a Scientist** Construct an Argument 84

Animal Senses 86

Light and Sight 88

Investigate How We See 90

NATIONAL GEOGRAPHIC LEARNING **Think Like a Scientist** Use a Model 92

NATIONAL GEOGRAPHIC LEARNING **Science Career** Dog Whisperer 94

Earth Science

Earth's Systems: Processes that Shape the Earth 98

Rainfall in the United States 100

Pacific Northwest Forest 102

Southwest Desert 104

Central Plains Grassland 106

Eastern Temperate Forest 108

Weathering 110

Erosion and Deposition 112

Wind Changes the Land 114

Water Changes the Land 116

Investigate Weathering and Erosion 118

Ice Changes the Land 120

Living Things Change the Land 122

Landslides Change Earth's Surface 124

NATIONAL GEOGRAPHIC LEARNING | Think Like an Engineer Make Observations 126

Natural Hazards 130

Earthquakes 132

Investigate Earthquakes 134

Tsunamis 136

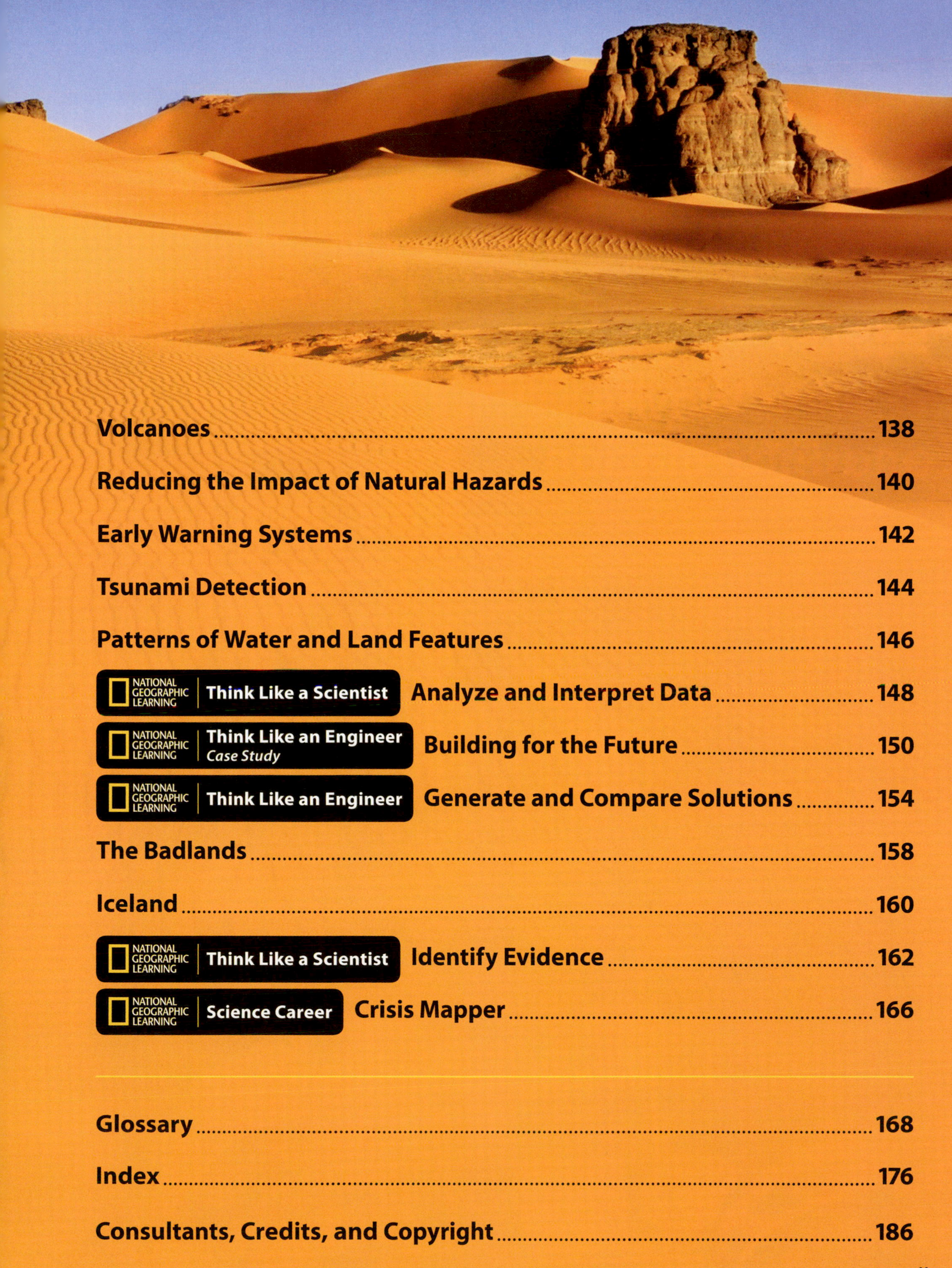

Volcanoes 138

Reducing the Impact of Natural Hazards 140

Early Warning Systems 142

Tsunami Detection 144

Patterns of Water and Land Features 146

NATIONAL GEOGRAPHIC LEARNING | Think Like a Scientist — Analyze and Interpret Data 148

NATIONAL GEOGRAPHIC LEARNING | Think Like an Engineer *Case Study* — Building for the Future 150

NATIONAL GEOGRAPHIC LEARNING | Think Like an Engineer — Generate and Compare Solutions 154

The Badlands 158

Iceland 160

NATIONAL GEOGRAPHIC LEARNING | Think Like a Scientist — Identify Evidence 162

NATIONAL GEOGRAPHIC LEARNING | Science Career — Crisis Mapper 166

Glossary 168

Index 176

Consultants, Credits, and Copyright 186

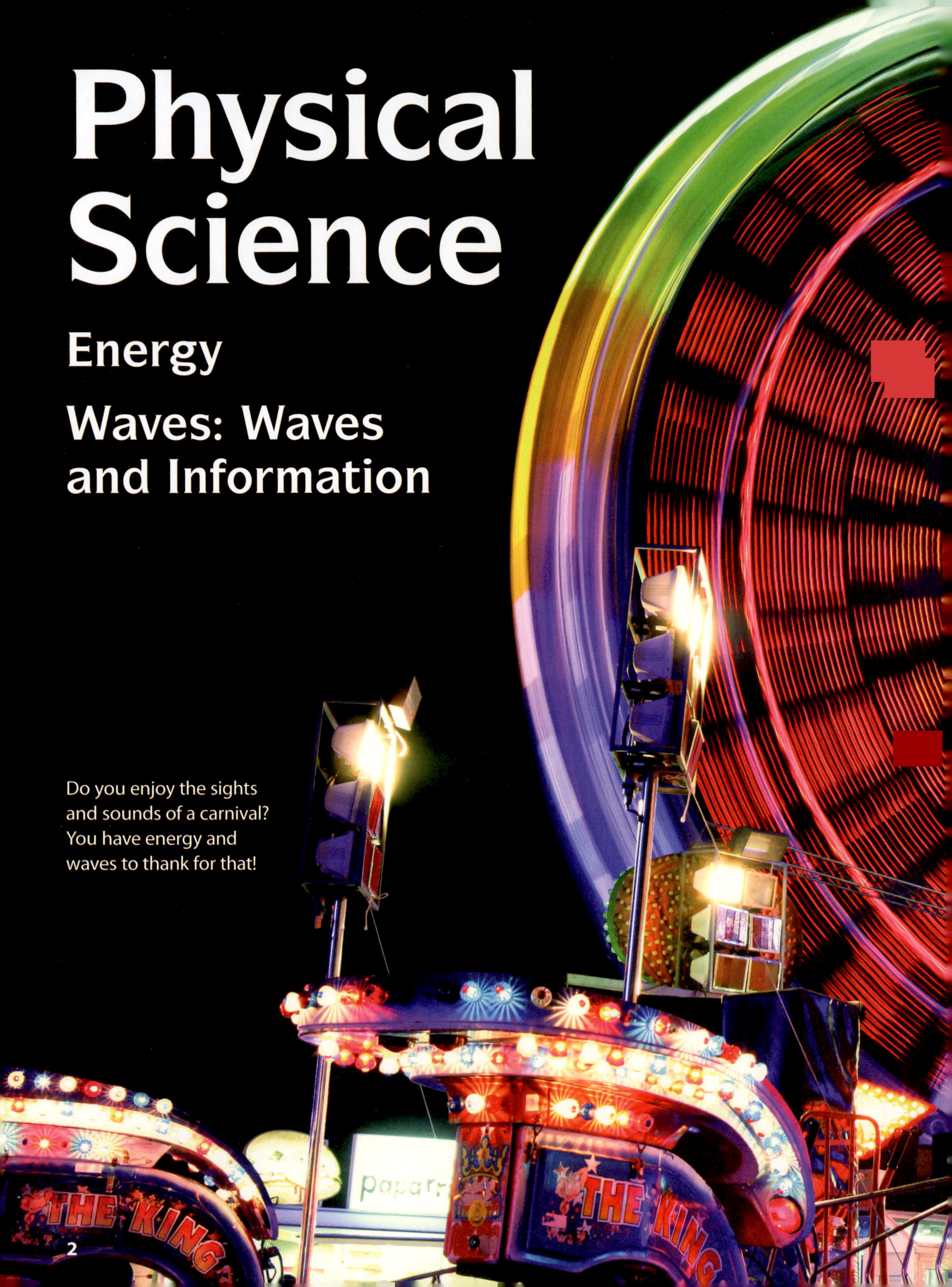

Physical Science

Energy

Waves: Waves and Information

Do you enjoy the sights and sounds of a carnival? You have energy and waves to thank for that!

SALIDA

Batter Up!

A baseball hurls toward you at a tremendous speed. You swing . . . and miss! The ball was moving faster than you thought!

When a ball flies through the air, **energy** is moved from one place to another. All moving objects have energy. The faster they are moving, the more energy they have.

The pitcher can affect the ball's energy of motion by how hard he throws.

NEXT GENERATION SCIENCE STANDARDS | DISCIPLINARY CORE IDEAS

PS3.A: Definitions of Energy

- The faster a given object is moving, the more energy it possesses. (4-PS3-1)
- Energy can be moved from place to place by moving objects or through sound, light, or electric currents. (4-PS3-2), (4-PS3-3)

Pitchers can decide how fast they want to throw a baseball. A pitcher can give the ball more energy to increase its speed, or less energy to decrease its speed. If a hitter thinks the ball is moving faster or slower than it is actually moving, the hitter will miss. And that is just what the pitcher wants!

Pitchers can adjust their grip on the ball to throw it at different speeds. A typical fastball can move more than 145 km/h (90 mph).

A typical changeup moves slower, up to 145 km/h (90 mph).

Wrap It Up!

1. **Explain** How can energy be moved from place to place?
2. **Compare** How does the energy of a fastball compare with the energy of a changeup? How do you know?

Investigate

Speed

? How is the speed of an object related to its energy?

The fastest baseballs thrown clock unbelievable speeds over 161 km/h (100 mph). Imagine the energy such a ball possesses! Pitchers who throw the fastest pitches can even sustain injuries from using so much force. In this investigation, you can compare the speeds of balls rolled with different amounts of force.

Materials

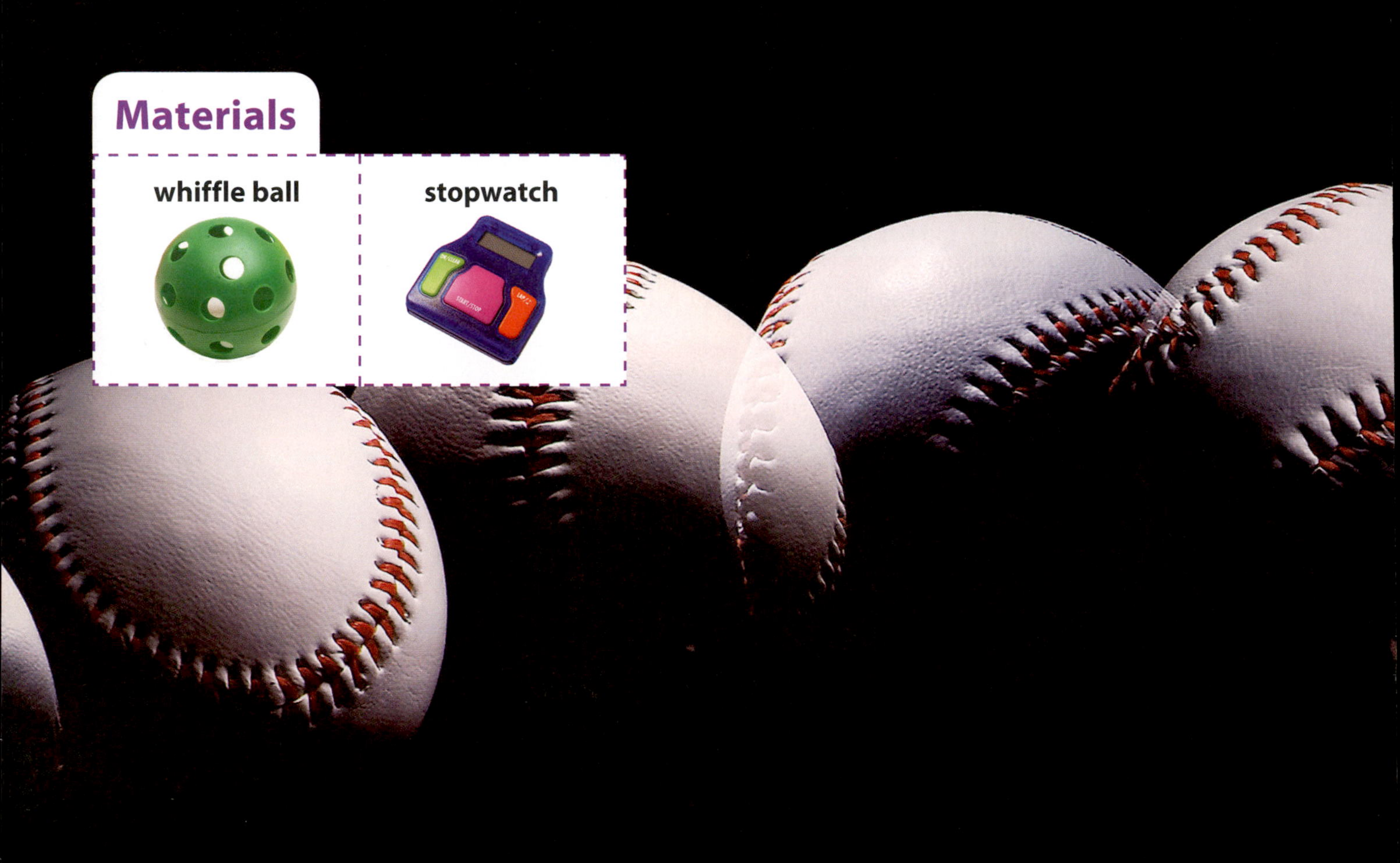

NEXT GENERATION SCIENCE STANDARDS | PERFORMANCE EXPECTATION
4-PS3-1. Use evidence to construct an explanation relating the speed of an object to the energy of that object.

1 With a partner, kneel about 3 meters from a wall or other surface. Place a ball on the starting line. Have your partner ready to use the stopwatch.

2 Say "Go" and gently release the ball so it moves very slowly. Have your partner time how long it takes for the ball to roll to the wall. Record the time.

3 Repeat steps 1 and 2, this time rolling the ball with slightly more force.

4 Repeat steps 1 and 2, this time rolling the ball with even more force.

Wrap It Up!

My science notebook

1. **Describe** The less time it takes the ball to roll to the wall, the greater its speed. Describe the speeds of the ball in your three trials.
2. **Explain** Use your results as evidence to explain how the speed of an object is related to its energy.

Hit the Ball

A fastball speeds toward a batter. He hits it! The ball is in **motion**—changing position. As the ball collides with the bat, energy is **transferred** from the bat to the ball. As a result, the ball's direction changes, and it soars toward the outfield. The hitter races to first base.

The batter's swing affects the bat's energy when it collides with the ball.

NEXT GENERATION SCIENCE STANDARDS | DISCIPLINARY CORE IDEAS
PS3.B: Conservation of Energy and Energy Transfer
Energy is present whenever there are moving objects, sound, light, or heat. When objects collide, energy can be transferred from one object to another, thereby changing their motion. (4-PS3-2), (4-PS3-3)

1) The ball has energy as it moves toward the bat. The bat also has energy as it moves toward the ball. **2)** When the bat and ball collide, energy is transferred between the bat and the ball. **3)** The ball's motion is changed. The ball is now moving in a new direction and may be moving at a new speed, too.

When objects collide, they can change each other's motion in different ways. For example, the direction of a ball's motion changes when the ball bounces off a wall. The speed of a ball's motion changes when the ball encounters air and wind and slows down. And the ball's speed of motion changes to a full stop when the ball collides with an outfielder's glove. In each of these collisions, energy is transferred and the motion of the ball is changed.

Wrap It Up!

1. **Name** What two features of motion can change when energy is transferred to an object?
2. **Apply** Describe what would happen if you transferred energy to a motionless soccer ball by kicking it.
3. **Infer** Does the ball's energy increase or decrease when its motion has stopped?

Investigate

Motion

? What happens to an object's energy when it collides with another object?

The pitcher, the teams, the fans . . . everyone breathlessly watches the motion of a baseball after it makes contact with the bat. They are all eager to know how the ball's energy and motion will change. Will it fly right past the outfielder or collide with his glove and then change direction? In this investigation, you can explore how a ball's energy changes when it collides with objects in different ways.

Materials

whiffle ball	whiffle bat

NEXT GENERATION SCIENCE STANDARDS | PERFORMANCE EXPECTATION

4-PS3-3. Ask questions and predict outcomes about the changes in energy that occur when objects collide.

1 Kneel about 3 meters away from a partner. Have your partner roll the ball to you. Observe and record how the ball's energy changes as you catch the ball. Roll the ball back to your partner.

2 Hold the bat flat on the floor. Have your partner roll the ball so that it bounces off the bat. Observe and record how the ball's energy changes. Roll the ball back to your partner.

3 Again, hold the bat flat on the floor. Have your partner roll the ball to you. While keeping the bat on the floor, gently swing the bat at the ball. Observe and record how the ball's energy changes. Roll the ball back to your partner.

4 Repeat step 3, but this time, swing the bat with slightly more force. Observe and record how the ball's energy changes.

Wrap It Up!

1. **Ask** Write your own question and answer about how the energy of a ball changes when it collides with an object.
2. **Predict** What might happen to the energy of the ball if you hit it harder?

Sounds of the Game

Whoosh . . . smack! When you attend a baseball game, energy is moving all around you in the form of sound. A ball whooshes through the air and smacks into the catcher's mitt. A roar erupts from the crowd as the umpire yells, "You're out!" All these sounds occur as energy is transferred through the air in rapid, back and forth movements called **vibrations.**

When the moving ball smacks into a catcher's mitt and comes to a stop, its energy isn't lost. Some of the energy is changed, or **transformed,** into heat. The surrounding air gets heated, vibrates, and sound is produced.

When the ball hits the catcher's mitt, the energy of its motion is transformed into heat and sound.

NEXT GENERATION SCIENCE STANDARDS | DISCIPLINARY CORE IDEAS

PS3.A: Definitions of Energy
Energy can be moved from place to place by moving objects or through sound, light, or electric currents. (4-PS3-2), (4-PS3-3)

PS3.B: Conservation of Energy and Energy Transfer
Energy is present whenever there are moving objects, sound, light, or heat. When objects collide, energy can be transferred from one object to another, thereby changing their motion. In such collisions, some energy is typically also transferred to the surrounding air; as a result, the air gets heated and sound is produced. (4-PS3-2), (4-PS3-3)

As the ball slams into the catcher's mitt, tiny air particles around the mitt are moved. These air particles push the air particles around them, and so on. As a result, waves of air vibrations quickly spread throughout the entire stadium. When the vibrations reach a person's eardrums, the person hears the sound.

When the player catches the ball, some of the ball's energy of motion will transform to sound energy. Some of the player's energy of motion will change to sound when he hits the ground, too.

My science notebook

Wrap It Up!

1. **Recall** How does energy of motion transform into sound?
2. **Infer** Explain why covering your ears with your hands reduces the sound you hear.
3. **Infer** Why might a catcher's mitt feel warm after he catches a ball?

Investigate

Sound

What evidence can you observe that sound transfers energy?

Your eardrums detect small vibrations, but some loud noises produce vibrations you can feel with the rest of your body. For example, when music at a baseball park plays over a loudspeaker, you might feel the vibrations. Sometimes you can see the vibrations produced by sound. In this investigation, you can observe the effect of sound on grains of salt.

NEXT GENERATION SCIENCE STANDARDS | PERFORMANCE EXPECTATION
4-PS3-2. Make observations to provide evidence that energy can be transferred from place to place by sound, light, heat, and electric currents.

1 Use a rubber band to attach plastic wrap over the opening of the cup. Sprinkle salt in the center of the plastic wrap.

2 Hold the paper towel tube so that it is directed at the salt but not touching it. Predict what will happen when you speak into the tube. Record your prediction.

3 Speak softly into the paper towel tube. Observe what happens to the salt grains. Record your observations.

4 Repeat step 3, but this time speak loudly into the tube. Observe what happens to the salt grains. Record your observations.

Wrap It Up!

1. **Observe** Did your observations match your prediction? Explain.
2. **Compare and Contrast** Describe what happened to the salt grains when you spoke into the tube loudly, then softly.
3. **Give Evidence** Use your observations to give evidence that energy can be transferred from place to place by sound.

The Sun's Light

What can transfer energy faster than the motion of a fastball or the vibrations of sound? Light. Light is energy that you can see. In about 8 minutes, or the time it takes you to find your seat in a baseball stadium and sit down, light energy has already traveled all the way from the sun to your eyes. That's about 149,600,000 kilometers (92,960,000 miles)!

Light transfers energy from the sun through space. When the sun's light energy reaches Earth, that energy is transferred to countless objects and transformed in different ways. Objects sitting in sunlight become warm as the light is transformed into heat. Light also bounces off objects in our surroundings. When the light enters our eyes, it allows us to see.

The immense energy of the sun comes from chemical reactions that occur deep inside the sun's core.

NEXT GENERATION SCIENCE STANDARDS | DISCIPLINARY CORE IDEAS

PS3.A: Definitions of Energy

Energy can be moved from place to place by moving objects or through sound, light, or electric currents. (4-PS3-2), (4-PS3-3)

PS3.B: Conservation of Energy and Energy Transfer

- Energy is present whenever there are moving objects, sound, light, or heat. (4-PS3-2), (4-PS3-3)
- Light also transfers energy from place to place. (4-PS3-2)

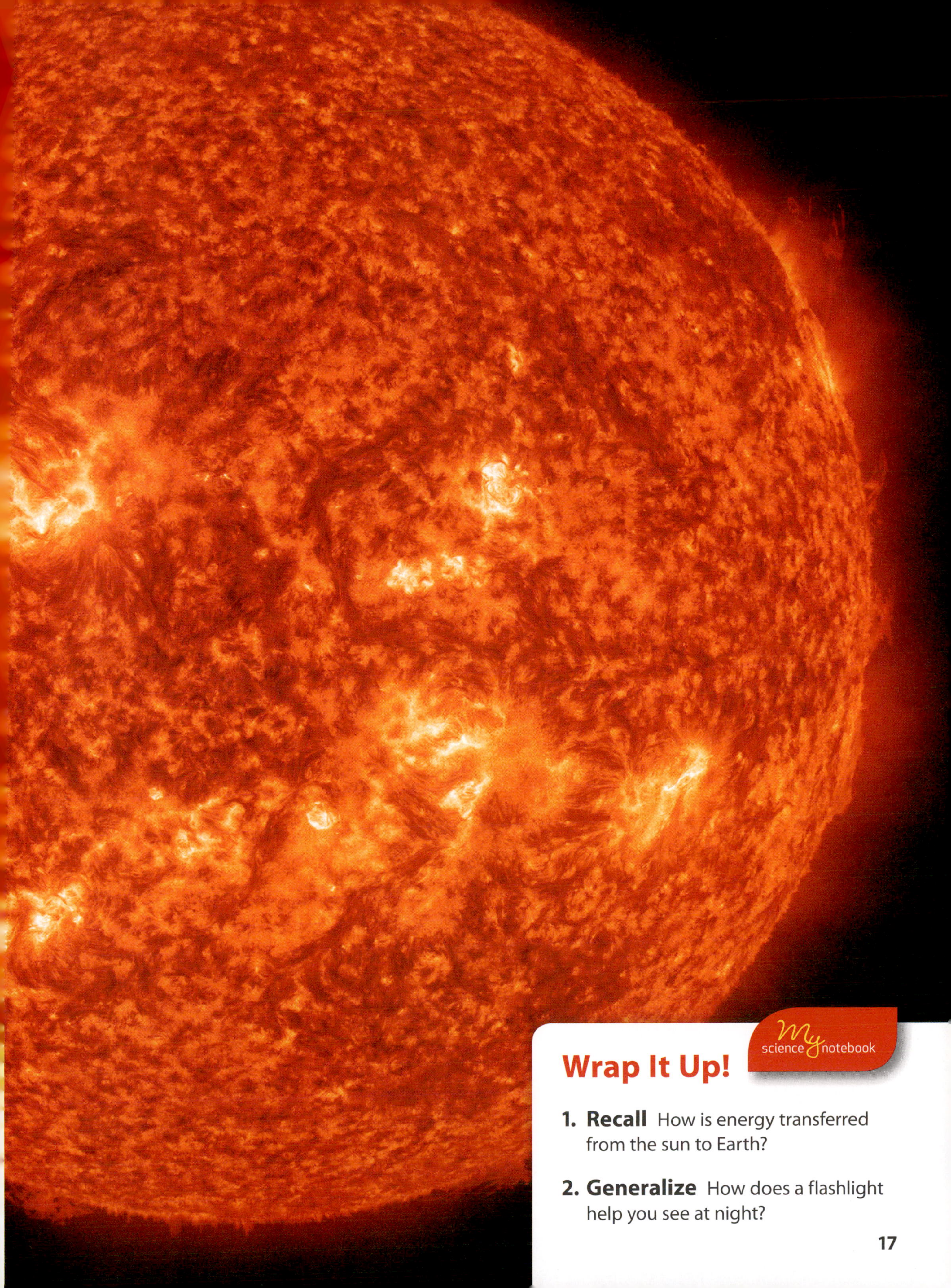

Wrap It Up!

1. **Recall** How is energy transferred from the sun to Earth?

2. **Generalize** How does a flashlight help you see at night?

Investigate

Light

? **What evidence can you observe that light transfers energy?**

Just like sound energy can change the motion of salt grains, light energy can change objects. When direct sunlight strikes a piece of construction paper, the light causes a change in the paper. In this investigation, you can observe the effect of light energy on the color of construction paper.

Materials

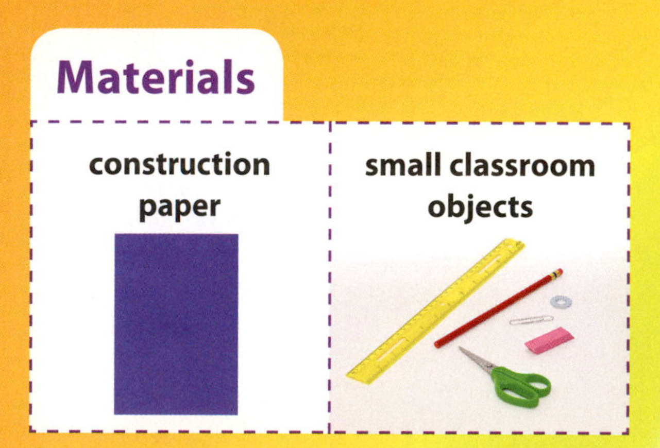

NEXT GENERATION SCIENCE STANDARDS | PERFORMANCE EXPECTATION
4-PS3-2. Make observations to provide evidence that energy can be transferred from place to place by sound, light, heat, and electric currents.

1 Place a piece of construction paper on a flat surface in a sunny spot.

2 Select a few objects to place on the construction paper.

My science notebook

3 Predict what will happen to the color of the paper after it has been exposed to sunlight. Record your prediction.

4 Leave the paper alone until it has received a full hour or more of sunlight. Remove the objects and record your observations.

Wrap It Up!

1. **Predict** Did your results support your prediction? Explain.

2. **Infer** What might be the result if a clear object were placed on the paper?

3. **Give Evidence** Use the observations you made to give evidence that energy can be transferred from place to place by light.

Heat It Up!

Heat is another way for energy to move from place to place. The particles that make up matter are always vibrating or moving. The energy of the moving particles is **thermal energy.** The faster the particles are moving, the more thermal energy the matter has. Heat is the transfer of thermal energy from a warmer object to a cooler object.

This camper can keep his drink warm using thermal energy from the campfire.

NEXT GENERATION SCIENCE STANDARDS | DISCIPLINARY CORE IDEAS
PS3.A: Definitions of Energy
The faster a given object is moving, the more energy it possesses. (4-PS3-1)
PS3.B: Conservation of Energy and Energy Transfer
Energy is present whenever there are moving objects, sound, light, or heat. (4-PS3-2), (4-PS3-3)

A campfire has a lot of thermal energy. The camper kneels next to the campfire on a cool, dark night. Some of the energy of the campfire is transferred to the camper and helps him keep warm.

Heat can be used in many other ways. Thermal energy in buildings keeps us warm. You probably dry your clothes using thermal energy from a clothes dryer.

This popcorn maker heats popcorn seeds until they pop out of the hard outer shells.

Warm air from a hair dryer causes water from wet hair to evaporate.

My science notebook

Wrap It Up!

1. **Define** What is heat?
2. **Apply** In what ways do you use thermal energy every day?
3. **Relate** Besides heat, what other forms of energy are transferred by the campfire?

Investigate

Heat

What evidence can you observe that heat transfers energy from place to place?

Warm air can transfer thermal energy to butter and melt it. That's why butter left out on a hot day turns into a liquid puddle! In this investigation, you can observe how a metal spoon can transfer thermal energy between water and butter.

Materials

3 cups with water of different temperatures	3 metal spoons	3 small dabs of butter

NEXT GENERATION SCIENCE STANDARDS | PERFORMANCE EXPECTATION

4-PS3-2. Make observations to provide evidence that energy can be transferred from place to place by sound, light, heat, and electric currents.

1 Place three cups of water on your desk. Record descriptions of the temperatures of the water in your science notebook.

2 Place a dab of butter on the stem of each spoon. Predict which dab of butter will melt first after the spoons are placed in the cups of water. Record your predictions.

3 Work with your group to place the spoons with butter in the cups of water. Be sure you put the spoons in the water at the same time.

4 Observe the butter every minute for 10 minutes. Record your observations.

Wrap It Up!

1. **Observe** List the cups of water in order from least to most thermal energy.
2. **Predict** Did your results support your predictions? Explain.
3. **Give Evidence** How did the amount of thermal energy in the cups affect the melting of the butter? Use evidence from your observations in your explanation.

It’s Electric

You probably know that to get the light bulb in a lamp to light, you need to plug the cord into an electrical outlet and turn on the switch. To produce light, the lamp uses electricity, or **electrical energy.** Electrical energy is the energy of moving charged particles. When electrical energy is transferred through wires, it is called **electric current.** Look at the photograph of the many carnival lights. What makes all these lights glow and the rides move? Electricity!

Electric current is a useful way to transfer energy from place to place. Energy from a faraway power plant can be moved through wires to all the buildings in a town. There it can be transformed into light, heat, sound, and motion.

At a carnival, electric current is transformed into colorful lights, festive sounds, and moving rides.

NEXT GENERATION SCIENCE STANDARDS | DISCIPLINARY CORE IDEAS

PS3.A: Definitions of Energy
Energy can be moved from place to place by moving objects or through sound, light, or electric currents. (4-PS3-2), (4-PS3-3)

PS3.B: Conservation of Energy and Energy Transfer
Energy can also be transferred from place to place by electric currents, which can then be used locally to produce motion, sound, heat, or light. The currents may have been produced to begin with by transforming the energy of motion into electrical energy. (4-PS3-2), (4-PS3-4)

Electric current transfers energy to a waffle iron, where it is transformed into heat.

The electric guitar is attached to an amplifier. The amplifier makes the sound from the guitar louder.

Wrap It Up!

My science notebook

1. **Define** What is electric current?
2. **Recall** Give examples of how electrical energy can be used to produce light, heat, sound, and motion.
3. **Apply** You have learned about some uses of electrical energy. What other ways do you use electrical energy every day?

Electric Circuits

The colorful lights in the picture run on electricity. The wires and light bulbs make up an **electric circuit.** An electric circuit is a complete path through which electric current can pass. A battery and a power plant are sources of electricity for circuits.

A battery has stored energy. The battery has a positive end and a negative end. If you attach a wire from one end of the battery to the other, you produce a complete path through which an electric current can pass. For the current to flow, the circuit must be complete. If the wire is not connected to both ends of the battery, the energy cannot move through the circuit.

For these lights to shine, they must be connected to a source of electricity in a complete circuit.

NEXT GENERATION SCIENCE STANDARDS | DISCIPLINARY CORE IDEAS

PS3.A: Definitions of Energy

Energy can be moved from place to place by moving objects or through sound, light, or electric currents. (4-PS3-2), (4-PS3-3)

PS3.B: Conservation of Energy and Energy Transfer

Energy can also be transferred from place to place by electric currents, which can then be used locally to produce motion, sound, heat, or light. (4-PS3-2), (4-PS3-4)

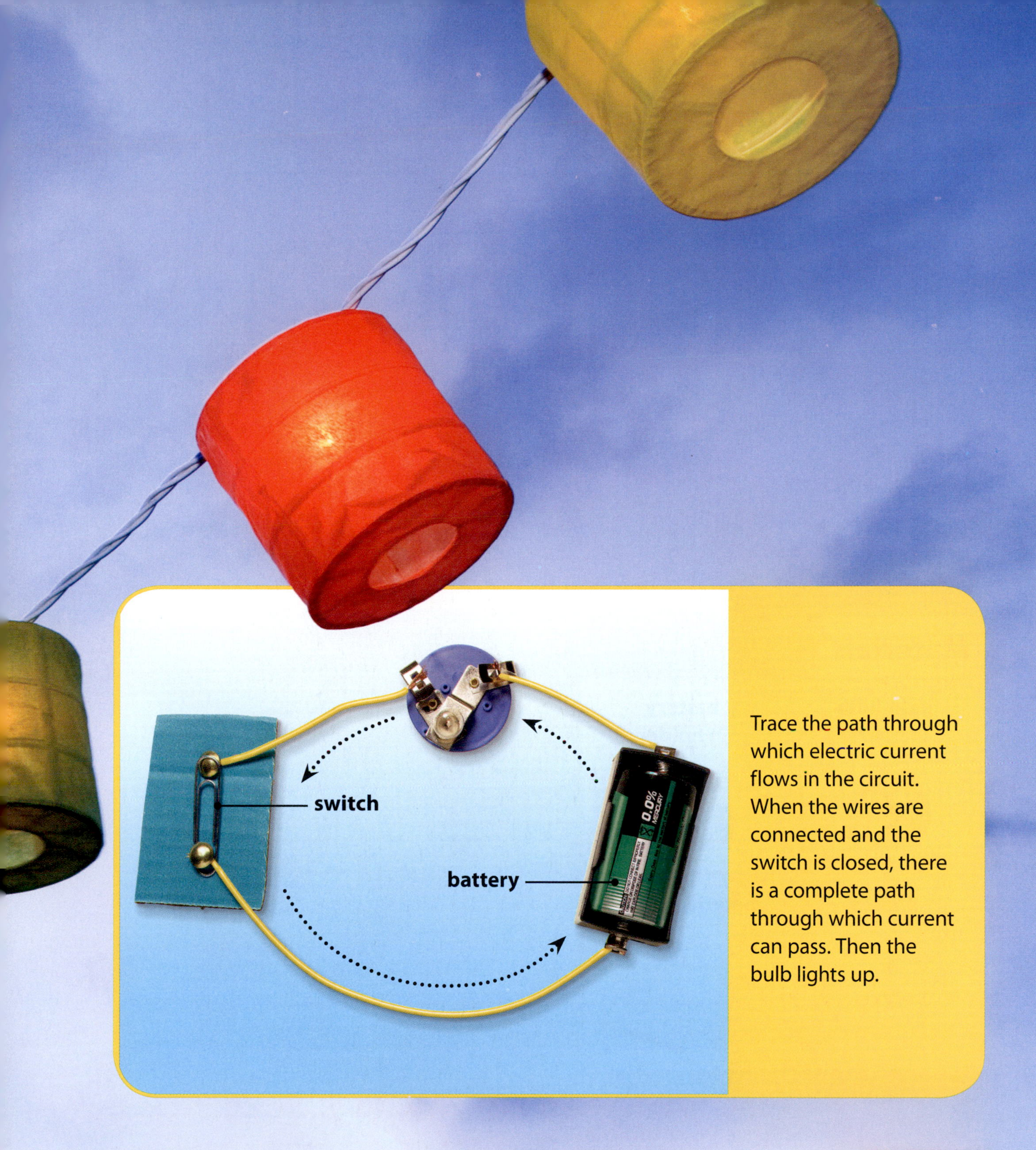

Trace the path through which electric current flows in the circuit. When the wires are connected and the switch is closed, there is a complete path through which current can pass. Then the bulb lights up.

Wrap It Up!

1. **Describe** How does an electric circuit transfer energy?
2. **Predict** What happens to the flow of electric current when you flip a light switch off? Explain.

Investigate

Electric Circuits

Which materials can complete an electric circuit?

For electricity to flow, a circuit must be complete. Then the electric current can flow and light a bulb. In this investigation, you can find out which materials can complete an electric circuit.

Materials

light bulb in holder

battery in holder

3 wires

materials to test

NEXT GENERATION SCIENCE STANDARDS | DISCIPLINARY CORE IDEAS

4-PS3-2. Make observations to provide evidence that energy can be transferred from place to place by sound, light, heat, and electric currents.

1 Attach wires to a battery holder and to a bulb holder as shown.

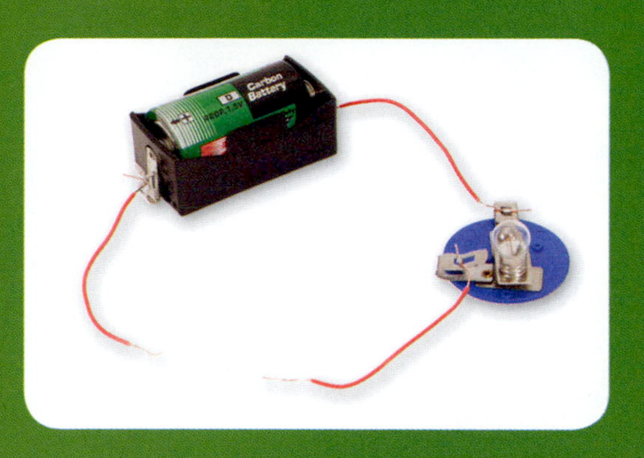

2 Notice that there are two wire ends that are not connected. Touch the ends of the wires together to make a complete circuit. Record your observations in your science notebook.

3 Predict whether a rubber band will complete the circuit if you touch the wire ends to it. Record your prediction. Touch the ends of the wires to the rubber band. Record your observations.

4 Predict whether other materials will complete the circuit. Record your predictions. Test the materials. Record your observations.

Wrap It Up!

1. **Predict** Did your results support your predictions? Explain.
2. **Describe** What materials are needed to build a complete circuit? How are the materials used to complete the circuit?
3. **Conclude** How can energy be transferred from place to place by electric currents? Use evidence from the activity in your answer.

Spin It!

Where does the electricity you use in your home come from? It must be generated using other forms of energy. Energy can't be created or destroyed. But energy can be changed, or transformed, from one form into another. One source of energy is wind. The moving air causes the wing-like blades of a wind turbine to turn around and around. The rotating blades have **energy of motion** that can be transformed into electricity.

NEXT GENERATION SCIENCE STANDARDS | DISCIPLINARY CORE IDEAS
PS3.B: Conservation of Energy and Energy Transfer
Energy can also be transferred from place to place by electric currents, which can then be used locally to produce motion, sound, heat, or light. The currents may have been produced to begin with by transforming the energy of motion into electrical energy. (4-PS3-2), (4-PS3-4)

The energy from the moving blades of a wind turbine runs a generator. The generator transforms the energy of motion into electricity. People can use this electricity in homes, schools, and offices.

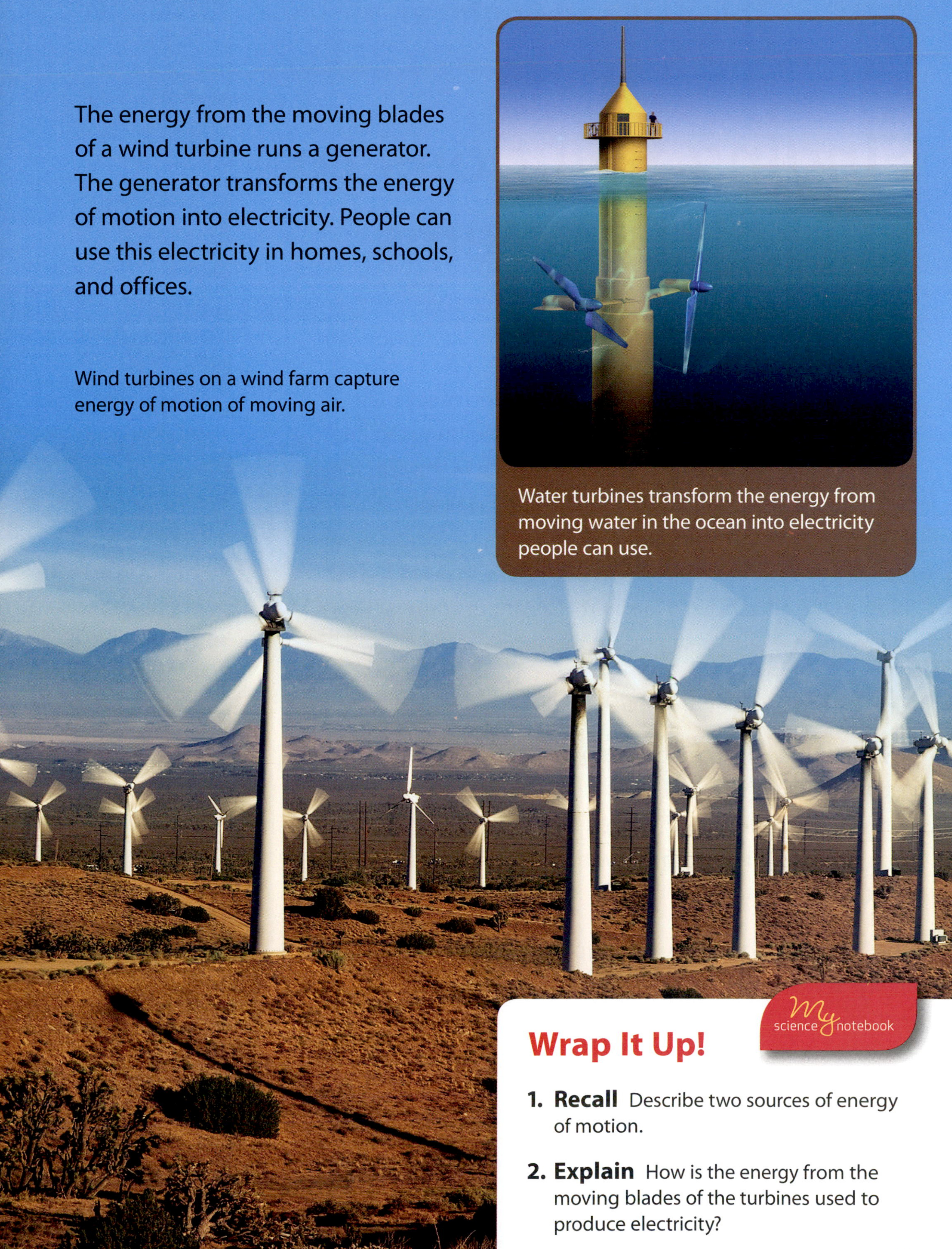

Wind turbines on a wind farm capture energy of motion of moving air.

Water turbines transform the energy from moving water in the ocean into electricity people can use.

Wrap It Up!

My science notebook

1. **Recall** Describe two sources of energy of motion.
2. **Explain** How is the energy from the moving blades of the turbines used to produce electricity?

Finding Solutions to Energy Problems

Problem

T.H. Culhane helps people all over the world meet their energy needs. T.H. is an expert in finding solutions to energy problems. He got his start studying people in the rain forests of Borneo. The people had few resources available, yet they thrived in their environment. T.H. thought, "How can we use the idea of working with limited resources to do the same thing in the cities?" Let's talk with T.H. about some of his environmental and health solutions.

NGL Science Why did you go to rain forests?

T.H. Culhane I learned that people who live in the rain forest get all of their energy, all their food, and all their ecosystem services—recycling their wastes back into energy and food—from the forest. I went to live with them to learn how they did it.

NGL Science Your work takes you all over the world. You always try to find sustainable energy solutions. What does that mean?

NEXT GENERATION SCIENCE STANDARDS | DISCIPLINARY CORE IDEAS
ETS1.A: Defining Engineering Problems
Possible solutions to a problem are limited by available materials and resources (constraints). The success of a designed solution is determined by considering the desired features of a solution (criteria). Different proposals for solutions can be compared on the basis of how well each one meets the specified criteria for success or how well each takes the constraints into account. (secondary to 4-PS3-4)

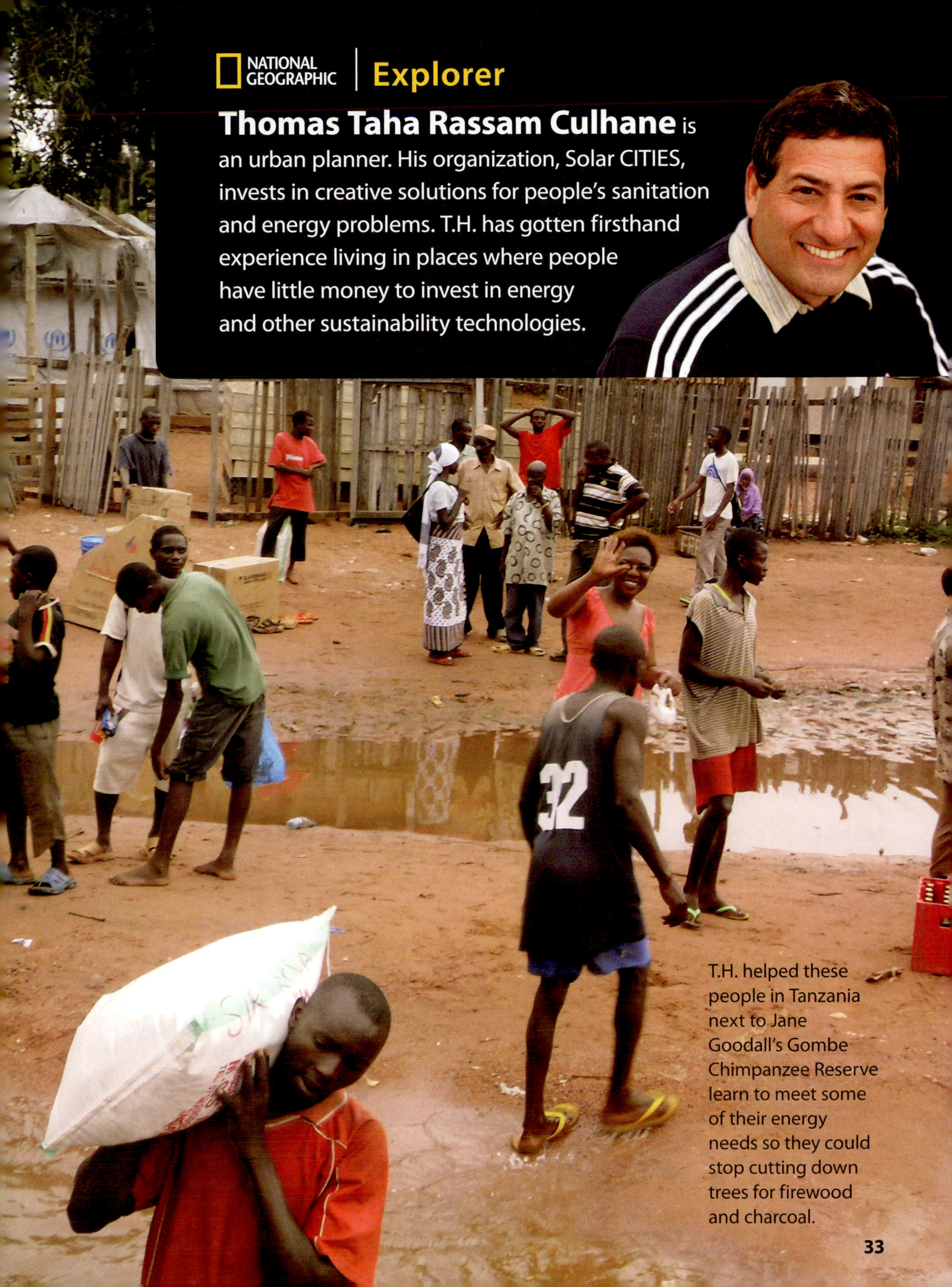

Explorer

Thomas Taha Rassam Culhane is an urban planner. His organization, Solar CITIES, invests in creative solutions for people's sanitation and energy problems. T.H. has gotten firsthand experience living in places where people have little money to invest in energy and other sustainability technologies.

T.H. helped these people in Tanzania next to Jane Goodall's Gombe Chimpanzee Reserve learn to meet some of their energy needs so they could stop cutting down trees for firewood and charcoal.

T.H. Culhane *Sustainable* means doing something in such a way that you can keep doing it. For example: You can save some of your garden seeds from this year. Then plant the seeds next year. You can do this year after year. You'll never run out of seeds. That's a sustainable way to grow a garden.

NGL Science What energy problems have you identified?

T.H. Culhane One: How can we get enough energy? Most energy people use originally comes from the sun. We have to transform energy from the sun into a form we can use. Two: How can we get rid of wastes without harming the environment?

Solution

NGL Science How have you helped people solve their energy problems in places with limited resources?

T.H. Culhane The idea is quite simple. Most energy people use originally comes from the sun. For example, the sun's energy makes plants grow. When animals eat the plants, they create waste from what they eat. We can use the waste for energy. We just need to transform the energy into a form we can use.

T.H. Culhane explains an energy solution to village leaders in Tanzania, Africa. **1)** A holding tank is installed. **2)** Animal wastes are added to the tank. **3)** The expandable tank that captures and holds the methane gas is installed.

T.H. Culhane and villagers add ground up food scraps, dead flowers, and other organic wastes to the biodigester tank. This dramatically increases gas production.

Think Like an Engineer
Case Study

NGL Science How do we transform energy into forms we can use?

T.H. Culhane We have developed a simple device called a "home-scale" biodigester. The biodigester can transform energy in household wastes into energy and nutritious fertilizer we can use. To minimize energy use, we frequently use recycled materials to build this device.

NGL Science How did you get useful energy from wastes?

T.H. Culhane The biodigester transforms organic wastes. We can grind up our food scraps. Then the chopped up food gets added to the biodigester. So does human waste from the toilet. Bacteria feed on the wastes. As they feed, the bacteria produce methane gas. We burn the methane gas on the stove for cooking. You can even burn the gas in a generator to produce electricity!

NGL Science Does the biodigester produce anything besides gas?

T.H. Culhane Yes, the process also produces a rich liquid fertilizer. We use it in gardens.

NGL Science How did you get the idea of turning garbage into energy?

T.H. Culhane It came from years spent working on my Ph.D. with the Zabaleen trash pickers community of Cairo, Egypt, who make their living from recycling garbage. Getting rid of wastes without harming the environment is important. Some cultures understand that everything has a use. Food wastes and toilet wastes are useful. They're forms of energy. So a problem is also a solution!

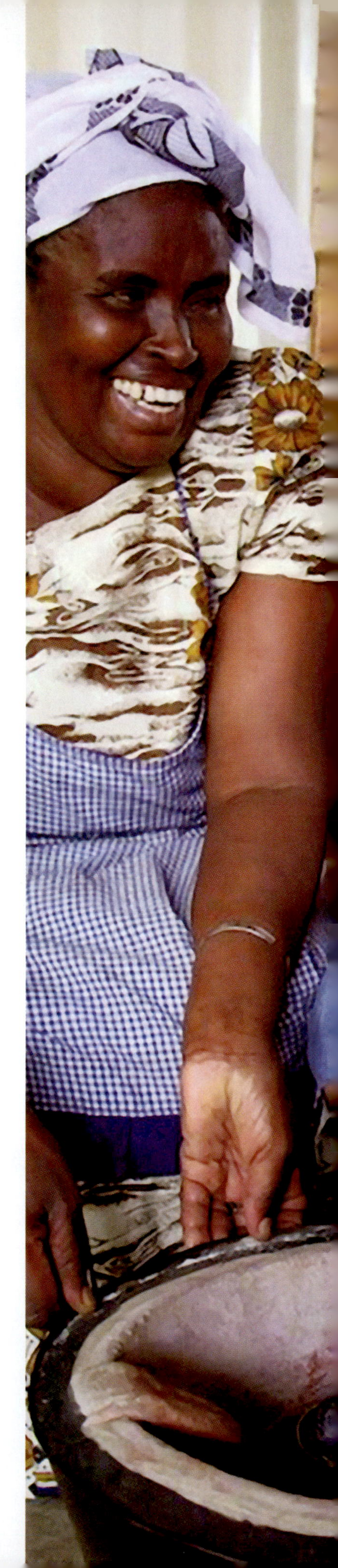

T.H. Culhane and the cook in a school in one of Nairobi's most crowded urban slums prepare to use a new stove fueled by biogas that eliminates the smoke, health, and environmental problems associated with charcoal.

Wrap It Up!

My science notebook

1. **Identify** What two problems did T.H. Culhane identify?
2. **Explain** How does T.H. Culhane's energy solution make use of available materials?

Design, Test, and Refine a Device

You've read how T.H. Culhane applied scientific ideas to design his biodigester. As he made and refined his design, he accounted for the lack of resources in the areas his device would be used. Now it's your turn. Imagine that you are working to find an energy solution for people living in remote mountain communities without access to electricity. The current solution of using wood fires to cook food is not sustainable. Wood fires use up forest resources and produce harmful smoke in people's homes.

1. **Define the problem.** My science notebook
   Solar ovens convert sunlight energy into heat. What problem would solar ovens solve for the mountain communities? List the criteria for a successful solution. What constraints must be taken into account? Remember, the people using this oven will need to build it far away from a hardware store. You must use simple, inexpensive materials. The oven must also be easy to construct.

2. **Find a solution.**
   Study several solar oven designs, and then draw and label your own design. Describe how your design will meet the criteria of simple, low cost, and easy to build. Create a prototype of your solar oven using available materials.

NEXT GENERATION SCIENCE STANDARDS | PERFORMANCE EXPECTATIONS

4-PS3-4. Apply scientific ideas to design, test, and refine a device that converts energy from one form to another.

3-5-ETS1-1. Define a simple design problem reflecting a need or a want that includes specified criteria for success and constraints on materials, time, or cost.

3-5-ETS1-3. Plan and carry out fair tests in which variables are controlled and failure points are considered to identify aspects of a model or prototype that can be improved.

This woman in Mali in Africa prepares food using a solar cooker.

3. **Test your solution.**
   Decide how you will test your prototype. What measurements will you take to provide evidence that your solar oven works? What tools will you need to take those measurements? When your teacher has approved your plan and you have what you need, carry out your test. Record your observations.

4. **Refine or change your solution.**
   Analyze your findings. How could your prototype be improved? Refine your design and test it again. Record your changes.

5. **Compare your data.**
   Did your change improve the performance of your solar oven design? Do you need to make more refinements? Continue to refine your prototype until you are happy with its performance.

6. **Analyze and explain your results.**
   Explain the scientific basis of your design solution.
   - What forms of energy are involved in your design?
   - How does your oven transform one type of energy into another?
   - Which criteria does your design meet?
   - What evidence do you have to show that your prototype is successful?
   - How could your design be further improved?

7. **Present your results.**
   Present your arguments and your prototype to the class. Evaluate your classmates' prototypes using the same criteria you used to evaluate your own.

This woman in Chad, in Africa, makes tea with water boiled by a solar oven. She did not have to gather firewood.

Design, Test, and Refine a Device

A new board game is coming out soon. The game developers have asked you, an engineer, to design a buzzer for the game. The buzzer will need to be battery-powered and controlled by a switch. Your buzzer needs to use materials from those provided by your teacher, and be quick and low-cost to build.

1. **Define the problem.** My science notebook
   You know you need to build a buzzer that can be turned on and off. You can use a circuit similar to the one you built in *Investigate Electric Circuits*, except a buzzer can be used in place of the light bulb. In order to turn the buzzer on and off, you will need to include a switch in your circuit. Look at the switch in the circuit on page 27 for hints on how to do that.

2. **Find a solution.**
   Study the materials available to you. Which ones will you use to build your prototype? Draw and label a diagram of your prototype. Explain how you think your prototype will work.

3. **Test your solution.**
   Plan how you will test your circuit design. What evidence will you provide to show that your circuit design works? After your teacher has approved your plan and you have what you need, carry out your test. Record your observations.

NEXT GENERATION SCIENCE STANDARDS | PERFORMANCE EXPECTATIONS

4-PS3-4. Apply scientific ideas to design, test, and refine a device that converts energy from one form to another.

3-5-ETS1-1. Define a simple design problem reflecting a need or a want that includes specified criteria for success and constraints on materials, time, or cost.

3-5-ETS1-3. Plan and carry out fair tests in which variables are controlled and failure points are considered to identify aspects of a model or prototype that can be improved.

4. **Refine or change your solution.**
   Analyze your results. Did your prototype work? Are there changes you could make that would improve your prototype? Make the changes and retest. Keep refining and testing your prototype until you are sure it is the best it can be.

5. **Analyze and explain your results.**
   Explain the scientific basis of your design solution.
   - What forms of energy are involved in your design?
   - How does it transform one type of energy into another?
   - Refer to the criteria you identified earlier for a successful device. Which criteria does your design meet?
   - What evidence do you have to show that your prototype is successful?

6. **Present your results.**
   Present your arguments and your design to the class. Evaluate your classmates' designs using the same criteria you used to evaluate your own.

This game buzzer converts electrical energy to light and sound energy.

Nonrenewable Energy Resources

Inside power plants, fuel such as coal, oil, or natural gas is burned to heat water and create steam. The steam turns the blades of turbines, providing energy of motion. Generators transform the energy of motion into electricity. Power plants do not "produce energy," though that is a common expression. They convert the stored energy from fuel into a form of energy for practical use.

Coal, oil, and natural gas are fossil fuels. A **fossil fuel** is a source of energy that formed from the remains of plants and animals that lived millions of years ago. Pressure and heat over time caused the remains to change into fossil fuels. Fossil fuels are considered **nonrenewable energy resources** because they will eventually run out.

Natural gas is a gas form of fossil fuel. Natural gas is drilled from the ground.

NEXT GENERATION SCIENCE STANDARDS | DISCIPLINARY CORE IDEAS

ESS3.A: Natural Resources
Energy and fuels that humans use are derived from natural sources, and their use affects the environment in multiple ways. Some resources are renewable over time, and others are not. (4-ESS3-1)

PS3.D: Energy in Chemical Processes and Everyday Life
The expression "produce energy" typically refers to the conversion of stored energy into a desired form for practical use. (4-PS3-4)

Coal is a fossil fuel that is mined from the ground using large machines.

Nuclear power plants use chemical energy from uranium reactions to produce electricity. Uranium is a nonrenewable energy resource mined from the ground.

Oil rigs remove oil and natural gas from under the ocean floor. These fossil fuels are being removed faster than nature can replace them.

My science notebook

Wrap It Up!

1. **List** Identify four nonrenewable energy resources.
2. **Identify** What type of energy is transformed into electricity in nuclear power plants?
3. **Interpret** What do people really mean when they say a power plant "produces" energy?

Renewable Energy Resources

Unlike fossil fuels and nuclear power, some sources of energy are **renewable energy resources.** Renewable energy sources will never run out. For example, **solar energy** and **wind energy** are renewable. No matter how much sunlight and wind we use, there will always be more.

Hydroelectric dams capture the energy of motion from water in rivers. When water flows through the dam, it spins turbines inside the dam and generates electricity. Hydroelectric dams are a renewable energy resource.

Solar panels transform the sun's light and heat energy into electricity.

NEXT GENERATION SCIENCE STANDARDS | DISCIPLINARY CORE IDEAS
ESS3.A: Natural Resources
Energy and fuels that humans use are derived from natural sources, and their use affects the environment in multiple ways. Some resources are renewable over time, and others are not. (4-ESS3-1)

In windy locations, wind turbines transform wind energy to electricity. Most wind turbines are tall to catch the strong, steady winds high above the ground.

My science notebook

Wrap It Up!

1. **List** List three renewable energy sources.
2. **Relate** How is electricity generation from wind and water similar to electricity generation from fossil fuels?

Energy Resources and the Environment

People cannot produce energy to meet their needs without converting it from other sources. Scientists work on developing energy resources that can support society and the environment over a long time. Solar, wind, and water resources are renewable, and they are cleaner than fossil fuels. Still, no option is perfect. Even renewable resources affect the environment. Scientists are always looking for new sources of energy that have few disadvantages.

The chart on the next page gives some advantages and disadvantages of different energy resources.

A worker climbs a wind turbine at an offshore wind farm.

NEXT GENERATION SCIENCE STANDARDS | DISCIPLINARY CORE IDEAS

ESS3.A: Natural Resources

Energy and fuels that humans use are derived from natural sources, and their use affects the environment in multiple ways. Some resources are renewable over time, and others are not. (4-ESS3-1)

PS3.D: Energy in Chemical Processes and Everyday Life

The expression "produce energy" typically refers to the conversion of stored energy into a desired form for practical use. (4-PS3-4)

ADVANTAGES		DISADVANTAGES
Oil is relatively inexpensive because we already have the machines to obtain and use it.		Oil is nonrenewable, and burning oil produces air pollution. Oil spills harm ecosystems.
Coal is relatively inexpensive because we already have the machines to obtain and use it.		Coal is nonrenewable, and burning coal produces air pollution. Coal mining damages the land and pollutes water in the ground.
Natural gas burns more cleanly than the other fossil fuels.		Natural gas is nonrenewable, and drilling for natural gas can pollute air and water.
Nuclear energy generates electricity from nuclear reactions without polluting the air.		Uranium is nonrenewable. Wastes from nuclear energy plants can cause harm to living things, including people.
Solar energy is renewable and does not cause pollution.		Solar fields require large amounts of land, and their use depends on the availability of sunlight.
Wind energy is renewable and does not cause pollution.		Wind farms require large amounts of land and may harm bird populations. Wind does not blow steadily in all places.
Hydroelectric energy uses moving water. It is renewable and does not cause pollution.		Hydroelectric dams destroy river habitats and may disrupt fish populations.

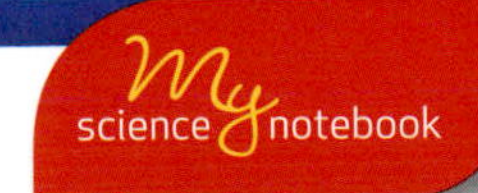

Wrap It Up!

1. **Evaluate** Which resource has the greatest disadvantage? Explain.
2. **Make Judgments** Is there such a thing as a "clean" fossil fuel? Explain.

Obtain and Combine Information

People's needs and wants change over time, as do their demands for new and improved technologies. More people live in the United States now than ever before, and our demand for electricity is greater than it ever has been. As we move toward the future, we need to look carefully at the technologies we use to produce energy. We must think about how well they meet our changing needs and how their use impacts our environment.

Study and compare the data in the graphs on the next page. The graphs contrast U.S. consumption of energy from fossil fuels and from renewable resources, such as wind, water, and sunlight. The graphs show energy consumption in units called British thermal units, BTU for short. One BTU is about equal to the amount of heat generated by a burning matchstick. A quadrillion BTUs is about equal to the energy contained in 183 million barrels of oil.

The United States uses over 18.5 million barrels of oil per day.

NEXT GENERATION SCIENCE STANDARDS | PERFORMANCE EXPECTATION
4-ESS3-1. Obtain and combine information to describe that energy and fuels are derived from natural resources and their uses affect the environment.

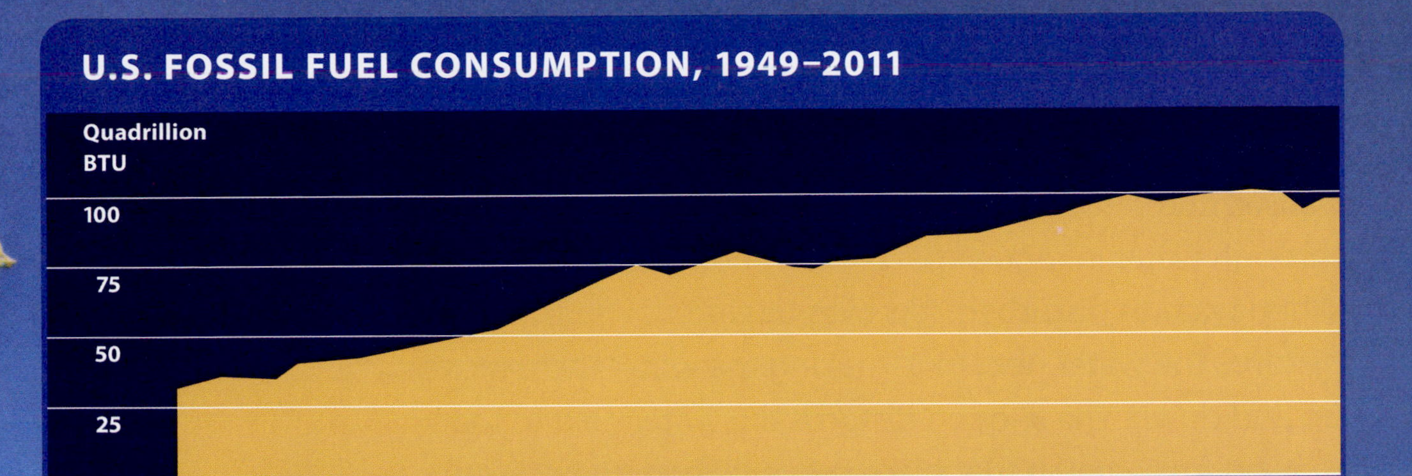

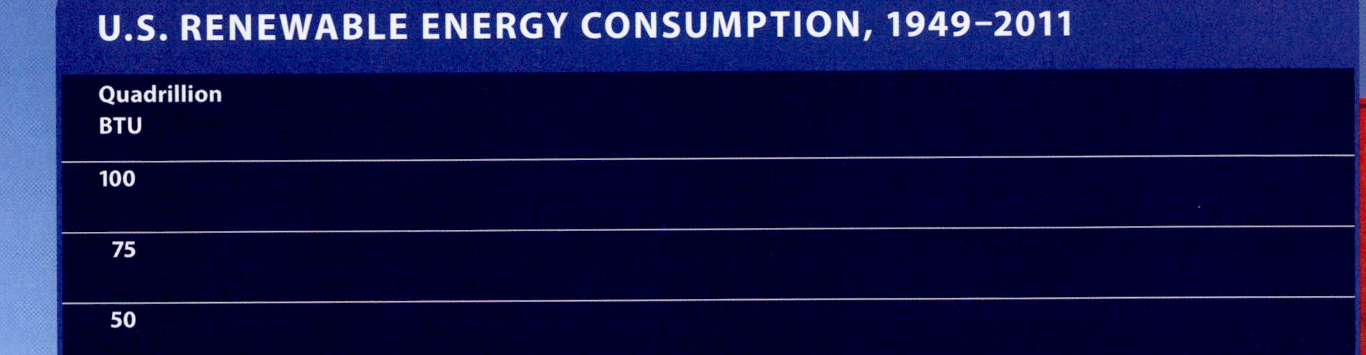

My science notebook

Wrap It Up!

1. **Explain** From where do we get energy and fuels?

2. **Infer** Reread *Energy Resources and the Environment* on the previous pages. Use that information and the graphs to infer how the current pattern of U.S. energy consumption affects the environment.

3. **Predict** What might happen if the United States continues on its current energy path? Cite evidence from the graphs and your reading in your answer.

Waves

Up and down, up and down, the surfer bobs. Energy is transferring through water beneath her. Energy travels through water in regular patterns of motion called **waves**. When water is disturbed, it moves up and down. This motion disturbs the water next to it, making the water there move up and down. This motion disturbs the water next to that, making the water in that place move up and down. Pretty soon the up-and-down motion has traveled far through the water. As each wave passes through the water beneath the surfer, the board she sits on rises and falls.

While the surfer waits for a wave that is tall enough to ride, smaller waves move her up and down without moving her forward.

NEXT GENERATION SCIENCE STANDARDS | DISCIPLINARY CORE IDEAS
PS4.A: Wave Properties
Waves, which are regular patterns of motion, can be made in water by disturbing the surface. When waves move across the surface of deep water, the water goes up and down in place; it does not move in the direction of the wave except when the water meets the beach. (4-PS4-1)

As a water wave gets close to the beach, it gets tall. Some of the water "breaks" or curls over and falls toward the front of the wave. The surfer catches a ride by sliding down the steep front of the wave.

Wrap It Up!

1. **Describe** Describe how a wave travels across the ocean.

2. **Explain** Why doesn't the surfer move forward until she is atop a tall wave?

Wave Properties

Waves spread out in all directions from their source. If you have ever dropped a pebble into a pond and watched the ripples, you have seen a wave pattern.

When you drop a pebble into a pond, it makes small waves. A larger rock makes larger waves. The sizes of waves differ in height and in the distance between their highest points.

These water waves are not very tall or far apart. In other words, they have a low amplitude and a short wavelength.

NEXT GENERATION SCIENCE STANDARDS | DISCIPLINARY CORE IDEAS
PS4.A: Wave Properties
Waves of the same type can differ in amplitude (height of the wave) and wavelength (spacing between wave peaks). (4-PS4-1)

The distance between the highest point, or crest, of a wave and the middle point of the wave is called **amplitude.** Amplitude can also be measured from the lowest point, or trough, of the wave and the middle point of the wave.

Amplitude is one property of waves. **Wavelength** is another. Wavelength is the distance from one crest to the next crest, or one trough to the next trough.

Amplitude measures the height of crests and depth of troughs. Wavelength measures how far apart they are.

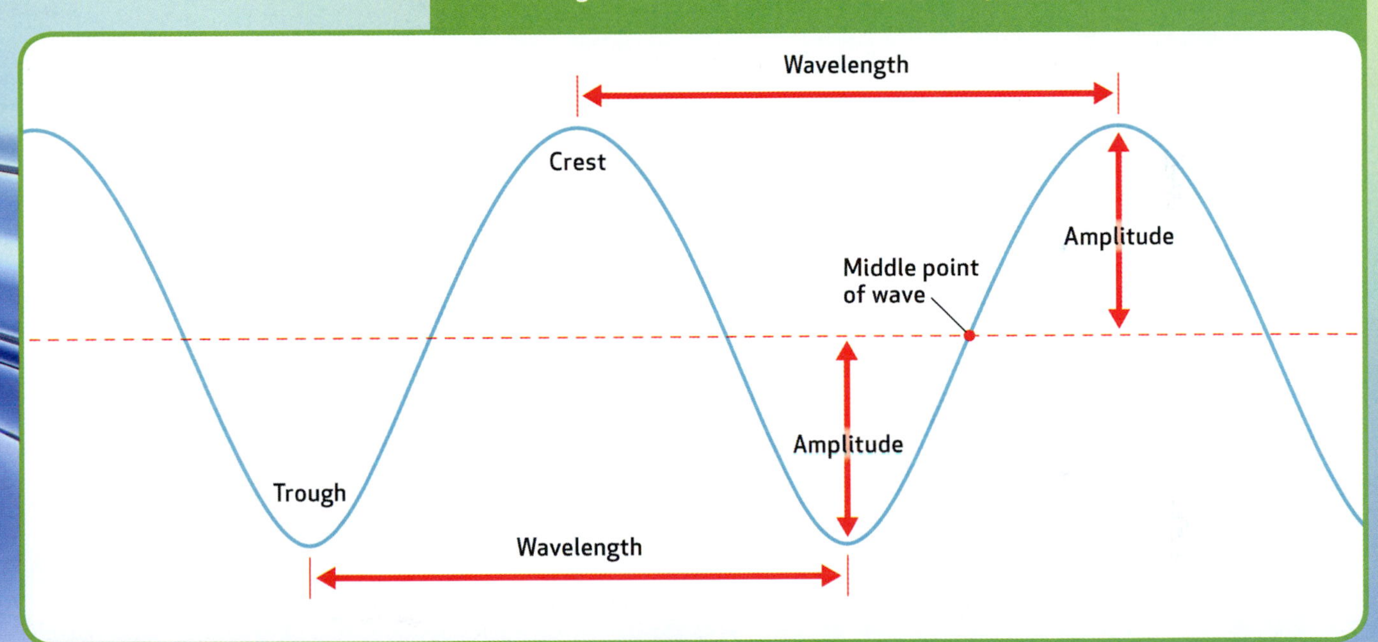

Wrap It Up!

1. **Recall** What are amplitude and wavelength?

2. **Infer** Waves transfer energy of motion. Which transfers more energy: a wave with a high amplitude or a wave with a low amplitude? Explain.

Investigate

Wavelength and Amplitude

? How can you model wavelength and amplitude?

Did you know that sound is a wave? Like all waves, sound waves have the properties of wavelength and amplitude. The wavelength of a sound wave determines how high or low it sounds. The amplitude of a sound wave determines its loudness.

Make these sounds as you draw the waves. A dog growl is a low-pitched, quiet sound. A diagram of a dog growl has a long wavelength and a low amplitude. A bird chirp is a high-pitched, loud sound. A diagram of a bird chirp has a short wavelength and a high amplitude.

In this investigation, you will use wire to model waves with different wavelengths and amplitudes.

Materials

chenille stem	marker

NEXT GENERATION SCIENCE STANDARDS | PERFORMANCE EXPECTATION
4-PS4-1. Develop a model of waves to describe patterns in terms of amplitude and wavelength and that waves can cause objects to move.

1 Use the chenille stem to make a model of a wave. Describe its wavelength and amplitude. For example, it may have a long wavelength and low amplitude.

2 Draw and label a picture of your wave.

3 Keeping the amplitude the same, change the wavelength of your wave. Draw and label a picture of your wave.

4 Make a new wave with different properties. Draw and label your wave.

Wrap It Up!

1. **Describe** Describe the properties of each wave you modeled.
2. **Analyze** How can two waves with the same wavelength differ?

Investigate

How Waves Move Objects

How can you model how waves cause objects to move?

For some people, the rise and fall of a boat's floor beneath their feet is enough to lull them to sleep. For others, it's enough to make them sick! In this investigation, you'll explore how waves transfer energy to objects, causing them to move.

NEXT GENERATION SCIENCE STANDARDS | PERFORMANCE EXPECTATION

4-PS4-1. Develop a model of waves to describe patterns in terms of amplitude and wavelength and that waves can cause objects to move.

1 Fill the pan half full with water. Dip the short edge of the card down into the water at an angle. Keeping the card in the water, gently wave the card up and down to produce a wave.

2 Select an object and place it in the water. Use the card to make a wave. Observe the wave and the motion of the object. Record your observations. Remove the object.

3 Now use the card to make a wave with a lower amplitude. Record your observations and what you did to make the wave.

4 Now use the card to make a wave with a shorter wavelength. Record your observations and what you did to make the wave.

Wrap It Up!

1. **Relate** How did your motions relate to the characteristics of each wave you made?
2. **Conclude** How did your model show how waves can cause objects to move?

Information Technology—GPS

When you look at a map on a phone, you can zoom in and out and even see views from the street. All of this is possible because the information has been put into digital code form, or **digitized.**

Digital maps usually have a little dot showing exactly where you are. How does the phone know where you are? The phone uses **GPS.** GPS stands for **Global Positioning System.** This system can locate anything with a GPS receiver.

Satellites in space **transmit,** or send, radio signals to a receiver on Earth, such as a smartphone. The phone receives the signals and uses them to pinpoint a location on a map.

NEXT GENERATION SCIENCE STANDARDS | DISCIPLINARY CORE IDEAS
PS4.C: Information Technologies and Instrumentation
Digitized information transmitted over long distances without significant degradation. High-tech devices, such as computers or cell phones, can receive and decode information—convert it from digitized form to voice—and vice versa. (4-PS4-3)

GPS can be very useful in navigation, or finding your way.

Wrap It Up!

My science notebook

1. **Restate** What is GPS?
2. **Apply** List as many ways as you can think of to transmit a message. Which of those do you think use digital code?

Information Technology—Cell Phones

Radio signals, such as the ones used in GPS, can transmit over long distances. But radio signals aren't always stable. If you have ever tried talking to a friend with a radio walkie-talkie, you have probably noticed your friend's voice wasn't very clear.

A better way to communicate over long distances is with digital code. When you talk into a cell phone, the phone converts the sound of your voice into digitized form before transmitting it as a radio signal.

The radio signal is picked up by the nearest cell phone tower in your network. The tower redirects the signal through the network to your friend's phone. Your friend's phone converts the digitized information back into the sound of your voice.

Cell phones use digital code to maintain the sound quality of conversations over long distances.

NEXT GENERATION SCIENCE STANDARDS | DISCIPLINARY CORE IDEAS
PS4.C: Information Technologies and Instrumentation
Digitized information transmitted over long distances without significant degradation. High-tech devices, such as computers or cell phones, can receive and decode information—convert it from digitized form to voice—and vice versa. (4-PS4-3)

Digital information remains stable over long distances. Digital information is coded using a pattern of 0s and 1s.

Wrap It Up!

My science notebook

1. **Sequence** Put the following sentences in the correct sequence: The digital information is transmitted through a radio network to the other phone. The phone converts the digital signal into sound. The phone converts sound into a digital signal.

2. **Compare** How do radio signals compare with digital signals for transmitting information over long distances?

Investigate

Use a Code

How can you use patterns to transmit information?

People have been using codes to transmit information over long distances for years. In the 1800s, Samuel Morse invented a code that uses long and short signals. Originally transmitted with a telegraph, the dots and dashes represented faster or slower clicking signals. Morse code can also be transmitted with light using faster or slower flashes.

In this investigation, you can transmit your own message to a partner using Morse code.

MORSE CODE

A	• —	U	• • —
B	— • • •	V	• • • —
C	— • — •	W	• — —
D	— • •	X	— • • —
E	•	Y	— • — —
F	• • — •	Z	— — • •
G	— — •		
H	• • • •		
I	• •		
J	• — — —		
K	— • —	1	• — — — —
L	• — • •	2	• • — — —
M	— —	3	• • • — —
N	— •	4	• • • • —
O	— — —	5	• • • • •
P	• — — •	6	— • • • •
Q	— — • —	7	— — • • •
R	• — •	8	— — — • •
S	• • •	9	— — — — •
T	—	0	— — — — —

Each character in a word is represented by a series of dots and dashes. This is Morse code for SOS:

• • • — — — • • •

It stands for "save our ship" and was used by boats to signal the need for help. Now its use is not limited to boats.

Materials

flashlight

NEXT GENERATION SCIENCE STANDARDS | PERFORMANCE EXPECTATION

4-PS4-3. Generate and compare multiple solutions that use patterns to transfer information.

1 Study the Morse code key. Think of a message that is three to five words long. Record your message. Do not show your partner.

2 Use the flashlight to transmit your coded message to a partner. Remember to leave "spaces" between letters, and longer spaces between words.

3 Have your partner record your code and use the key to decode the message.

4 Compare the message you sent to the message your partner received. Record your observations.

To use a telegraph to communicate, both the sender and receiver must understand the code.

Wrap It Up!

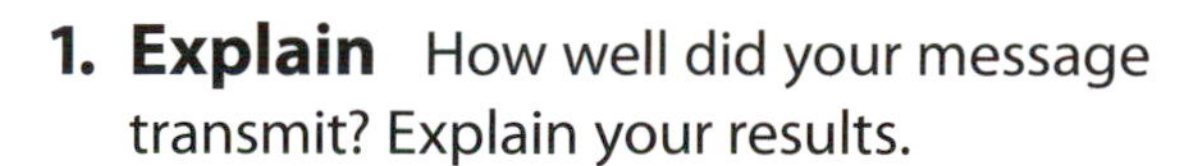

1. **Explain** How well did your message transmit? Explain your results.
2. **Compare** How does Morse code compare with digital code as a way of transmitting information?

Compare Multiple Solutions

You can find a bar code on the package of nearly every grocery item in your kitchen. Many smartphones have apps that translate bar codes into price information. Codes are all around you! But all codes don't require electronic devices to decode them.

You can invent your own code. Imagine that you are at summer camp and want to communicate with your friend in the next cabin, only a few feet away. Cell phones are not allowed. You both know how to use Morse code, but flashlights won't work because you can't see into each other's cabin. It's up to you to devise a way to communicate with each other.

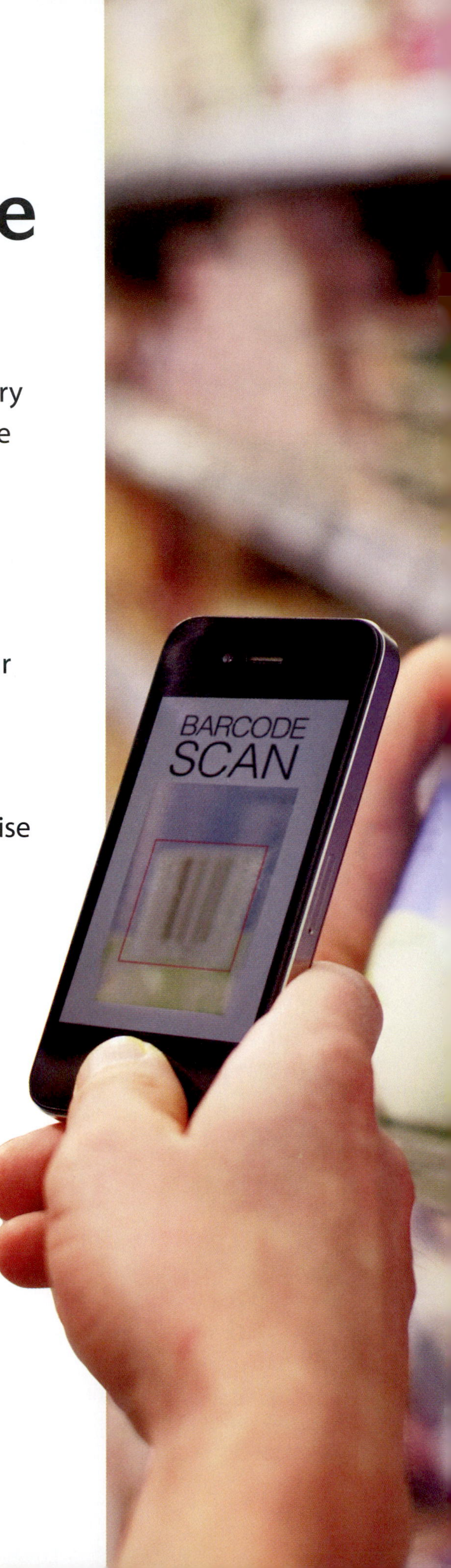

Bar codes contain information that help stores keep track of the products they sell.

NEXT GENERATION SCIENCE STANDARDS | DISCIPLINARY CORE IDEAS

ETS1.C: Optimizing The Design Solution

Different solutions need to be tested in order to determine which of them best solves the problem, given the criteria and the constraints. (secondary to 4-PS4-3)

PERFORMANCE EXPECTATION

4-PS4-3. Generate and compare multiple solutions that use patterns to transfer information.

1. **Define the problem.** 

   What problem needs to be solved? List the criteria for a successful solution. What constraints limit your choices?

2. **Find a solution.**
   Study the materials available to you. Select the materials you think would work best for transmitting Morse code. Plan how you will test your idea. How will you simulate being in separate cabins? What message will you use to test your solution? What observations will you make? Record all your ideas in your science notebook.

3. **Test your solution.**
   What observations will you make? Once your teacher has approved your plan and you have what you need, carry out your test. Record your observations.

4. **Refine or change your solution.**
   Analyze your results. Did your solution work? Does it allow you to communicate clearly? How could your solution be further improved? Revise your solution and test it again.

5. **Share your solution.**
   Present your information technology solution to the class. Evaluate your classmates' solutions using the same criteria you used to evaluate your own. How many different ways did your class come up with to communicate? Which do you think is most effective?

Animal Tracker

Have you ever seen a group of birds flying together in the same direction and wondered where they were going? Martin Wikelski investigates how and where animals migrate. Sometimes animals travel thousands of miles to meet their basic needs. Wikelski believes that people could learn a lot from studying these patterns of mass movement.

How does Wikelski observe animal movement? He follows them! Wikelski tracks animals either in person or by using communication technology. In one study, he attached tiny radio transmitters to individual sparrows. Using receivers on airplanes, Wikelski and his team tracked the sparrows' movements. They found that the adult migrating birds can find their way even after winds blow them thousands of miles off course.

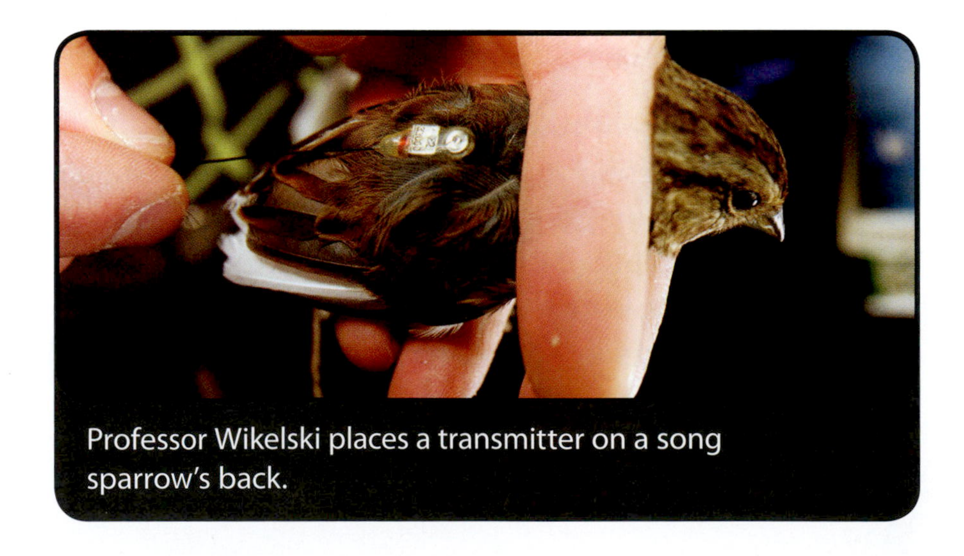

Professor Wikelski places a transmitter on a song sparrow's back.

This radio transmitter is so tiny it can be glued to the back of a tropical orchid bee.

NEXT GENERATION SCIENCE STANDARDS | CONNECTIONS TO NATURE OF SCIENCE
Scientific Knowledge Is Based on Empirical Evidence
Scientists look for patterns and order when making observations about the world.

NATIONAL GEOGRAPHIC | **Explorer**

Martin Wikelski is a behavioral ecologist who studies patterns of animal movement. Wikelski has followed birds from Europe to Africa. He has flown with ducks from Sweden to Germany. And he has ridden a motorcycle across the Alps on the trail of songbirds. In addition to birds, he has tracked bats, giant tortoises, bumblebees, and dragonflies.

Life Science

Structure, Function, and Information Processing

The Borneo orangutan finds fruit for its next meal.

External Structures of a Wild Rose

Have you ever seen a wild rose plant like the one in the photo? The rose is a type of plant that produces flowers. Its flowers are beautiful, but they are also important to the plant. Like all plants, the wild rose is made up of different kinds of structures. Its external structures are the parts that you can see on the outside of the plant. They include leaves, stems, roots, and flowers.

Leaves, stems, and roots have important functions in the growth and survival of the plant. The wild rose plant also has structures that allow it to reproduce—its flowers. Flowers produce seeds, which can grow into new plants.

Both the color and scent of the wild rose plant's flowers attract insects.

NEXT GENERATION SCIENCE STANDARDS | DISCIPLINARY CORE IDEAS
LS1.A: Structure and Function
Plants and animals have both internal and external structures that serve various functions in growth, survival, behavior, and reproduction. (4-LS1-1)

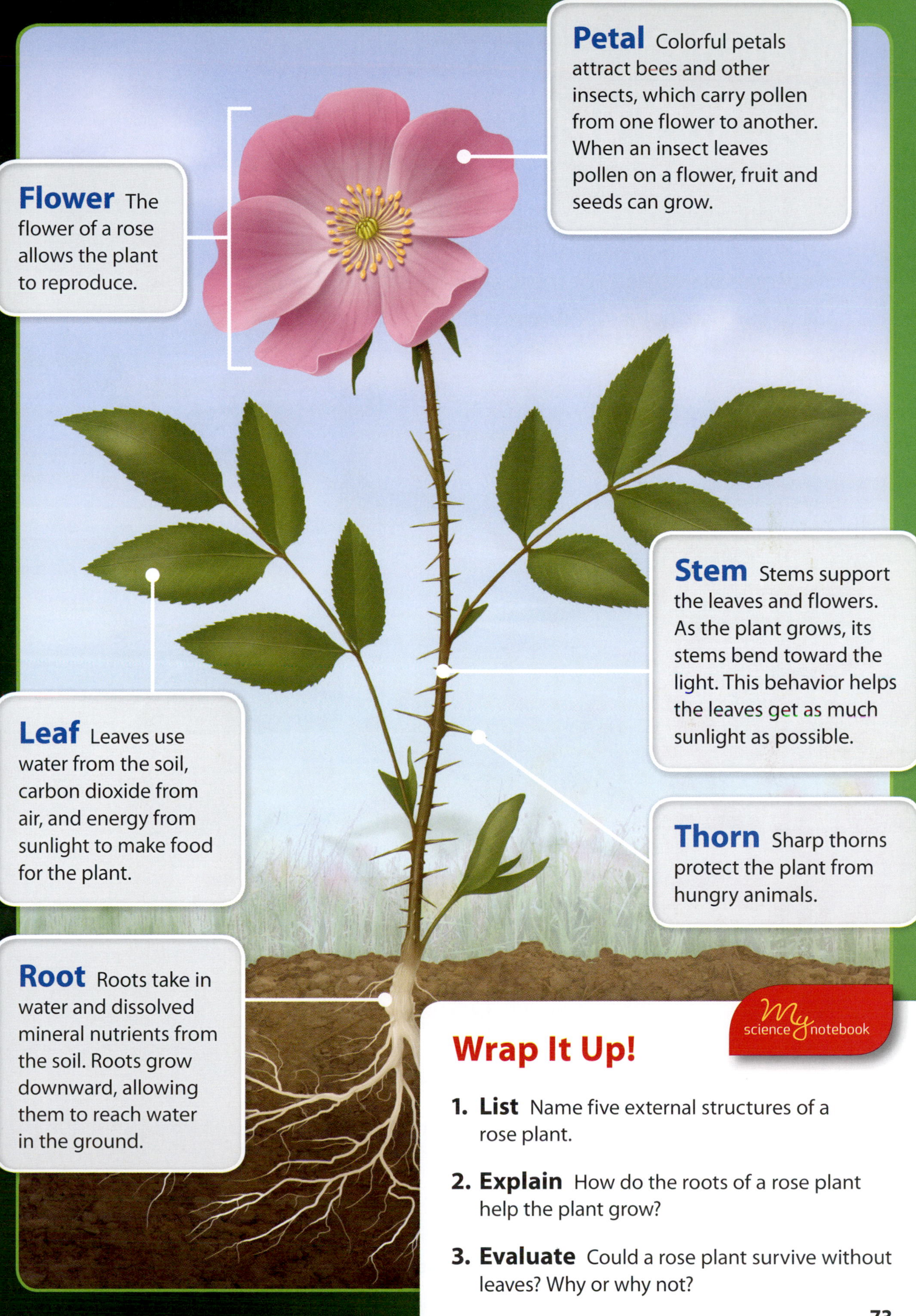

My science notebook

Wrap It Up!

1. **List** Name five external structures of a rose plant.
2. **Explain** How do the roots of a rose plant help the plant grow?
3. **Evaluate** Could a rose plant survive without leaves? Why or why not?

Internal Structures of a Wild Rose

Plants have internal structures that help them grow, survive, and reproduce. These structures exist inside the plant. Many of these structures are hard to see without a magnifying glass or microscope.

The many stamens of this wild rose flower surround the pistil in the center.

NEXT GENERATION SCIENCE STANDARDS | DISCIPLINARY CORE IDEAS
LS1.A: Structure and Function
Plants and animals have both internal and external structures that serve various functions in growth, survival, behavior, and reproduction. (4-LS1-1)

Flower In the center of the flower are the stamens and pistil. **Stamens** make pollen. For fruit to develop, pollen must be transferred to the pistil. Then the **pistil** develops into a fruit with seeds inside. Each seed can grow into a new plant.

Stem Inside each stem are bundles of tiny tubes. Some tubes carry water from the roots up to the leaves and flowers. Other tubes carry food from the leaves to the rest of the plant.

Roots Tiny hairs on the roots take in water and mineral nutrients from the soil.

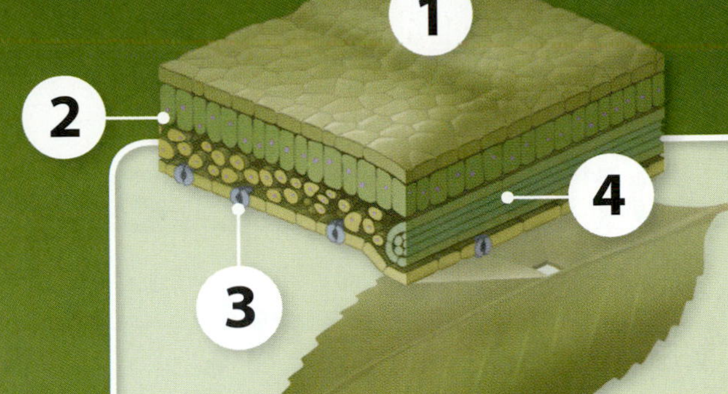

Leaf Leaves use water from the soil, carbon dioxide from the air, and energy from sunlight to make food for the plant. Leaves are made up of several parts.

1 The outer layer protects the leaf and keeps it from drying out.
2 In the middle is the food-making layer.
3 Openings in the bottom of the leaf let air into the food-making layer.
4 Veins are made up of tiny tubes. Some tubes carry water to the leaf. Other tubes carry food from the leaves to the rest of the plant.

Wrap It Up!

1. **Identify** What are the structures of a leaf? What are their functions?
2. **Cause and Effect** How do the stamens of a flower help a plant reproduce?
3. **Analyze** Are stems necessary to the survival of a plant? Why?

Construct an Argument

A buttercup plant and a wild rose look very different. But a buttercup, like a wild rose, has external and internal structures that help it survive.

Use sticky notes to label the different external and internal structures of the buttercup plant. Then answer the questions below.

1. **List.** 

What external structures did you label? What internal structures did you label?

2. **Compare.**
Work with a group. Compare your labels. Work together until everyone in your group has all the buttercup plant's structures labeled. Then compare the buttercup plant's structures to the wild rose's structures on pages 72–75.

3. **Construct an argument.**
Have each person in the group choose one labeled external structure and one labeled internal structure. Write an explanation arguing how, as with the wild rose, these structures help the buttercup plant grow, survive, or reproduce.

4. **Generalize.**
Come back together as a group. Present your arguments for the structures you labeled. Together, write a summary explaining how the structures of the buttercup plant and the wild rose help them survive.

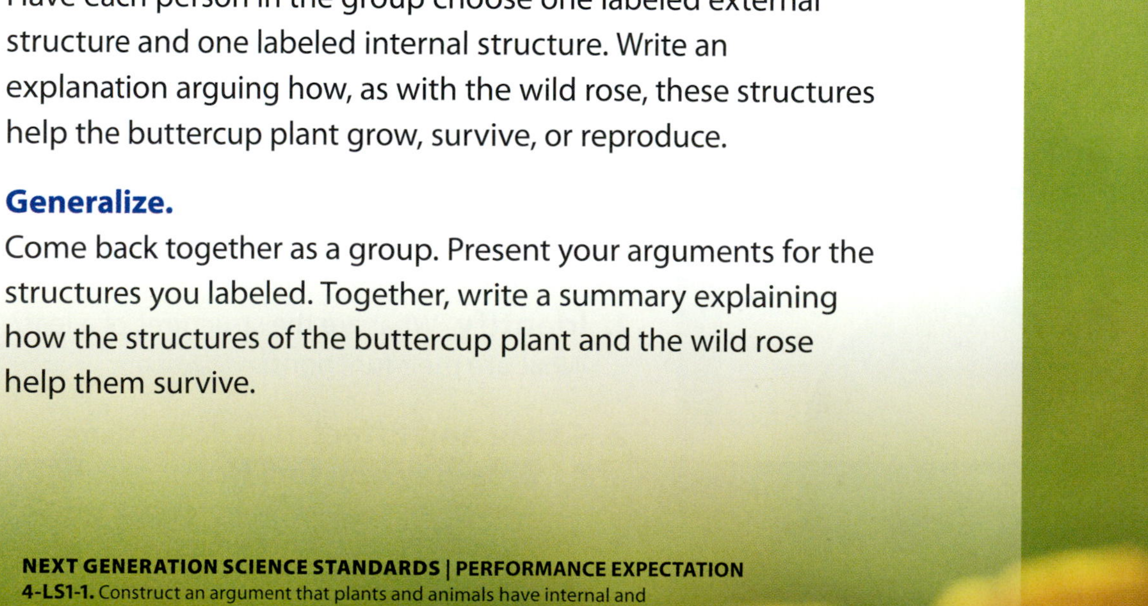

NEXT GENERATION SCIENCE STANDARDS | PERFORMANCE EXPECTATION
4-LS1-1. Construct an argument that plants and animals have internal and external structures that function to support survival, growth, behavior, and reproduction.

External Structures of an Elephant

What animal uses its nose to put food in its mouth and its feet to sense sound? An elephant! An elephant's body is made up of many different structures that allow it to grow, survive, and respond to its surroundings. The photo shows some of the external structures of an Asian elephant.

The Asian elephant is the biggest animal in Asia. To get enough energy to survive, an elephant needs to eat an enormous amount of food. No wonder elephants spend most of their time looking for food. What do they eat? Grasses, leaves, roots, bark, and fruit.

Mother elephants and their young travel in herds. These animals use a wide variety of sounds to communicate—trumpeting, roaring, snorting, grunting, and barking. They also make rumbling sounds that are too low for humans to hear. These low sounds can travel through the ground for long distances—as far as 32 kilometers (20 miles).

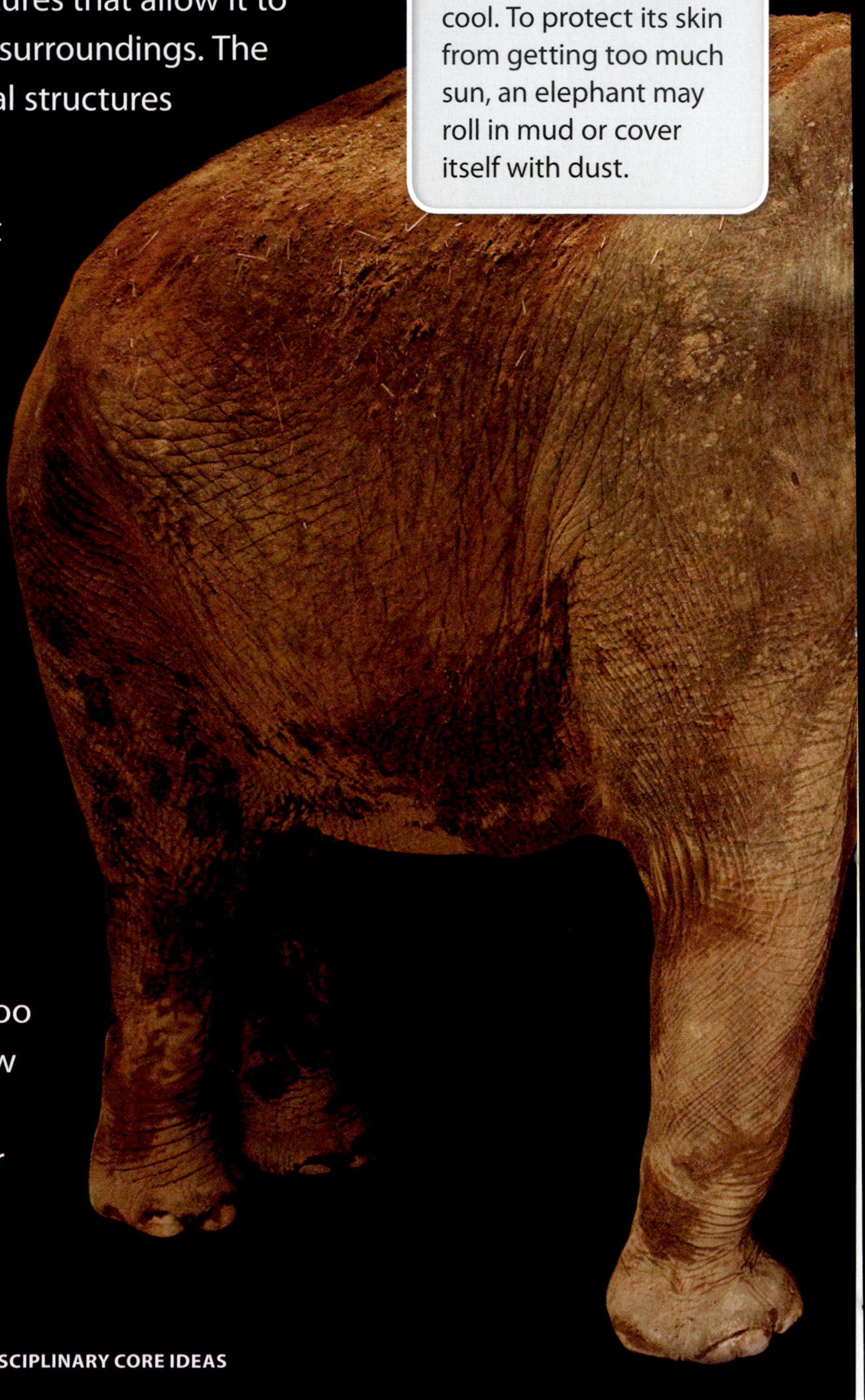

Skin Tough, wrinkled skin protects the elephant's internal organs. The skin also keeps the elephant cool. To protect its skin from getting too much sun, an elephant may roll in mud or cover itself with dust.

NEXT GENERATION SCIENCE STANDARDS | DISCIPLINARY CORE IDEAS
LS1.A: Structure and Function
Plants and animals have both internal and external structures that serve various functions in growth, survival, behavior, and reproduction. (4-LS1-1)

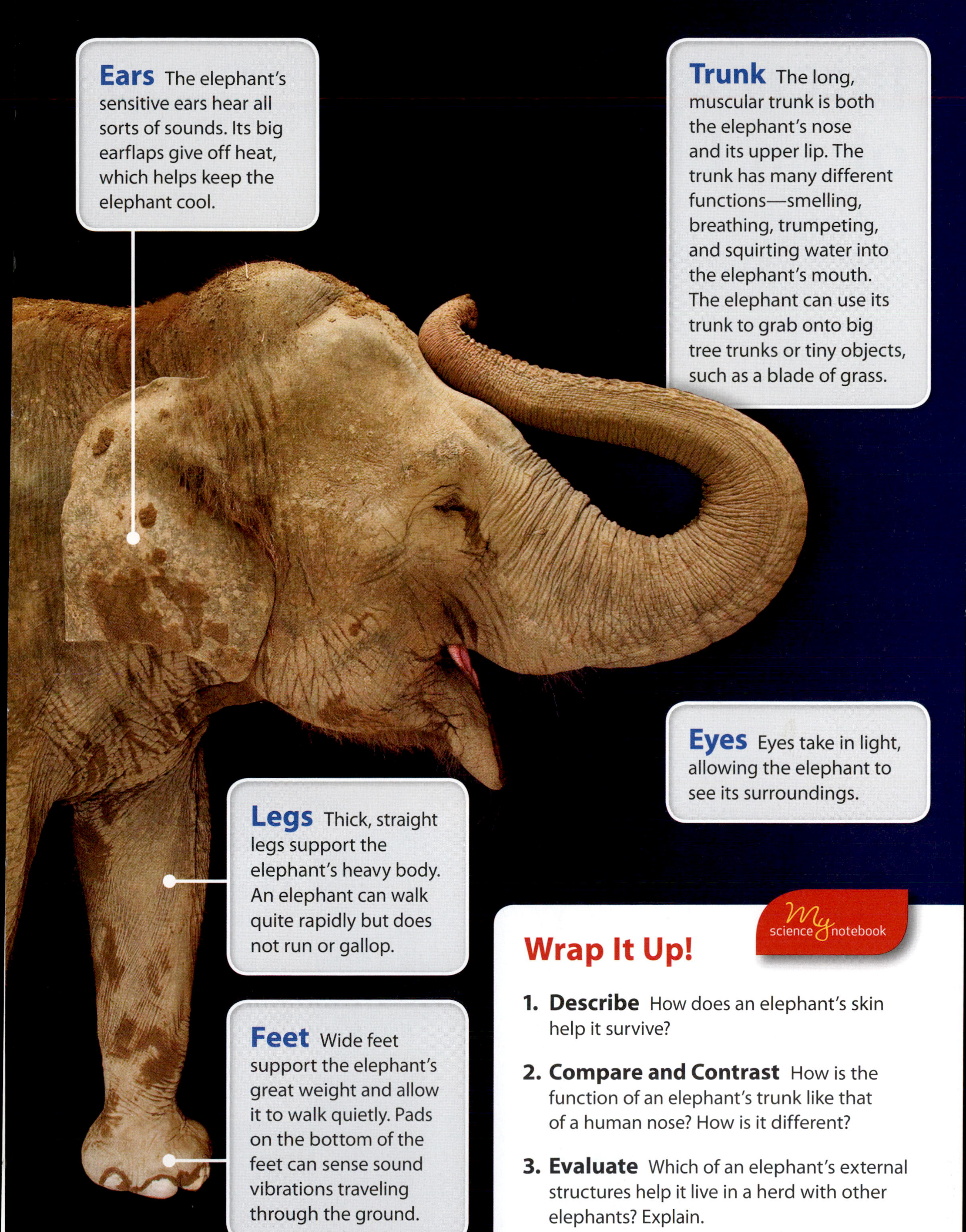

Wrap It Up!

My science notebook

1. **Describe** How does an elephant's skin help it survive?
2. **Compare and Contrast** How is the function of an elephant's trunk like that of a human nose? How is it different?
3. **Evaluate** Which of an elephant's external structures help it live in a herd with other elephants? Explain.

Internal Organs of an Elephant

An elephant's internal organs serve various functions. They allow the elephant to grow and survive. These functions include providing the elephant's body with food, water, and oxygen. All of the living parts of the elephant's body require these materials in order to survive.

Lungs The lungs take in oxygen from the air and release carbon dioxide. Blood traveling through the lungs picks up oxygen.

Stomach The large stomach stores food and begins the process of breaking it down. Food then travels to the intestine.

Intestines Most of an elephant's food is digested in the small intestine. Bacteria at the end of the small intestine help break down the food. Sugars and other chemicals from food are taken up by blood in the intestine walls. Undigested food moves from the small intestine to the large intestine.

Liver The liver produces many chemicals that are necessary for the functions of an elephant's body. For example, the liver produces bile. Bile helps break down fats during the process of digestion.

NEXT GENERATION SCIENCE STANDARDS | DISCIPLINARY CORE IDEAS
LS1.A: Structure and Function
Plants and animals have both internal and external structures that serve various functions in growth, survival, behavior, and reproduction. (4-LS1-1)

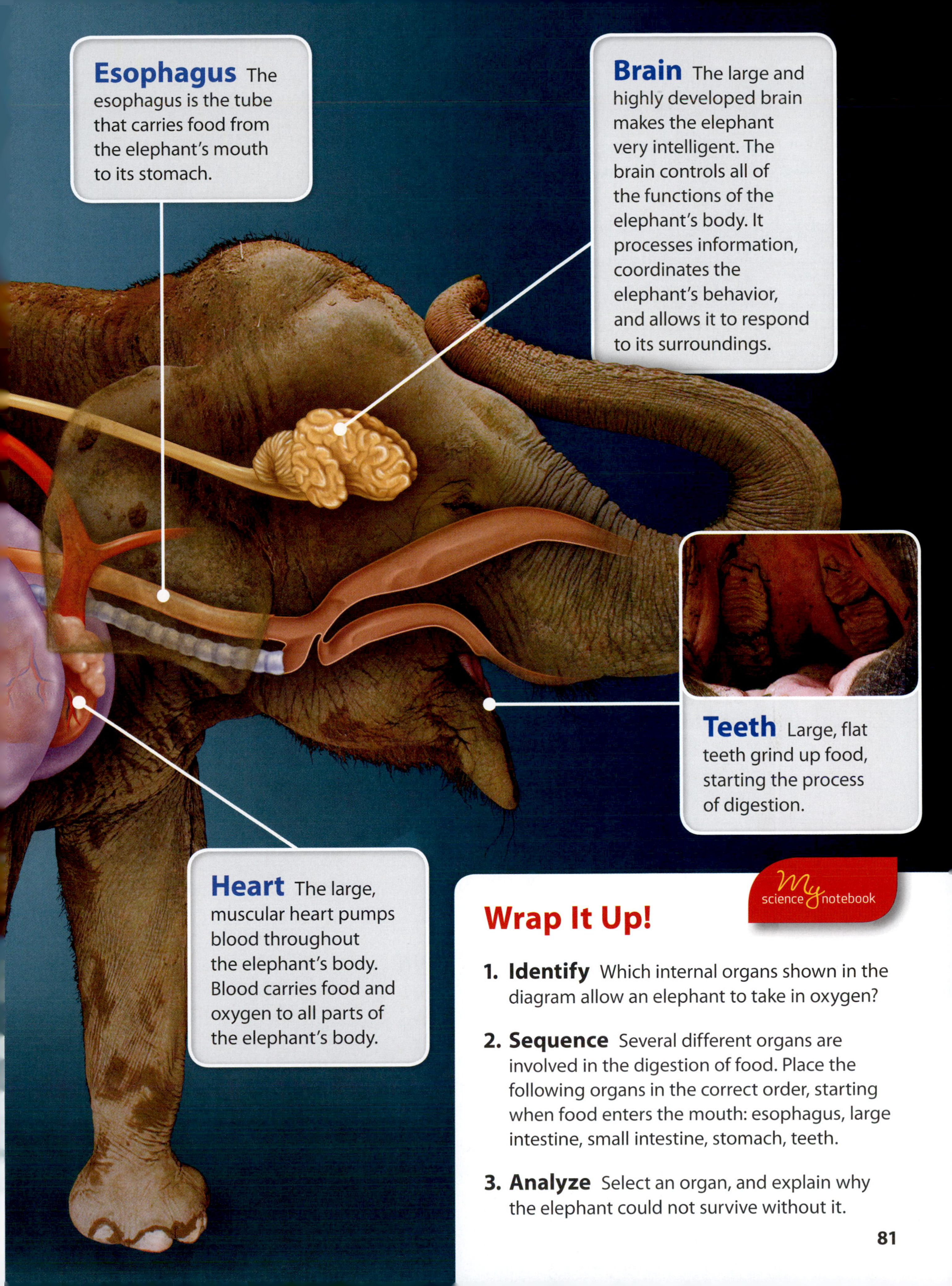

Wrap It Up!

1. **Identify** Which internal organs shown in the diagram allow an elephant to take in oxygen?

2. **Sequence** Several different organs are involved in the digestion of food. Place the following organs in the correct order, starting when food enters the mouth: esophagus, large intestine, small intestine, stomach, teeth.

3. **Analyze** Select an organ, and explain why the elephant could not survive without it.

Bones and Muscles of an Elephant

The internal structures of an elephant include its bones and muscles. Bones support the elephant's body and protect its internal organs. All of the bones in an elephant make up its skeleton. Muscles are attached to bones and work with the bones to move parts of the elephant's body.

Ribs The ribs protect the heart and lungs.

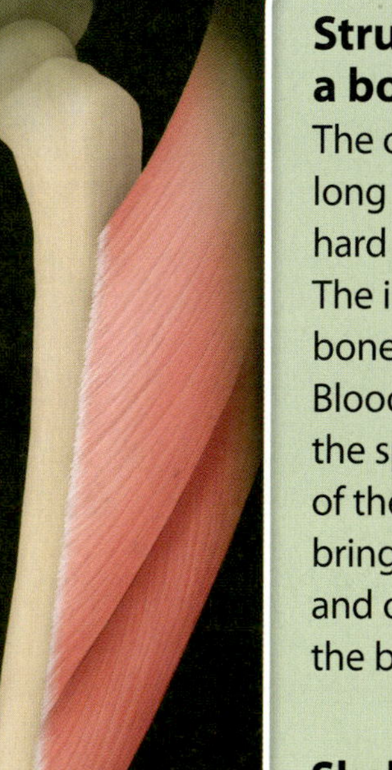

Structure of a bone
The outside of a long leg bone is hard and compact. The inside of the bone is spongy. Blood vessels in the spongy part of the bone bring nutrients and oxygen to the bone.

Skeletal muscles
Skeletal muscles are attached to bones. When a skeletal muscle contracts, it pulls on the bone and makes it move.

Pelvis The bones of the pelvis are beneath the muscles. The pelvis provides a frame that supports the back legs. Joints in the pelvis allow the elephant to move its legs so it can walk or swim.

NEXT GENERATION SCIENCE STANDARDS | DISCIPLINARY CORE IDEAS
LS1.A: Structure and Function
Plants and animals have both internal and external structures that serve various functions in growth, survival, behavior, and reproduction. (4-LS1-1)

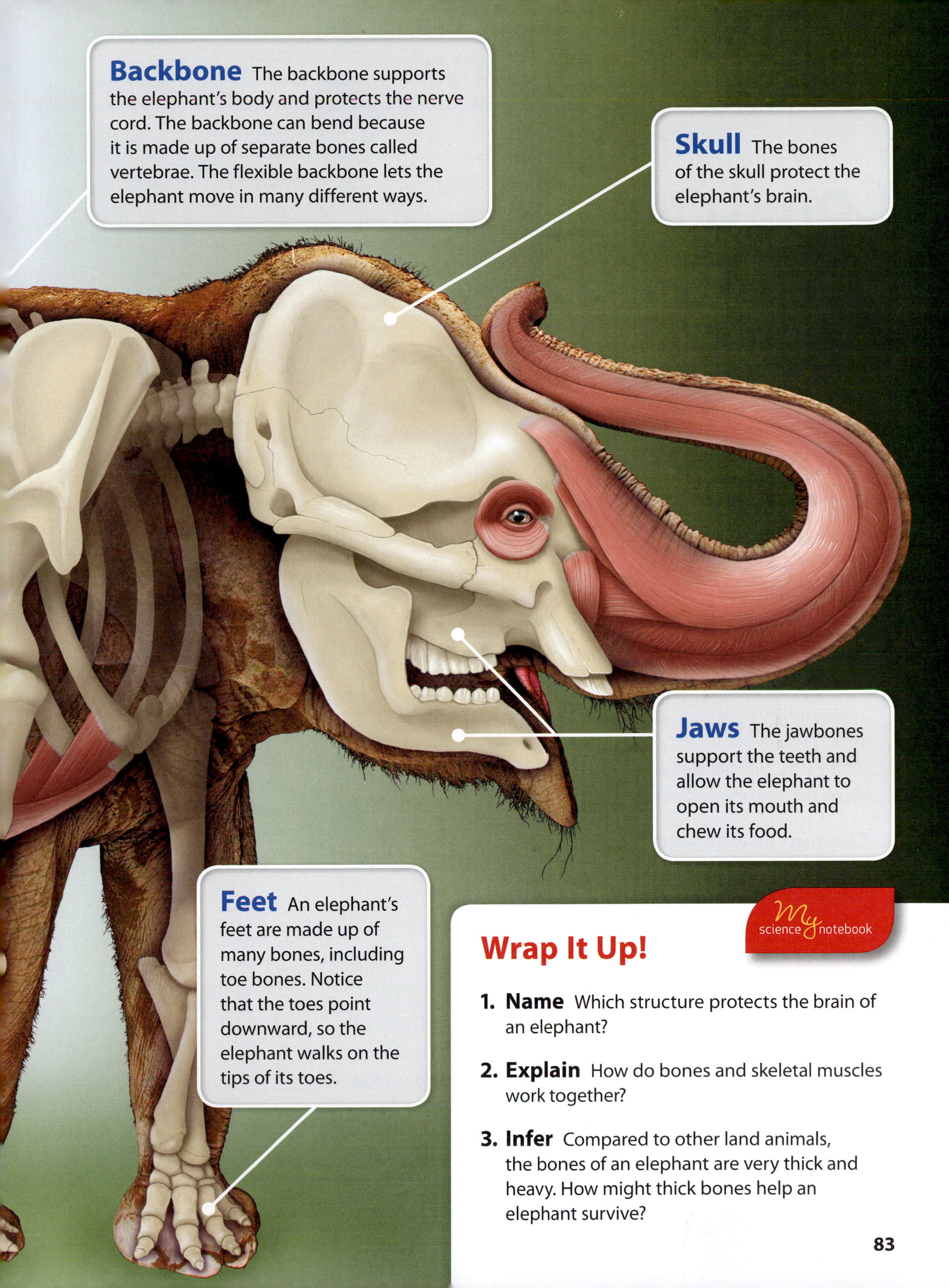

Wrap It Up!

1. **Name** Which structure protects the brain of an elephant?
2. **Explain** How do bones and skeletal muscles work together?
3. **Infer** Compared to other land animals, the bones of an elephant are very thick and heavy. How might thick bones help an elephant survive?

Construct an Argument

A wolf and an elephant look very different. But a wolf, like an elephant, has external and internal structures that help it survive.

Use sticky notes to label the different internal and external structures of the wolf. Then answer the questions below.

1. **List.** 

   What external structures did you label? What internal structures did you label?

2. **Compare.**
   Work with a group. Compare your labels. Did you all label the same structures? Work together until everyone in your group has all the wolf's structures labeled. Then compare the wolf's structures to the elephant's structures on pages 78–83.

3. **Construct an argument.**
   Have each person in the group choose one labeled external structure and one labeled internal structure. Write an explanation arguing how, as with an elephant, these structures help the animal grow, survive, behave in certain ways, or reproduce.

4. **Generalize.**
   Come back together as a group. Present your arguments for the structures you labeled. Together, write a summary explaining how the structures of the wolf and the elephant help them survive.

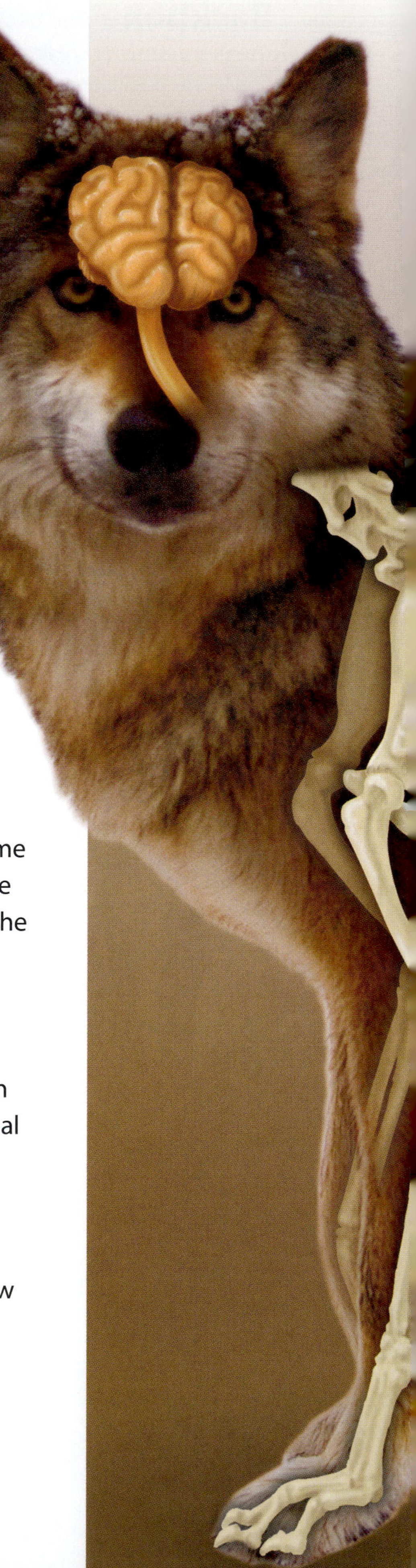

NEXT GENERATION SCIENCE STANDARDS | PERFORMANCE EXPECTATION
4-LS1-1. Construct an argument that plants and animals have internal and external structures that function to support survival, growth, behavior, and reproduction.

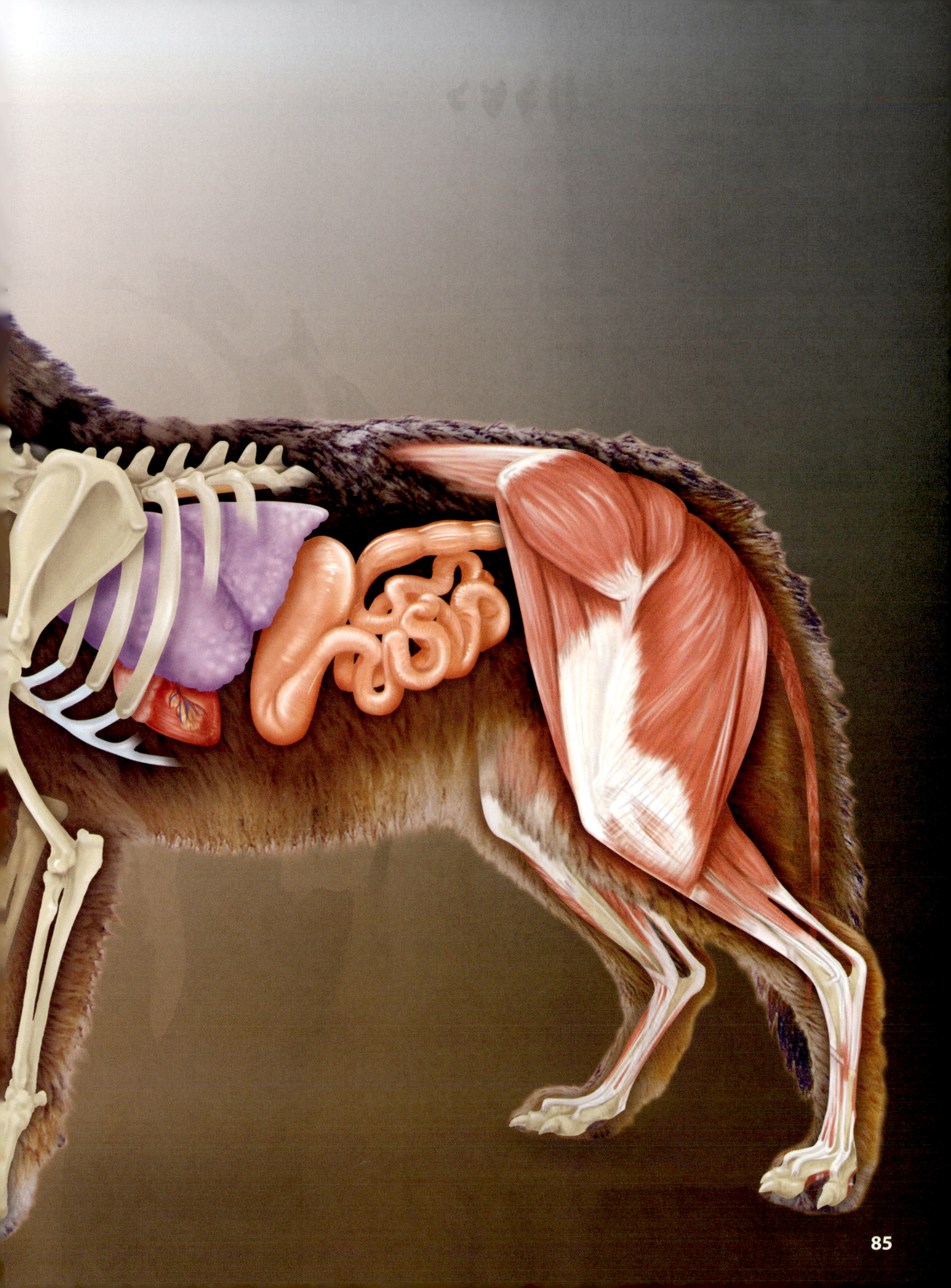

Animal Senses

The clouded leopard is a fierce cat that lives in the forests of Southeast Asia. It uses its senses to learn about its surroundings. Each sense receptor responds to a particular kind of information and sends signals to the brain. The brain processes those signals so they have meaning for the cat.

The clouded leopard uses its keen sense of hearing to learn when its predators and prey are nearby. For example, the sounds of a ground squirrel's movements travel through the air as vibrations. When those vibrations enter the cat's ears, sound receptors send signals to the cat's brain. The brain interprets those signals as the noises made by a ground squirrel. The cat uses that perception and its memories of hunting other animals to catch the ground squirrel. Read the captions to find out how the clouded leopard's other senses aid it in processing information.

NEXT GENERATION SCIENCE STANDARDS | DISCIPLINARY CORE IDEAS
LS1.D: Information Processing
Different sense receptors are specialized for particular kinds of information, which may be then processed by the animal's brain. Animals are able to use their perceptions and memories to guide their actions. (4-LS1-2)

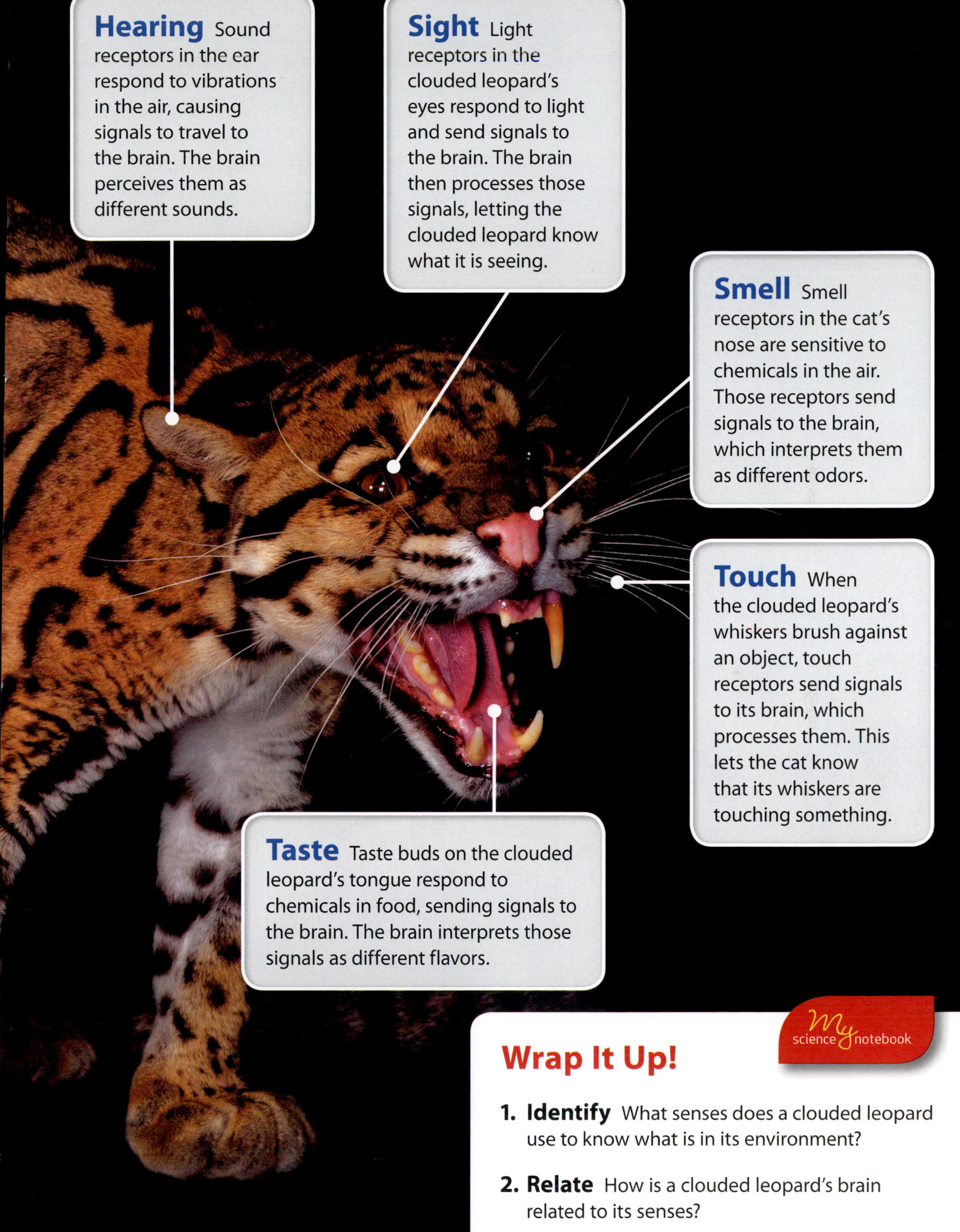

Wrap It Up!

My science notebook

1. **Identify** What senses does a clouded leopard use to know what is in its environment?

2. **Relate** How is a clouded leopard's brain related to its senses?

Light and Sight

How does a clouded leopard see a ground squirrel? First, sunlight **reflects** from, or bounces off, the squirrel. The light travels through the air. When the light enters the cat's eye, it hits light receptors at the back of the eyeball. Those receptors send signals to the brain. The brain processes the signals, so the clouded leopard understands that it is seeing a ground squirrel.

NEXT GENERATION SCIENCE STANDARDS | DISCIPLINARY CORE IDEAS
PS4.B: Electromagnetic Radiation
An object can be seen when light reflected from its surface enters the eyes. (4-PS4-2)

Trace the path of light from the sun to the leopard's eyes. The leopard's brain interprets signals from its eyes. Then the cat can pounce on its prey.

Clouded leopards have good eyesight. Their forward-facing eyes allow them to judge distances as they climb trees and hunt prey.

Wrap It Up!

1. **Define** What does the word *reflect* mean?
2. **Explain** How does light from the sun allow an animal to see?
3. **Infer** Clouded leopards often hunt at night. How do you think they are able to see objects at night?

Investigate

How We See

How can you model the idea that light allows objects to be seen?

Most objects do not give off their own light. You can only see such objects when light from another source bounces off of them and enters your eyes. Buildings are full of lamps and other light sources that enable us to see objects in spaces that sunlight does not reach. In this activity, you will explore how you see objects using a light source.

Materials

flashlight

classroom objects

shoebox

NEXT GENERATION SCIENCE STANDARDS | PERFORMANCE EXPECTATION

4-PS4-2. Develop a model to describe that light reflecting from objects and entering the eye allows objects to be seen.

1 Your teacher will give your group a shoebox with an object inside it. Then your teacher will turn out the lights in the room.

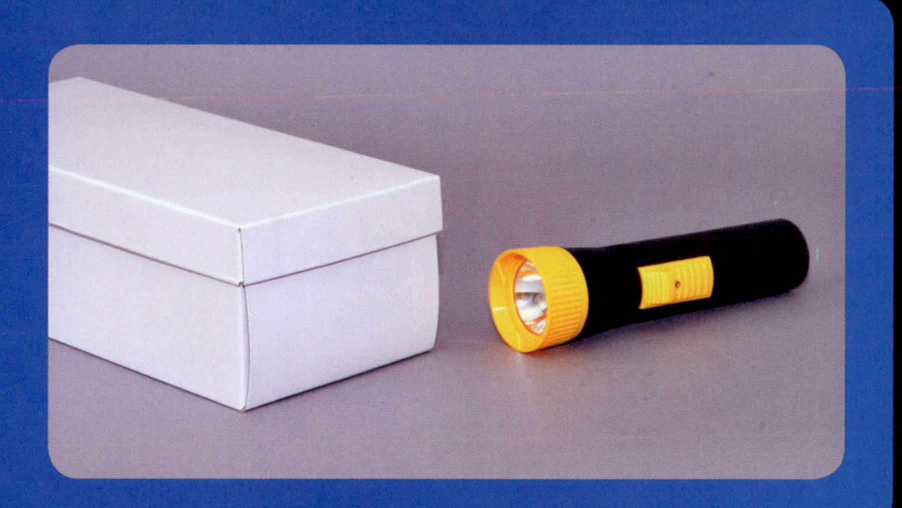

2 Open the shoebox. Shine a flashlight on the object in the box. Look at the object in the box. Then turn on the lights in the room.

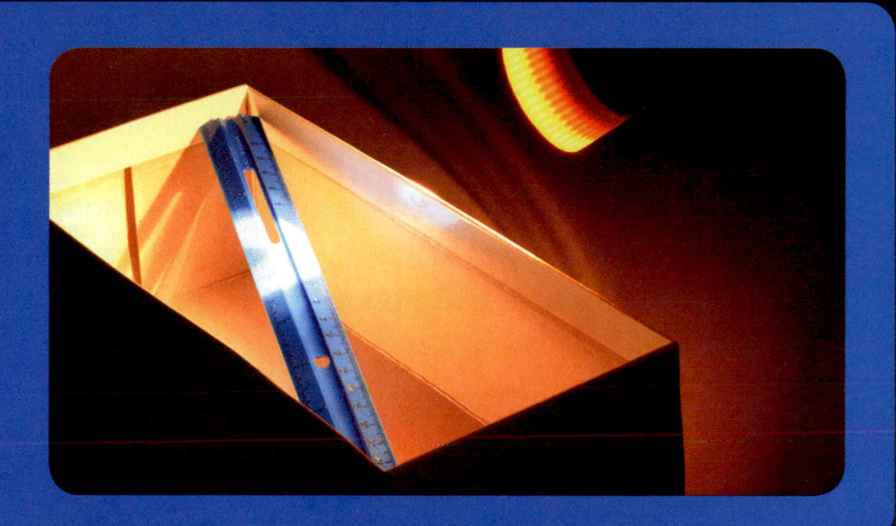

3 Draw a model that shows how light reflecting from the object and entering your eye allows you to see the object. Write captions that explain how light allows the object to be seen.

4 Explain your model to a classmate. Does your classmate have ideas for improving your model? Make changes to your model if needed.

Wrap It Up!

1. **Explain** Use information from your model to explain how light from the flashlight reached your eye.

2. **Apply** Is it possible to see an object when there is no light? Why or why not?

Use a Model

This is a beach mouse. It uses its senses to survive in the grassy sand dunes of Florida. It digs burrows into the sand. It uses the burrows to keep safe, raise young, and store seeds for food. The mice look for food mainly at night. Darkness helps them hide from other animals that could eat them. Imagine that this mouse has just come out of its burrow. It is late at night, and it is very dark.

1. **Make a model.** 

   Design a model that shows how the mouse receives information through several of its senses as it searches for food. Include how the mouse processes the information and responds to the information.

2. **Discuss your model.**
   Work with a partner. Compare your models. Talk about different ways the mouse receives, processes, and responds to information. Combine your ideas on one mouse model.

3. **Revise your model.**
   Imagine that an eastern diamondback rattlesnake is hunting the mouse. Research how snakes gather information from the environment. Then work with your partner to revise your model. Show what information the mouse receives and processes. Show what information the snake receives and processes.

4. **Share your model.**
   Meet with another pair of students. Discuss how the models you made are the same and different. Work together to make a generalization as to how different animals receive, process, and respond to information in different ways.

NEXT GENERATION SCIENCE STANDARDS | PERFORMANCE EXPECTATION
4-LS1-2. Use a model to describe that animals' receive different types of information through their senses, process the information in their brain, and respond to the information in different ways.

The beach mouse spends little time outside its burrow during daylight hours.

The eastern diamondback rattlesnake preys on rats, mice, squirrels, and birds.

Dog Whisperer

What can you do if your dog won't behave? Maybe it barks all the time or chews up shoes. Maybe it fights with other dogs or jumps up on strangers.

Cesar Millan knows how to fix all these problems. How? By using dog psychology. Psychology is the science of the mind and behavior.

Cesar uses his understanding of dogs to correct the behavior of pet dogs that are out of control. But mostly, Cesar shows dog owners how to change the way they treat their pets. Here's the surprising fact: When owners correct the way they treat their pets, their pets almost always stop misbehaving!

Cesar is called the "dog whisperer" because he has a special talent for interacting with dogs in ways that improve their behavior.

NEXT GENERATION SCIENCE STANDARDS | CONNECTIONS TO NATURE OF SCIENCE
Scientific Knowledge is Based on Empirical Evidence
Science findings are based on recognizing patterns.

Cesar Millan is a dog trainer, author, and star of the National Geographic Channel program "The Dog Whisperer." Originally from Mexico, Cesar now lives in the United States. Cesar founded and runs the Dog Psychology Center in Los Angeles, where he rehabilitates dogs with severe behavior problems. Cesar has won many awards, including one from the Humane Society for his work with dogs from shelters.

Science Career

NGL Science How do you rehabilitate dogs?

Cesar Millan I don't train dogs to respond to commands such as "sit" or "stay." Instead I try to connect with the dog's mind and natural instincts to help correct unwanted behaviors.

NGL Science How did you learn about dogs?

Cesar Millan As a child, I spent a lot of time on my grandfather's farm in Mexico. The working dogs on the farm were my true teachers in the art and science of dog psychology. I loved to watch dogs play with one another. The more hours I spent watching them, the more questions came into my mind. How did they coordinate their activities? How did they communicate with one another?

NGL Science How did you develop your methods of training dogs?

Cesar Millan My way of training came directly from my observations of dogs on my grandfather's farm. I interact with the dogs the way they interact with one another.

NGL Science Do you also work with the dog owners?

Cesar Millan The dog owner often thinks that the problem lies with the dog. But the problem is usually with the way the owner treats his dog. I often say, "I rehabilitate dogs, but I train people." My formula is simple: For a balanced, healthy dog, a human must share exercise, discipline, and affection, in that order!

Cesar plays with this dog as if he were a dog himself.

Earth Science

Earth's Systems: Processes that Shape the Earth

The Sakurajima volcano in southern Japan has erupted almost continuously since 1955.

The amount of rain that falls during the year varies from region to region. The map shows areas that receive different amounts of rain. Rainfall helps to shape the land and affects the types of plants and animals found in a region.

Pacific Northwest Forest

The Pacific Northwest coast is in a very rainy region. Some places within that region receive more than 250 centimeters (100 inches) of rain each year! This wet weather supports temperate rain forests with tall trees and many different kinds of plants and animals.

Southwest Desert

The deserts of the southwest receive less than 30 centimeters (12 inches) of rain a year. The living things that live here can survive with little water.

NEXT GENERATION SCIENCE STANDARDS | DISCIPLINARY CORE IDEAS
ESS2.A: Earth Materials and Systems
Rainfall helps to shape the land and affects the types of living things found in a region. Water, ice, wind, living organisms, and gravity break rocks, soils, and sediments into smaller particles and move them around. (4-ESS2-1)

Central Plains Grassland

The middle part of the United States receives moderate rain—about 51 centimeters (20 inches) each year. This much rainfall supports grasslands. It provides enough water for grasses, but not enough for most trees.

Eastern Temperate Forest

The eastern United States receives an average of around 127 centimeters (50 inches) of rain throughout the year. The plentiful rainfall supports temperate forests with many plants and animals.

Average Annual Precipitation
centimeters (inches)

- Over 180 (70)
- 100–180 (40–70)
- 50–100 (20–40)
- 0–50 (0–20)

Wrap It Up

My science notebook

1. **Define** About how much rainfall does the southwest desert region receive?
2. **Contrast** How is a grassland different from a temperate forest? Contrast the amount of rain and the kinds of plants in the two places.
3. **Generalize** How does the amount of rainfall affect the living things that can live in a region?

Pacific Northwest Forest

If you visit the Pacific Northwest, you'll see many tall trees. In fact, this region is home to the world's tallest trees! The rainy weather and mild temperatures support the growth of lush forests. Most of the trees are evergreen, which means that they keep their needle-shaped leaves all year long.

Many flowering shrubs and trees, such as currants and dogwoods, also live in these forests. Wildflowers, mosses, and ferns grow in the moist soil. This forest is home to many different kinds of animals, from insects and amphibians to birds and mammals.

When it is not raining in the Pacific Northwest, the trees are still often wrapped in moisture in the form of fog.

NEXT GENERATION SCIENCE STANDARDS | DISCIPLINARY CORE IDEAS
ESS2.A: Earth Materials and Systems
Rainfall helps to shape the land and affects the types of living things found in a region. (4-ESS2-1)

The sideband snail slides over the wet ground and decaying leaves on the forest floor.

The colorful western tanager eats insects, fruits, and seeds. This male tanager is sitting in a spruce tree in an Oregon forest.

Many wildflowers, such as these iris, grow in the damp soil.

Many long growing seasons with plentiful rain have allowed this Douglas fir tree to grow very tall.

Wrap It Up!

1. **Describe** What is the weather like in the Pacific Northwest?

2. **Identify** What kinds of trees are most common in the Pacific Northwest?

3. **Cause and Effect** How does the rainfall of the Pacific Northwest affect the variety of plants and animals that live there?

Southwest Desert

The Sonoran Desert in the southwestern United States looks very different from that of the Pacific Northwest. Why? This region receives very little rain. Plants that can survive with little water live in this dry region. Desert plants grow farther apart from each other than those in a forest. The animals that live in the desert must also be able to survive in harsh conditions without much water.

The tough, scaly skin of reptiles, such as this iguana, slows the loss of water from their bodies. This allows them to survive in the desert.

The saguaro cactus provides shelter around the nest of these Harris's hawks.

NEXT GENERATION SCIENCE STANDARDS | DISCIPLINARY CORE IDEAS
ESS2.A: Earth Materials and Systems
Rainfall helps to shape the land and affects the types of living things found in a region. (4-ESS2-1)

Cacti such as the tall, forked saguaro grow in the Sonoran Desert.

Plants that live in deserts must have ways to save water. The barrel cactus stores water in its thick stem.

Some plants, such as this purple mat, survive the dry weather as seeds. When there is enough rain, these plants quickly sprout, grow, flower, and set seeds.

My science notebook

Wrap It Up!

1. **Recall** What characteristics do animals have that help them survive in the Sonoran Desert?
2. **Explain** What characteristics do plants have that help them survive in the Sonoran Desert?
3. **Infer** Most amphibians have moist skin and lay their eggs in water. Would you expect there to be many kinds of amphibians in the Sonoran Desert? Explain.

Central Plains Grassland

The central portion of the United States receives more rain than the desert but less rain than the forest. The region supports grasslands, also called prairies. This large area of land is covered in grasses and many kinds of wildflowers, but few trees.

Bison adapt to the changing seasons by shedding their thick winter coat when the weather becomes warm.

This prairie chicken feeds on the grasses growing in a Nebraska prairie.

NEXT GENERATION SCIENCE STANDARDS | DISCIPLINARY CORE IDEAS
ESS2.A: Earth Materials and Systems
Rainfall helps to shape the land and affects the types of living things found in a region. (4-ESS2-1)

The climate in the grasslands varies each season. Winters are cold and summers are hot and dry. Prairie plants grow long roots that can absorb and store water. During extremely dry periods, fires are common. Many prairie plants regrow after a fire because their roots are protected below the ground.

Few tree species grow here because there is not enough rainfall and fires destroy saplings.

These colorful flowers attract bees and wasps that help the plant reproduce.

Prairie grasses have adaptations that help them survive the long, hot summers.

Wrap It Up!

My science notebook

1. **Describe** What is the weather like in the prairie during the summer?
2. **Summarize** What is one way in which prairie plants survive the extreme changes in weather?
3. **Infer** When do the grasslands get most of their rainfall?

Eastern Temperate Forest

The mid-Atlantic region receives rain and snow throughout the year that supports temperate forests. Temperate forests are wet environments. These forests are similar to those found in the Pacific Northwest, with one major difference. Most trees in temperate forests are deciduous, which means they lose their leaves in fall and produce new ones in the spring.

Abundant rainfall in temperate forests encourages the growth of many trees, shrubs, and other plants.

NEXT GENERATION SCIENCE STANDARDS | DISCIPLINARY CORE IDEAS
ESS2.A: Earth Materials and Systems
Rainfall helps to shape the land and affects the types of living things found in a region. (4-ESS2-1)

Deciduous trees drop their leaves before winter to help preserve energy when it is cold. During warmer months, large leaves help low-growing forest plants absorb sunlight.

White-tailed deer thrive in temperate forests. They find plenty of plants to eat here.

The Northern Saw-whet Owl makes its home in the eastern temperate forest. Owls live in the tall trees and hunt mice, chipmunks, and other small animals that are common there.

The large leaves of the trillium help it absorb the small amount of sunlight that reaches the forest floor.

Flowering trees, such as this dogwood, bloom in the eastern temperate forest in spring.

Wrap It Up!

1. **Define** What are deciduous trees?

2. **Compare and Contrast** How do temperate forests differ from forests found in the Pacific Northwest?

3. **Infer** What characteristics of the temperate forest make it a good home for many animals?

Weathering

Landforms such as mountains, valleys, and plains are shaped and changed over millions of years. These changes to Earth's surface include three processes. The first process is weathering. **Weathering** is the breaking apart, wearing away, or dissolving of rock into smaller particles. The smaller particles are called **sediment.** Wind, water, ice, chemicals, and even plants can cause weathering.

NEXT GENERATION SCIENCE STANDARDS | DISCIPLINARY CORE IDEAS
ESS2.A: Earth Materials and Systems
Rainfall helps to shape the land and affects the types of living things found in a region. Water, ice, wind, living organisms, and gravity break rocks, soils, and sediments into smaller particles and move them around. (4-ESS2-1)

Weathering by water and wind helped shape this rock into its arched shape. Eventually more weathering could weaken it until gravity causes it to fall.

Wrap It Up!

1. **Recall** What can cause weathering?
2. **Explain** Describe how weathering might affect rocks.

Erosion and Deposition

Weathering produces sediment. Sediment is loose material. It can be large boulders or tiny grains of sand. What happens to these weathered pieces of rock? Often they are moved to a new place. The moving of sediment from one place to another is the second step in the process, **erosion.** Wind, water, ice, living things, and gravity all can move sediment from one place to another.

NEXT GENERATION SCIENCE STANDARDS | DISCIPLINARY CORE IDEAS
ESS2.A: Earth Materials and Systems
Rainfall helps to shape the land and affects the types of living things found in a region. Water, ice, wind, living organisms, and gravity break rocks, soils, and sediments into smaller particles and move them around. (4-ESS2-1)

What happens to the sediment that is carried away? It is deposited or dropped in a new place. This action is another process called **deposition.** Moving water can deposit pebbles and sand downstream. Or wind can deposit sand on a beach to form a sand dune. Together the three processes of weathering, erosion, and deposition can shape and change the land.

Long ago, on what is now Deer Isle, Maine, huge moving masses of ice carried these boulders from far away. When the ice melted, the boulders were left in a new place.

My science notebook

Wrap It Up!

1. **Recall** What are some causes of erosion?
2. **Generalize** How do erosion and deposition change Earth's surface?
3. **Conclude** Why does erosion occur before deposition?

Wind Changes the Land

Wind can shape land through weathering, erosion, and deposition. Wind picks up and moves small pieces of sediment, such as soil and sand. When sand slams into rocks, tiny particles of the rocks break off. The moving sand acts like sandpaper, rubbing the rocks smooth.

Sand dunes form in windy places where sand is plentiful, such as on a beach or in a desert. Steady winds push grains

If conditions are right, grains of sand collect and form a sand dune. A sand dune is a landform caused when wind deposits sand.

NEXT GENERATION SCIENCE STANDARDS | DISCIPLINARY CORE IDEAS
ESS2.A: Earth Materials and Systems
Water, ice, wind, living organisms, and gravity break rocks, soils, and sediments into smaller particles and move them around. (4-ESS2-1)

of sand along the ground. Large objects such as rocks and plants block the sand's movement. A sand dune forms as more and more grains of sand are blocked by an object and become trapped. Over time, a sand dune may grow very tall. How tall? A few sand dunes reach heights of more than 500 meters (1640 feet)!

These landforms are called toadstool caprocks. The rock at the top has resisted weathering more than the rock beneath. The softer, lower rocks have worn away, leaving the unusual shapes.

Wrap It Up!

My science notebook

1. **Describe** How can wind change the shape of rocks?

2. **Cause and Effect** Explain how sand dunes form.

3. **Classify** Identify these two processes as erosion or deposition: Sand dunes form. Wind picks up particles of soil and sand.

Water Changes the Land

Moving water can change Earth's surface in a big way. For example, the Virgin River formed Zion Canyon in Utah. How does moving water change Earth's surface? Like wind, moving water carries **sediment,** such as sand and pebbles. Sediment carried by the Virgin River

NEXT GENERATION SCIENCE STANDARDS | DISCIPLINARY CORE IDEAS
ESS2.A: Earth Materials and Systems
Water, ice, wind, living organisms, and gravity break rocks, soils, and sediments into smaller particles and move them around. (4-ESS2-1)

scraped and chipped away rock on the sides and bottom of the river. It took many years for the river to carve the canyon.

If you hiked through the canyon, you might see that many of the rocks are smooth and rounded. Sediment carried along by the water bumps and rubs against larger rocks. Eventually the surface of the rocks wears away. The rocks become smoother, more rounded, and smaller. **Weathering, erosion,** and **deposition** are mostly slow processes, but over time they can make huge changes to Earth's surface.

Over time, the Virgin River eroded enough rock to form this canyon.

Rushing water in a stream in Zion Canyon weathers and erodes rocks, making them smoother.

Wrap It Up!

My science notebook

1. **Explain** How can water cause weathering, erosion, and deposition of rock?

2. **Generalize** How are some canyons a result of changes to Earth's surface by water?

Investigate

Weathering and Erosion

? How can you model the processes of weathering and erosion?

Scientists use various kinds of models to investigate how natural processes work. In this activity you'll explore two ways that weathering and erosion can change a rock called sandstone.

Materials

sandstone	paper towel	jar with lid
water	hand lens	stopwatch

NEXT GENERATION SCIENCE STANDARDS | DISCIPLINARY CORE IDEAS
ESS2.A: Earth Materials and Systems
Water, ice, wind, living organisms, and gravity break rocks, soils, and sediments into smaller particles and move them around. (4-ESS2-1)

1 Predict what will happen when you rub 2 pieces of sandstone together. Record your prediction in your science notebook. Hold the pieces of sandstone over the paper towel. Rub them together for a few seconds. Observe what happens. Record your observations.

2 Place 5 pieces of sandstone in a jar. Pour water into the jar to fill it about half full. Put the lid on the jar securely. Use the hand lens to observe the sandstone and the bottom of the jar. Record your observations.

3 Predict what will happen if you shake the jar. Then shake the jar for 3 minutes. Use the stopwatch to time yourself.

4 Observe the sandstone and the bottom of the jar. Record your observations.

Look at the photo. How might weathering and erosion have changed these rocks?

Wrap It Up!

1. **Compare** How do your results compare to your predictions? Explain.
2. **Analyze** What processes did you model in Steps 1 and 3? Explain.
3. **Apply** Use what you learned in this investigation to explain how the rocks in the picture were changed.

Ice Changes the Land

What can move rocks that are as big as houses or as small as silt or sand? Ice! A **glacier** is a huge area of slow-moving ice. In some places, massive glaciers changed Earth's surface as they slowly moved over the land. As glaciers moved, they scraped against rock, carving it into new shapes.

Long ago, a glacier scraped this rock smooth.

Ice can change Earth's surface in another way, too. Water seeps into cracks in rock. The water freezes and expands. The ice inside the cracks acts like a wedge. It makes the cracks wider. The repeated freezing and thawing of ice will eventually widen the crack so much that the rock splits apart.

NEXT GENERATION SCIENCE STANDARDS | DISCIPLINARY CORE IDEAS
ESS2.A: Earth Materials and Systems
Water, ice, wind, living organisms, and gravity break rocks, soils, and sediments into smaller particles and move them around. (4-ESS2-1)

Glaciers carved wide, U-shaped valleys in Yosemite National Park in California.

This hanging valley formed when a small, shallow glacier flowed into a larger glacier.

Glaciers carved steep cliffs in Yosemite Valley.

My science notebook

Wrap It Up!

1. **Compare** How do water and ice affect rock in similar ways?
2. **Explain** How does Yosemite Valley show that ice changes Earth's surface?

Living Things Change the Land

Organisms, or living things, can change the shape of the land. Over time, plants, animals, and other kinds of organisms can break rocks into smaller pieces and move them from one place to another.

Meerkats are excellent diggers, often making huge underground burrows. This meerkat is digging for insects in the Kalahari Desert of South Africa.

NEXT GENERATION SCIENCE STANDARDS | DISCIPLINARY CORE IDEAS

ESS2.A: Earth Materials and Systems

Water, ice, wind, living organisms and gravity break rocks, soils, and sediments into smaller particles and move them around. (4-ESS2-1)

ESS2.E: Biogeology

Living things affect the physical characteristics of their regions. (4-ESS2-1)

Some organisms move rocks and sediments to new places. The roots of most plants grow in soil. As they grow, they push the particles of soil aside.

Animals such as moles, gophers, worms, and many kinds of insects burrow through the ground. To make their burrows, they move particles of soil. A large animal such as a woodchuck can even move rocks!

The roots of plants can grow in the cracks of rocks. As the roots of this tree became larger, they broke the rock apart.

My science notebook

Wrap It Up!

1. **Define** What is an organism?
2. **Cause and Effect** How can plants break down rocks?

Landslides Change Earth's Surface

Gravity is the force that pulls objects toward Earth's center. Gravity plays a part in the erosion of land. Have you ever seen a road sign that says *Caution: Falling Rock*? Rock on a ledge or steep hill may suddenly break loose and fall. Often it's just a rock or two. But at other times, it's a landslide.

This landslide in El Salvador destroyed many buildings in the neighborhood at the bottom of the hill.

NEXT GENERATION SCIENCE STANDARDS | DISCIPLINARY CORE IDEAS
ESS2.A: Earth Materials and Systems
Rainfall helps to shape the land and affects the types of living things found in a region. Water, ice, wind, living organisms, and gravity break rocks, soils, and sediments into smaller particles and move them around. (4-ESS2-1)

A **landslide** is a rapid movement of rock, soil, and other material down a hill or mountain. A landslide can change Earth's surface. If enough material falls, the shape of the hill or mountain may change.

What causes landslides? Sometimes heavy rains can loosen soil and rock on slopes. Volcanoes and earthquakes can also start the motion. But it is the force of gravity that pulls the rocks and other loosened material downhill.

Landslides can leave large, bare areas on mountains and hills.

My science notebook

Wrap It Up!

1. **Explain** How do landslides cause rapid changes to Earth's surface?
2. **Cause and Effect** What makes material in any landslide move downhill?
3. **Infer** How might a landslide affect people?

Make Observations

You have read how running water can change the shape of the land. Now you'll use what you've learned to help solve a problem.

After a heavy rainstorm, a farmer discovered water making gullies in the soil. Water was eroding the soil as it made its way from a field to the nearby stream. If this continued, much of the soil would be washed away. Then the farmer would not be able to grow crops in the widening gully.

You have been asked to solve the problem of soil eroding from the farmer's field. To test your solution, you'll use small-scale models to investigate the rate, or speed, of erosion.

1. Define the problem.
How can you slow or prevent erosion of a farmer's field?

2. Find a solution. 

- To solve this problem, you will need to make a model of a hillside. You need to decide on the materials you will need for this model.
- After you build your hillside, you need to decide how you will measure the rate of erosion. What are some ways you could change how water flows down your model hillside? Which variables will you change when you test your model? Which variables will you keep the same? How will you know the tests you run are fair?
- Write out a plan for your investigation. Include a diagram of your hillside and the solutions to reduce erosion that you will test. Then collect the materials you need and build your model.

NEXT GENERATION SCIENCE STANDARDS | PERFORMANCE EXPECTATIONS
4-ESS2-1. Make observations and/or measurements to provide evidence of the effects of weathering or the rate of erosion by water, ice, wind, or vegetation.

3-5-ETS1-3. Plan and carry out fair tests in which variables are controlled and failure points are considered to identify aspects of a model or prototype that can be improved.

Too much rain can erode the soil in which crops are grown.

3. Test your solution.

- Carry out the test you planned. Collect data from your test.
- Repeat your test several times, changing the variables you identified in your plan. Record your observations.

4. Refine or change your solution.

- Analyze your data. Did your solution change the rate of erosion? Did each trial produce similar results? Did changing the variables you identified in your plan produce a better result? Could you make your solution work better? Test your new ideas.
- When you are satisfied with your solution, get ready to explain your plan to others. Draw a detailed diagram or make a poster of your solution. Include clear labels that explain how your solution will work. Be prepared to defend your solution as others question or challenge your plan. Plan your argument in detail. Be sure to provide evidence from your tests to support your solution.
- Share your solution with a partner. Record his or her comments and suggestions.
- Evaluate his or her comments. Which of the suggestions would improve your plan? Revise your plan to include the suggestions.

5. Present your solution.

- When you are satisfied that your solution is the best you can make it, revise your presentation. Then share your solution with the class.

Farmers prepare their fields in certain ways to prevent soil erosion. They plough the crop rows in directions that block water from running straight downhill. That gives water more time to soak into the soil. They also alternate sections of crops. Rows of sturdy plants, less likely to be washed away by erosion, protect more delicate plants.

Natural Hazards

Natural processes are constantly changing Earth's surface. In some cases, these processes are harmful to humans and other living things. Something that is harmful or dangerous is called a **hazard.** Earthquakes, tsunamis, and volcanoes are three natural processes that can be hazardous to humans.

Earthquakes

An **earthquake** is the shaking of Earth's surface caused by sudden movement of rock beneath the surface. Earthquakes can cause buildings to collapse and roads and bridges to buckle. They may also break power lines and water pipes.

Volcanoes

Volcanoes form when molten rock from deep inside Earth rises to the surface. Volcanic eruptions can spew hot ash and molten rock high into the air. When these materials come down, they can bury buildings and roads and damage crops. Volcanic ash in the air can also disrupt airline traffic.

NEXT GENERATION SCIENCE STANDARDS | DISCIPLINARY CORE IDEAS
ESS3.B: Natural Hazards
A variety of hazards result from natural processes (e.g., earthquakes, tsunamis, volcanic eruptions). (4-ESS3-2)

Tsunamis

A **tsunami** is a series of ocean waves caused by an underwater earthquake, an underwater volcanic eruption, or a landslide. Anything that causes Earth's surface to move beneath the water also moves the water. When large tsunamis come ashore, they can destroy buildings, roads, or even entire villages.

Wrap It Up!

My science notebook

1. **Define** What does the word *hazard* mean?
2. **List** What are three of the natural processes that can be hazardous to humans?
3. **Infer** Why might a volcanic eruption be dangerous for airline traffic?

Earthquakes

An earthquake is a natural process caused by the movement of parts of Earth's surface. Earthquakes start along a fault boundary. A **fault** is a break in Earth's surface where huge slabs of rock slip past, move away from, or push against each other. The slabs of rock often become locked together along a fault line. If the slabs break free, energy is released and moves through the rocks. This makes the ground shake.

Several million earthquakes occur every year. Most of them happen far from people or are too weak for people to notice.

Strong earthquakes can be very dangerous. Their violent shaking can raise and lower the land and change the course of rivers. Powerful earthquakes can damage buildings and other structures. Roads buckle, railroad tracks twist, and bridges collapse. Water pipes and electric power lines break. It can take years for people to repair the damage caused by a strong earthquake.

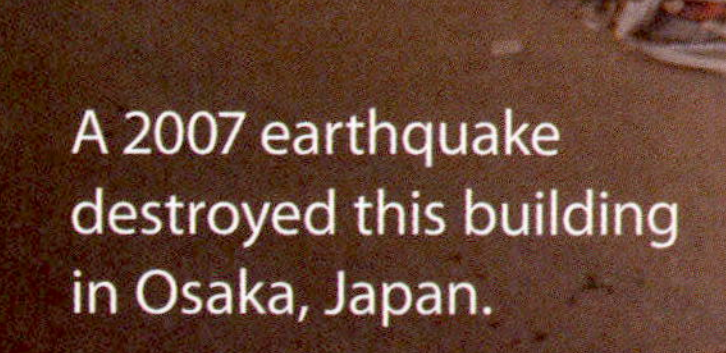

A 2007 earthquake destroyed this building in Osaka, Japan.

NEXT GENERATION SCIENCE STANDARDS | DISCIPLINARY CORE IDEAS

ESS3.B: Natural Hazards

A variety of hazards result from natural processes (e.g., earthquakes, tsunamis, volcanic eruptions). Humans cannot eliminate the hazards but can take steps to reduce their impacts. (4-ESS3-2)

Wrap It Up!

1. **Define** What is a fault?
2. **Cause and Effect** How can an earthquake affect structures built by humans? Give three examples.
3. **Apply** How might an earthquake affect people's ability to get from place to place?

Investigate

Earthquakes

? How can you demonstrate liquefaction?

During earthquakes, many buildings have collapsed or tilted and sunk into the ground. Why? Often the buildings were built on loose materials such as rocky soil, sand, or mud. There is also a lot of water in the ground. When the earthquake shook the ground, these materials became liquid-like. The change of a solid area of ground such as sand or mud into a less stable liquid-like condition is called **liquefaction.** Liquefaction is one of the hazards caused by earthquakes.

n this investigation, you will use a model to learn how earthquakes affect structures built on sand or mud.

Materials

NEXT GENERATION SCIENCE STANDARDS | DISCIPLINARY CORE IDEAS
ESS3.B: Natural Hazards
A variety of hazards result from natural processes (e.g., earthquakes, tsunamis, volcanic eruptions). Humans cannot eliminate the hazards but can take steps to reduce their impacts. (4-ESS3-2)

1 Fill the pan with sand, leaving about 9 cm at the top.

2 Place the pan on a table. Then pour in water to just below the surface of the sand. Record your observations.

3 Push one end of the block down into the wet sand so it stands up like a building. Predict what will happen when you repeatedly tap the mallet against the pan. Record your prediction.

4 Hold the pan in place. Very gently tap the side of the pan repeatedly with the mallet. Observe what happens to the sand and the block. Record your observations.

The water-filled holes in this photograph were caused as water bubbled up through the soil as a result of an earthquake. Imagine how such a change in the soil would affect a building!

Wrap It Up!

1. **Explain** What did the block represent? What did hitting the pan with the mallet represent?
2. **Analyze** How did hitting the pan with the mallet affect the block?
3. **Revise** How could you change the model to test ways to make buildings constructed on sand or mud more stable? Write your plan. Include a diagram of how your test would work.

Tsunamis

A tsunami is a series of fast-moving ocean waves caused by an earthquake, or an underwater volcanic eruption, or a landslide. In places where the ocean is deep, tsunami waves may be only a few centimeters high. As the waves approach shore, they increase in height. Some tsunami waves come ashore gently. Others become huge walls of water.

The Great Tōhoku Earthquake of 2011 in Japan set off a series of tsunamis. A few minutes after the earthquake, one tsunami crashed over this seawall in Miyako City. Tsunamis up to 12 meters (39 feet) tall destroyed coastal areas across northeastern Japan.

NEXT GENERATION SCIENCE STANDARDS | DISCIPLINARY CORE IDEAS
ESS3.B: Natural Hazards
A variety of hazards result from natural processes (e.g., earthquakes, tsunamis, volcanic eruptions). (4-ESS3-2)

When a large tsunami smashes into land, its turbulent water can destroy houses, roads, and other structures. People may be swept away. Very large tsunamis can destroy entire villages.

This tsunami was caused by a strong earthquake that struck Japan in March of 2011.

My science notebook

Wrap It Up!

1. **Recall** What are three events that cause tsunamis?
2. **Explain** How do some tsunamis change when they reach land?
3. **Summarize** Why are tsunamis dangerous?

Volcanoes

A volcanic eruption is a natural process on Earth's surface. Volcanoes form when **magma,** or melted rock inside Earth, rises to the surface. In all volcanoes, molten rock **erupts** or flows through an opening. Magma that erupts onto Earth's surface is called **lava.** Sometimes the lava flows down the side of the volcano and hardens into rock.

Mt. Sakurajima on the main island of Japan erupted on August 18, 2013.

NEXT GENERATION SCIENCE STANDARDS | DISCIPLINARY CORE IDEAS
ESS3.B: Natural Hazards
A variety of hazards result from natural processes (e.g., earthquakes, tsunamis, volcanic eruptions). Humans cannot eliminate the hazards but can take steps to reduce their impacts. (4-ESS3-2)

Hot, expanding gases in magma can cause explosive eruptions. Hot ash and lava released during an eruption may quickly bury or destroy nearby forests, fields, and towns. Such eruptions are a hazard to the people living nearby. The ash in the air makes it difficult to breathe. Volcanoes may also release poisonous gases.

There is little people can do to limit the dangers of a volcano. The safest solution is to not live close to one.

Ash covered these rental cars in Kagoshima, Japan, after Mt. Sakurajima erupted.

My science notebook

Wrap It Up!

1. **Identify** What are some of the hazards caused by erupting volcanoes?
2. **Compare and Contrast** What is the difference between magma and lava?
3. **Research** What were the effects of the most recent volcanic eruption closest to where you live?

Reducing the Impact of Natural Hazards

Natural hazards cannot be eliminated, but people can take steps to reduce their impact. For example, engineers can design buildings and bridges to withstand the violent shaking of earthquakes. This reduces damage and saves lives.

NEXT GENERATION SCIENCE STANDARDS | DISCIPLINARY CORE IDEAS
ESS3.B: Natural Hazards
A variety of hazards result from natural processes (e.g., earthquakes, tsunamis, volcanic eruptions). Humans cannot eliminate the hazards but can take steps to reduce their impacts. (4-ESS3-2)

The suspension bridge contains a motion-detection system, which warns drivers of earthquake danger.

The San Francisco-Oakland Bay Bridge in California is designed to withstand a major earthquake.

Wrap It Up!

My science notebook

1. **Review** What features can help protect the bridge from earthquake damage?

2. **Evaluate** The features that help keep bridges safe during an earthquake also make the bridges more expensive. Do you think all bridges should have these features?

Early Warning Systems

If people know that a natural hazard is likely to occur, they can take steps to reduce its impact. But how can people know when they are in danger? Scientists use many methods to help predict when natural hazards are likely to happen.

Seismometers measure earthquake activity. An increase in small earthquakes can mean a larger earthquake is about to happen or a volcano is becoming more active. If a volcanic eruption is predicted, people can **evacuate,** or move to safer areas. Governments can prepare emergency supplies, such as food and water. Emergency workers can be ready to help with injuries. All of these steps reduce the impact and help save lives.

Seismometers detect earthquake waves. An increase in earthquakes near a volcano can be a sign of increasing volcanic activity.

Ground motions sensed by seismometers are converted into electronic signals. The signals are transmitted by radio and recorded on **seismographs**.

NEXT GENERATION SCIENCE STANDARDS | DISCIPLINARY CORE IDEAS
ESS3.B: Natural Hazards
A variety of hazards result from natural processes (e.g., earthquakes, tsunamis, volcanic eruptions). Humans cannot eliminate the hazards but can take steps to reduce their impacts. (4-ESS3-2)

Scientists use several devices and methods to constantly collect information about a volcano's activity. Radio transmissions allow the data to be sent instantly to a monitoring system. Scientists can tell what is happening with a volcano almost as soon as it happens. They can also use that information to predict what is likely to happen next.

In addition to tracking seismic activity, scientists also monitor gases, temperature, and water at a volcano site. A change in the amount of carbon dioxide and other gases given off by the volcano can signal a coming eruption. So can temperature changes in rocks at the surface or rocks underground. Scientists monitor water levels and look for changes in chemistry in water near volcanoes, too.

Swelling, sinking, or cracking of the ground near a volcano can mean magma is moving beneath the surface. Tiltmeters detect any movement of magma close to the surface.

Scientists continually monitor the gases coming out of active volcanoes.

Wrap It Up!

1. **Describe** What information do scientists collect to monitor volcanoes?

2. **Explain** How is the information used to reduce the impact of volcanic eruptions?

Tsunami Detection

Most of the events that cause tsunamis occur on the floor of the ocean. It's not easy to collect seismic activity data there. It's also not possible to monitor the entire ocean floor. A sudden earthquake or landslide on the sea floor can cause a tsunami that may strike land hundreds or even thousands of kilometers away.

Scientists use a system of floating devices called buoys to measure changes in the depth of ocean water. A pattern of depth changes can indicate that a tsunami has formed. If scientists can detect a tsunami as soon as it happens, they can alert people on land that a tsunami may be on the way. People can then leave areas near the coast and move inland or to higher ground.

The tsunami warning buoy is floating on the ocean.

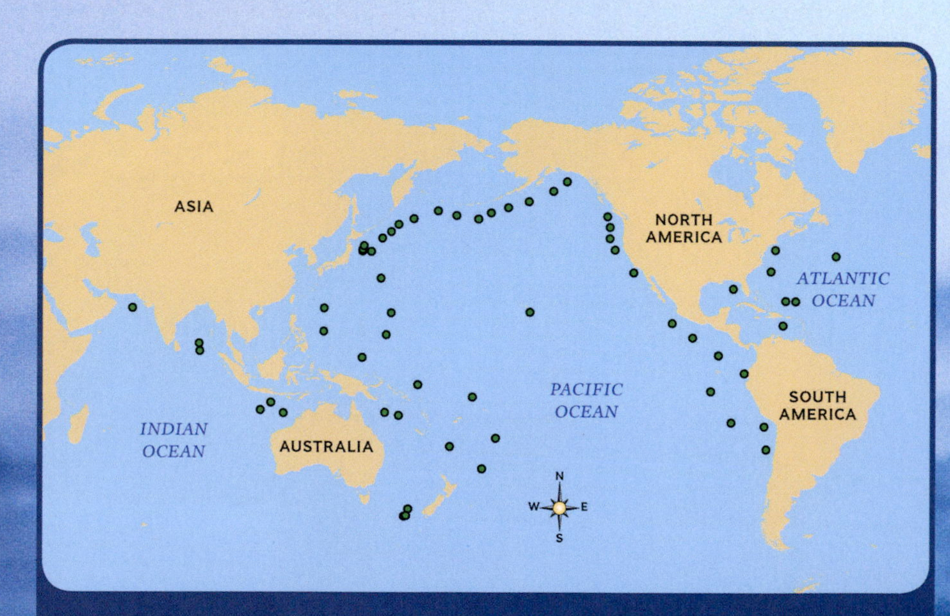

The map shows the ring of tsunami warning buoys that encircles the Pacific Ocean.

NEXT GENERATION SCIENCE STANDARDS | DISCIPLINARY CORE IDEAS
ESS3.B: Natural Hazards
A variety of hazards result from natural processes (e.g., earthquakes, tsunamis, volcanic eruptions). Humans cannot eliminate the hazards but can take steps to reduce their impacts. (4-ESS3-2)

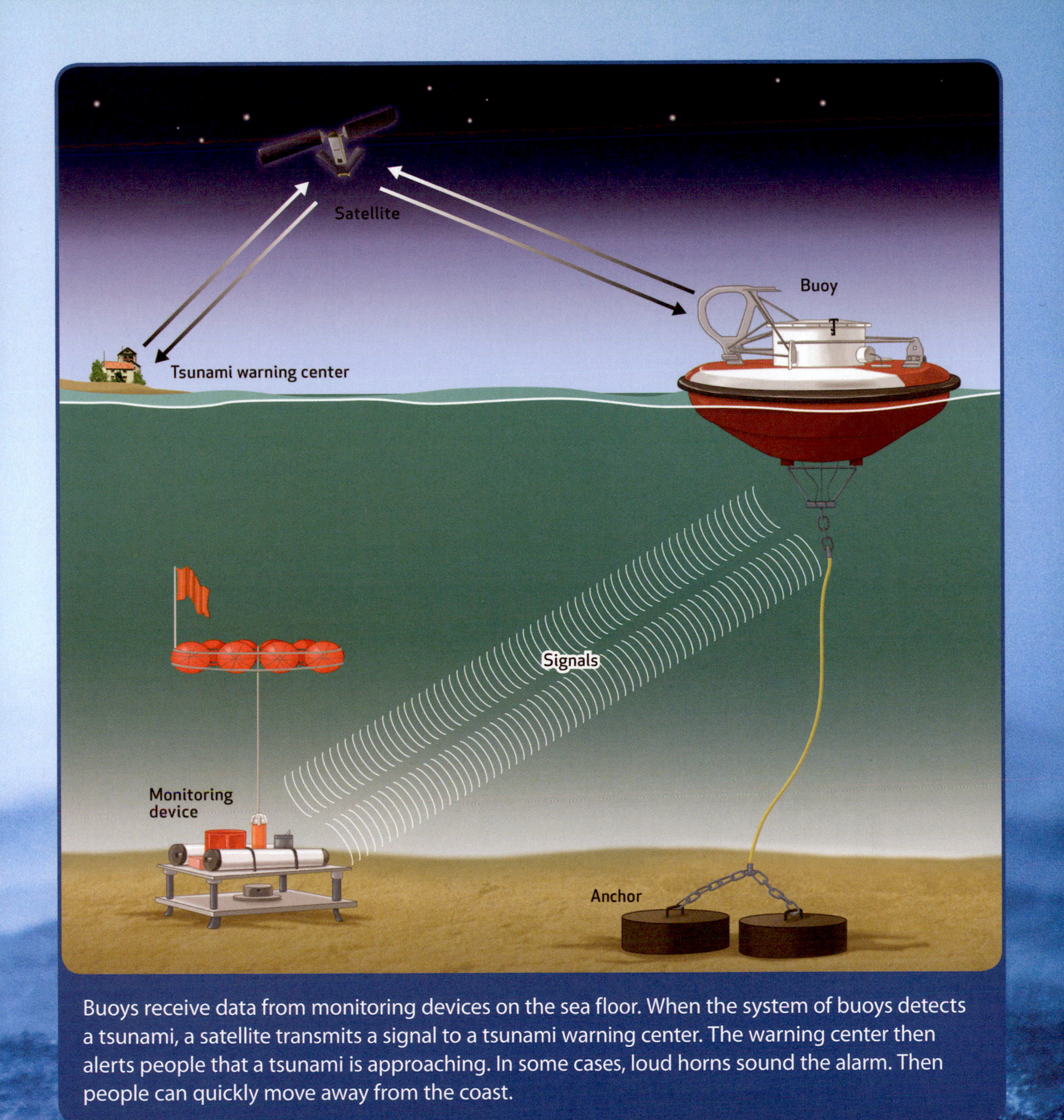

Buoys receive data from monitoring devices on the sea floor. When the system of buoys detects a tsunami, a satellite transmits a signal to a tsunami warning center. The warning center then alerts people that a tsunami is approaching. In some cases, loud horns sound the alarm. Then people can quickly move away from the coast.

Wrap It Up!

1. **Describe** How do scientists detect tsunamis?
2. **Explain** Why are satellites useful in predicting tsunamis?

Patterns of Water and Land Features

Maps can be used to find the location of land and water features on Earth. This map shows some of the major features on land, such as mountain ranges.

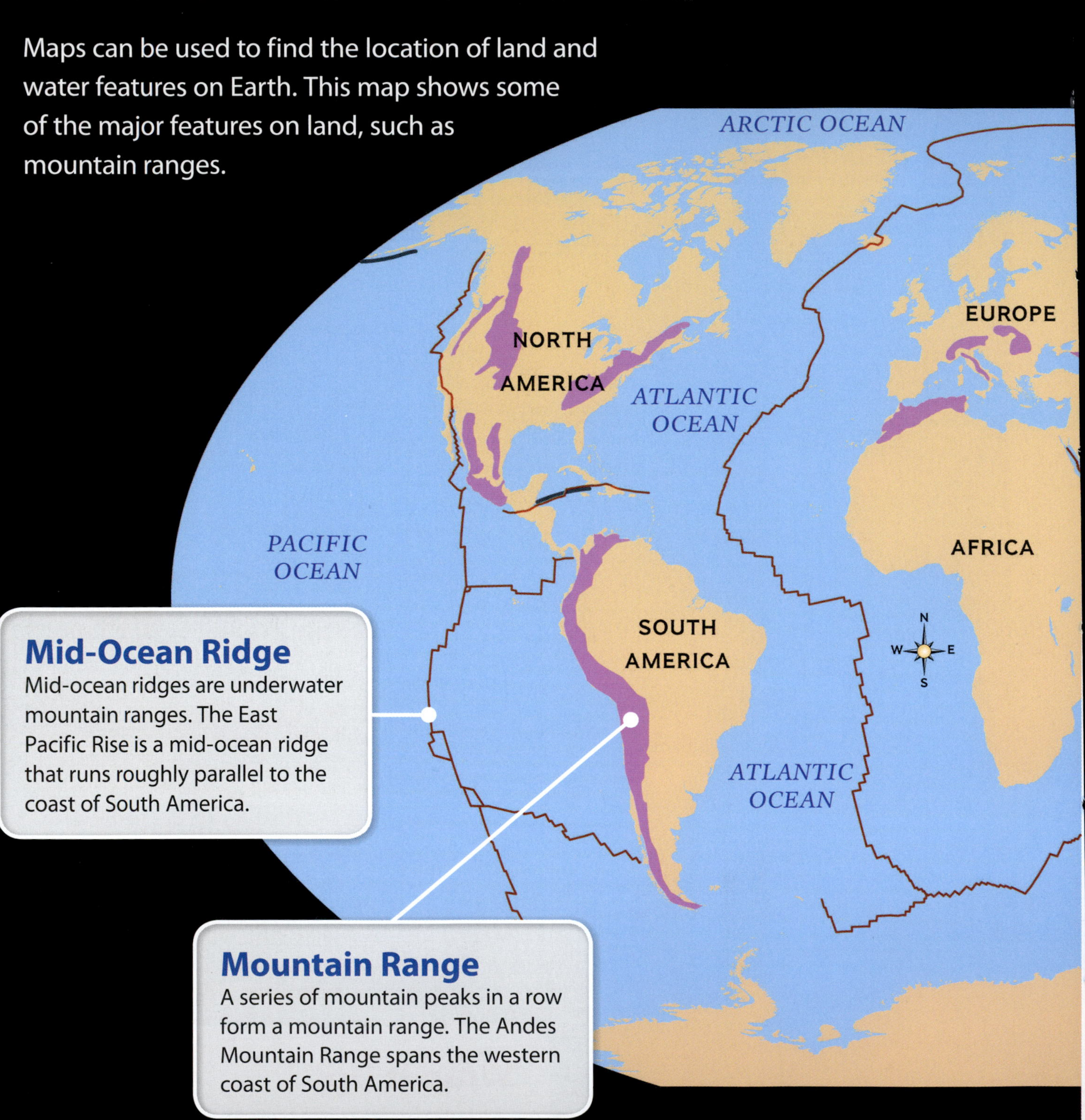

Mid-Ocean Ridge

Mid-ocean ridges are underwater mountain ranges. The East Pacific Rise is a mid-ocean ridge that runs roughly parallel to the coast of South America.

Mountain Range

A series of mountain peaks in a row form a mountain range. The Andes Mountain Range spans the western coast of South America.

NEXT GENERATION SCIENCE STANDARDS | DISCIPLINARY CORE IDEAS
ESS2.B: Plate Tectonics and Large-Scale System Interactions
The locations of mountain ranges, deep ocean trenches, ocean floor structures, earthquakes, and volcanoes occur in patterns. Major mountain chains form inside continents or near their edges. Maps can help locate the different land and water feature areas of Earth. (4-ESS2-2)

The map also shows two major features found on the ocean floor—mid-ocean ridges and deep ocean trenches. A **mid-ocean ridge** is an underwater mountain chain made up of thousands of volcanic peaks. A **deep ocean trench** is a steep underwater canyon.

My science notebook

Wrap It Up!

1. **Analyze** What symbol on the map represents a mountain range? How does the location of the mountain range in South America differ from that of the main mountain range in Asia?
2. **Analyze** What symbol represents a deep ocean trench? The Mariana Trench is the deepest trench on Earth. In which ocean is it located? Which continent is located to the west of this trench? *Hint*: Use the compass rose to find directions.
3. **Interpret** Describe the location of the East Pacific Rise.

NATIONAL GEOGRAPHIC LEARNING | Think Like a Scientist

Analyze and Interpret Data

This is the same map as on the previous pages, but two features have been added to it. One of these is the location of Earth's active volcanoes. The other is the location of some of the major earthquakes that have occurred since 1900. Study the map to see what patterns you can find.

Hawai'i

The Hawaiian Islands consist of many volcanoes. Kīlauea, Mauna Loa, and Hualalai have erupted within the past 200 years.

San Andreas Fault

The San Andreas Fault extends through California for about 1,300 kilometers (about 800 miles). It is the site of frequent earthquakes.

NEXT GENERATION SCIENCE STANDARDS | PERFORMANCE EXPECTATION
4-ESS2-2. Analyze and interpret data from maps to describe patterns of Earth's features.

Wrap It Up!

1. **Analyze** Which symbol on the map represents an active volcano? Which symbol represents a major earthquake?

2. **Interpret** Describe the pattern in the location of the active volcanoes.

3. **Analyze** Where do most earthquakes take place?

4. **Interpret** Compare the locations of earthquakes and volcanoes. What pattern do you see?

NATIONAL GEOGRAPHIC LEARNING

Think Like an Engineer
Case Study

Building for the Future

Problem

How can engineers make buildings more earthquake resistant?

In March 2011 a powerful earthquake just off the coast of Japan shook the ground and set off an enormous tsunami. The earthquake and tsunami damaged more than one million buildings and killed or injured thousands of people. Yet most buildings survived the earthquake, and many lives were saved. Why was the damage in Japan less than might have been expected? Because earthquakes are common in Japan, the government requires all new buildings to be designed and built to withstand earthquakes.

The recent disaster has inspired an engineer named Masaaki Saruta and his team to develop structures that are even more resilient. The engineers know they cannot prevent earthquakes, but they can develop solutions that will lessen the impact of these disasters.

This building was destroyed by an earthquake in Japan.

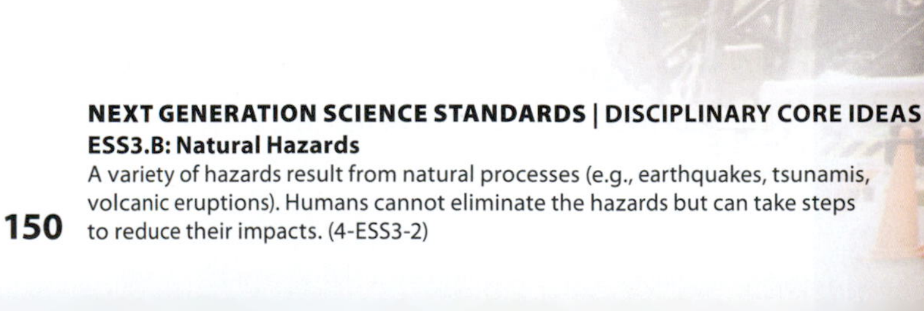

NEXT GENERATION SCIENCE STANDARDS | DISCIPLINARY CORE IDEAS
ESS3.B: Natural Hazards
A variety of hazards result from natural processes (e.g., earthquakes, tsunamis, volcanic eruptions). Humans cannot eliminate the hazards but can take steps to reduce their impacts. (4-ESS3-2)

Masaaki Saruta works at the Shimizu Corporation in Tokyo, Japan. He is the group leader of the Vibration Control Engineering Group. Saruta and his team of engineers design buildings to withstand the destructive forces of earthquakes.

Solution

Over the years, the engineers at Shimizu have developed a variety of earthquake-resistant designs. In some of the designs, buildings are separated from their foundations by a system of pads or bearings. This is called base isolation, and it separates the frame of a building from the violent shaking of an earthquake.

Masaaki and his team invented a new kind of base isolation called the core-suspended isolation system (CSI). In CSI, the core of a building is a large pillar of reinforced concrete. The floors of the building hang from the top of the core. The core is isolated from the rest of the building by large rubber bearings. When an earthquake strikes, the core absorbs the vibrations. The core may sway, but the floors of the building do not sway. Instead they remain upright.

The team has also designed a building that partially floats on water! The building stands on rubber bearings and its foundation rests in a pool of water. To test their design, they built a large model of the building. Their tests showed that water reduces the movement of a building. But the real test came during the March 2011 earthquake, when the engineers found that the features of their building cut the effects of shaking by more than half.

Masaaki says, "We want to come up with technologies that save people's lives." His team's engineering solutions are helping to accomplish this goal.

The Safety and Security Center was constructed using the core-suspended isolation system.

Wrap It Up!

1. **Summarize** Describe how a building is constructed using the core-suspended isolation system.
2. **Explain** How does CSI lessen the impact of an earthquake?
3. **Relate** How are base isolation and core-suspended isolation systems related?

NATIONAL GEOGRAPHIC LEARNING | Think Like an Engineer

Generate and Compare Solutions

You've read about Masaaki Saruta and his team's work to design and build earthquake-resistant buildings. Now it's your turn to use some of the same techniques engineers use. Working with a team, you'll design and test model houses to see if you can design and build a more damage-resistant structure.

1. **Define the problem.**
   How can you make a house more earthquake-resistant?

2. **Find a solution.** 

   - Work with a team. Think about how earthquakes damage buildings. You may need to do some research to find out more about earthquake damage to different kinds of structures. Then you can decide what kind of house you are going to build, how you will build it, and what kind of damage you want to prevent. Record your ideas as you develop your plan.
   - Now think about the materials you'll need and the methods you'll use to make your house earthquake resistant. How will you determine whether your house is resistant to earthquakes? How will you simulate an earthquake? Write out a plan for your investigation.

3. **Test your solution.**
   - Collect the materials you need and build your model house. Subject the house to an earthquake. What damage did it cause? How might you protect the house from this damage? Build a new house and test your ideas. Record your observations.

NEXT GENERATION SCIENCE STANDARDS | PERFORMANCE EXPECTATION
4-ESS3-2. Generate and compare multiple solutions to reduce the impacts of natural Earth processes on humans.

DISCIPLINARY CORE IDEAS
ETS1.B: Designing Solutions to Engineering Problems
Testing a solution involves investigating how well it performs under a range of likely conditions. (secondary to 4-ESS3-2)

This model house has been built with a set of braces inside its structure to make it more stable in earthquakes. The design team uses a large moving platform to reproduce an earthquake. Then they check to see how well the model house has survived being shaken.

- Adjust your design, and test again. Be sure that you conduct a fair test. In a fair test you change only one variable each time you carry out the test. Record your observations each time you carry out a new test.

4. Refine or change your solution.

- Analyze your results. Which solutions worked best? How do you know?
- Once you have at least two solutions for making your house resistant to earthquake damage, develop a document or presentation that clearly describes your findings. Use diagrams with labels to explain your solutions, and provide evidence from your tests to support your selection of the best designs.
- Share your designs with another team. Explain your evidence showing why you selected these two designs as the best ones. Record comments and suggestions from your classmates. Which of their ideas would improve your plan? After you have discussed your design solutions with your classmates, revise the design of your house to make it even more earthquake resistant.

5. Present your solution.

When you are satisfied that your solution is the best you can make it, revise your presentation. Then share your solutions with the class.

The team examines cracks in the model to evaluate how well the construction resisted damage during the shaking.

The Badlands

These rock formations are found in Badlands National Park in South Dakota. The colorful horizontal bands are layers of different kinds of sedimentary rock. **Sedimentary rock** is formed from sediment on Earth's surface. The sediment is deposited in layers on land or on the sea floor. Over millions of years, the layers harden into rock. The oldest layers are on the bottom, and the newer layers on top.

Each band of rock started out as a flat layer of sediments. The sediments included the remains of plants and animals. Some of those remains became **fossils,** or traces of plants and animals that lived long ago. Scientists use fossils to tell what kinds of organisms lived at the time the different layers of sediment were laid down. The location of certain fossils also indicates the order in which rock layers formed.

NEXT GENERATION SCIENCE STANDARDS | DISCIPLINARY CORE IDEAS
ESS1.C: The History of Planet Earth
Local, regional, and global patterns of rock formations reveal changes over time due to earth forces, such as earthquakes. The presence and location of certain fossil types indicate the order in which rock layers were formed. (4-ESS1-1)

30 MILLION YEARS AGO: The light-colored layers include sandstone and layers of ash from volcanoes. Fossils of desert plants show that the land was very dry.

33 MILLION YEARS AGO: The most colorful rocks were deposited when the climate was changing the land from a wet floodplain to drier grassland. Fossils found in these rock layers include tortoises.

37 MILLION YEARS AGO: The pale gray and green layers were deposited when there were forests, rivers, and shallow lakes on land. Fossils include enormous mammals that lived on savannahs.

65 MILLION YEARS AGO: The orange and yellow layers formed from mudflats and forest soils. These layers include clay-filled holes, which show that big trees grew in the soil.

75 MILLION YEARS AGO: Ammonite fossils are found in layers of limestone or shale and show that this area was once at the bottom of the ocean.

30 MILLION

33 MILLION

37 MILLION

65 MILLION

75 MILLION

Wrap It Up!

My science notebook

1. **Define** What is a fossil?
2. **Explain** How do the layers of sedimentary rock form? Where are the oldest layers usually found?
3. **Draw Conclusions** Scientists find two layers of sedimentary rock. One layer contains a fossil of a palm tree. The other layer contains an ammonite fossil. Which layer is older? Explain.

Iceland

Iceland is an amazing place. Although it is located in the cold North Atlantic Ocean, it has many hot springs, geysers, and 35 active volcanoes! Why does Iceland have so much volcanic activity? Because the island lies on top of the Mid-Atlantic Ridge, an underwater mountain chain made up of thousands of volcanic peaks. Iceland is the only place on Earth where the Mid-Ocean Ridge rises above the surface of the ocean.

Over millions of years, volcanoes released thick flows of lava. The lava hardened into the dark rocks found all over Iceland. Volcanic rock is called igneous rock, and it does not form in layers. You won't find fossils in these rock formations, but their patterns do reveal how powerful forces have built up the land.

A large break in Earth's crust, called a **rift** or fault, runs through the middle of Iceland. It is part of the Mid-Atlantic Ridge. Land on each side of the rift pulls apart. Over time, this activity creates a new landform called a rift valley. Other land features can form, too. The lake in the photograph formed when volcanic activity caused the land to sink. Water then collected in the depression forming a lake.

NEXT GENERATION SCIENCE STANDARDS | DISCIPLINARY CORE IDEAS
ESS1.C: The History of Planet Earth
Local, regional, and global patterns of rock formations reveal changes over time due to earth forces, such as earthquakes. The presence and location of certain fossil types indicate the order in which rock layers were formed. (4-ESS1-1)

Lake Thingvallavatn is located in Iceland's rift valley. The dark ridge and split in the land to the left of the lake show the fault where huge slabs of land are being pulled in opposite directions.

My science notebook

Wrap It Up!

1. **Describe** How are volcanic forces changing Iceland?
2. **Cause and Effect** Why does Iceland have so much volcanic activity?
3. **Infer** What does Iceland's rift valley reveal about the earth's forces?

Identify Evidence

The Grand Canyon is an awesome wonder. Standing on its South Rim, you can see its plateaus and canyons spreading northward for about 29 kilometers (18 miles). In some places you can see the Colorado River at the bottom of the canyon, about 1,500 meters (5,000 feet) below.

The walls of the canyon are made up of horizontal layers of sedimentary rock. If you know what to look for, these layers reveal the history of the region. This is not the history of hundreds or thousands of years, but the history of how the landscape has changed over hundreds of millions of years!

NEXT GENERATION SCIENCE STANDARDS | PERFORMANCE EXPECTATION

4-ESS1-1. Identify evidence from patterns in rock formations and fossils in rock layers to support an explanation for changes in a landscape over time.

The Colorado River carved the deep canyon over millions of years, exposing layer after layer of rock.

The different colored layers in the walls of the Grand Canyon contain different types of rock. The oldest layers are at the bottom. The newest layers are at the top.

Think Like a Scientist

Limestone, sandstone, and shale are all sedimentary rocks. Look at the patterns in the rock formations for evidence of how the landscape in this region has changed over time. Then cite evidence from the diagram to answer the questions.

SANDSTONE This layer formed when a desert with sand dunes covered the land. The tracks of spiders and scorpions were preserved in the sandstone.

REDWALL LIMESTONE These rocks formed in a wide, shallow sea. Sea animals such as corals and crinoids lived in the calm waters.

SHALE This layer formed when the land was covered by a warm, muddy sea. Animals such as trilobites, brachiopods, and crinoids lived here.

fossil trilobite

SANDSTONE This layer formed in an ancient sea where trilobites and brachiopods lived. There are traces of waves in the sands!

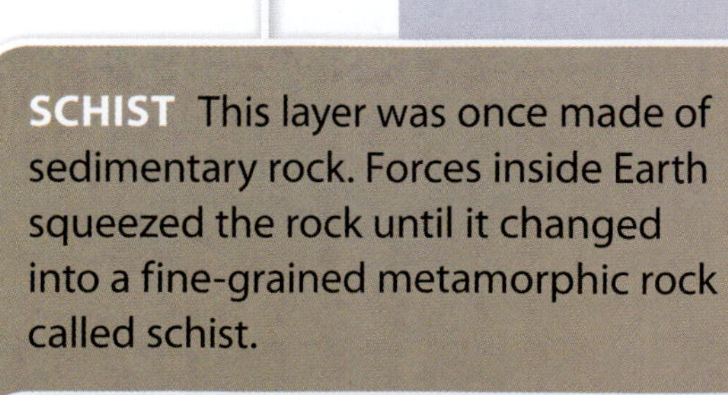

SCHIST This layer was once made of sedimentary rock. Forces inside Earth squeezed the rock until it changed into a fine-grained metamorphic rock called schist.

my science notebook

Wrap It Up!

1. **Interpreting Diagrams** What kind of rocks make up the newest layer shown in the diagram? Which rocks make up the oldest layer?
2. **Sequence** Use evidence from the diagram and what you know about sedimentary rocks to place these steps in order from first to last:
   - **a.** Sand dunes cover the land.
   - **b.** The Colorado River cuts through the layers of the land, forming a deep canyon.
   - **c.** Sedimentary rocks are squeezed together, forming schist.
   - **d.** A warm muddy sea covers the land.
3. **Infer** Use evidence from the diagram to explain how the landscape has changed over time.

Crisis Mapper

What can help speed medical aid to areas destroyed by tsunamis? What can help direct rescue helicopters to areas struck by earthquakes? What can help the United Nations deliver food and water to people suffering from droughts?

A map! Not just any map, but an online map with accurate information. In **crisis mapping,** Patrick Meier combines information from government and international agencies with tweets and text messages sent by volunteers. He and his team use this information to update an online map of the areas where disasters have struck. Crisis mapping saves lives by providing rescue workers with an up-to-the-minute picture of what is going on and where help is needed.

After the 2010 earthquake in Haiti, Patrick's efforts assisted citizens, aid workers, and the U.S. Coast Guard. Patrick's team of volunteers mapped the impact of the earthquake in near real time, providing professionals with the most up-to-date information available.

NEXT GENERATION SCIENCE STANDARDS | CONNECTION TO THE NATURE OF SCIENCE
Science is a Way of Knowing
Science is both a body of knowledge and processes that add new knowledge.

Explorer

Patrick Meier is a leader in the field of crisis mapping. Patrick didn't always think he would be working with new types of maps. In school he liked the subjects computer science and philosophy. Later in college he studied humanitarian affairs. In his work he combines his love of geography and technology in a way that helps people around the world.

NGL Science How did you get into mapping?

Patrick Meier When I was 12, the first Gulf War broke out. I had a big map of the Middle East and started physically mapping the updates with crayons and pens and markers.

NGL Science What have you and your team tracked so far?

Patrick Meier Haiti started it all. A month later there was an earthquake in Chile. Then the floods in Pakistan that summer. Russian fires in July. Floods in Brisbane in January. A major earthquake in Christchurch, New Zealand, that February.

NGL Science What are people saying about your crisis-mapping technology?

Patrick Meier Many humanitarian organizations say our crisis-mapping technology is revolutionizing disaster relief efforts. . . . Now we can pinpoint urgent needs instantly, saving time and lives.

Electricity travels through **electric circuits** to light up the night.

Glossary

A

amplitude (AMP-li-tüd)
Amplitude is the distance between the crest or trough and the middle point of a wave. (p. 55)

C

crisis mapping (CRĪ-sis MAP-ing)
Crisis mapping is the updating of an online map that shows where disasters have struck. (p. 166)

D

deep ocean trench (DĒP Ō-shun TRENCH)
A deep ocean trench is a steep underwater canyon. (p. 147)

deposition (de-pe-ZI-shun)
Deposition is the laying down of sediment and rock in a new place. (pp. 113, 117)

digitize (DIJ-i-tīz)
To digitize means to put information in digital code form. (p. 60)

E

earthquake (URTH-kwāk)
An earthquake is the shaking of the ground caused by the movement of Earth's crust. (p. 130)

electric current (i-LEK-trik KUR-ent)
Electric current is the transfer of electrical energy through a material. (p. 24)

electric circuit (i-LEK-trik SUR-kit)
An electric circuit is a complete path through which an electric current can pass. (p. 26)

electrical energy (i-LEK-trik-ul EN-er-jē)
Electrical energy is the energy of charged particles. (p. 24)

energy (EN-er-jē)
Energy is the ability to do work. (p. 4)

energy of motion (EN-er-jē uv MŌ-shun)
Energy of motion is the energy that is present when an object moves. (p. 30)

erosion (i-rō-zhun)
Erosion is the moving of sediment from one place to another. (pp. 112, 117)

erupt (ē-RUPT)
To erupt is to release melted rock, ash, and gases up through a volcano onto Earth's surface. (p. 138)

evacuate (ē-VAK-yū-āt)
To evacuate means to move from a place of danger to a safe place. (p. 142)

F

fault (FAWLT)
A fault is a break in Earth's surface where huge slabs of rock come together. (p. 132)

fossil (FO-sil)
A fossil is a trace of a plant or animal that lived long ago. (p. 158)

fossil fuel (FO-sil FYŪL)
A fossil fuel is a source of energy that formed from the remains of plants and animals that lived millions of years ago. (p. 44)

G

glacier (GLĀ-shur)
A glacier is a huge area of slow-moving ice. (pp. 120)

Global Positioning System (GPS)
(GLŌ-bul pe-ZI-shun-ing SIS-tem)
A Global Positioning System (GPS) is a tool that uses satellites to locate position. (pp. 60,)

gravity (GRA-vi-tē)
Gravity is a force that pulls things toward the center of Earth. (p. 124)

Gravity pulls Earth and the skydiver toward each other.

H

hazard (HA-zurd)
A hazard is something that is harmful or dangerous. (p. 130)

L

landslide (LAND-slīd)
A landslide is a rapid movement of rock, soil, and other material down a hill or mountain. (p. 125)

lava (LAH-vah)
Lava is melted rock that flows from a volcano onto Earth's surface. (p. 138)

liquefaction (li-qui-FAK-shun)
Liquefaction is the change of a solid area of ground into a liquid-like surface. (p. 134)

M

magma (MAG-mah)
Magma is melted rock inside Earth. (p. 138)

mid-ocean ridge (MID-ō-shun RIDJ)
A mid-ocean ridge is an underwater mountain chain made up of thousands of volcanic peaks. (p. 147)

motion (MŌ-shun)
When an object is moving, it is in motion. (p. 8)

N

nonrenewable energy source (non-rē-NŪ-i-bel EN-er-jē SORS)
A nonrenewable energy source is an energy source that will eventually run out. (p. 44)

O

organism (OR-ga-niz-um)
An organism is a living thing. (p. 122)

P

pistil (PIS-til)
A pistil is the part of the flower that forms fruit. (p. 75)

R

reflect (rē-FLEKT)
To reflect is to bounce light off of an object. (p. 88)

renewable energy source
(rē-NŪ-i-bel EN-er-jē SORS)
A renewable energy source is an energy source that will never run out. (p. 46)

rift (RIFT)
A rift is an opening where pieces of Earth's crust are being pulled apart. (160)

S

sand dune (SAND DŪN)
A sand dune is a landform caused when wind deposits sand. (p. 114)

sediment (SE-dah-ment)
Sediment is material that comes from the weathering of rock. (p. 110)

sedimentary rock (se-di-MENT-u-rē ROK)
Sedimentary rock is rock formed from sediment from the weathering and erosion of rock on Earth's surface. (p. 158)

seismograph (SĪZ-mō-graf)
A seismograph is a device that records earthquake activity. (p. 142)

seismometer (sīz-MO-mi-ter)
A seismometer is a tool that measures earthquake activity. (p. 142)

solar energy (SOL-er EN-er-jē)
Solar energy is heat and light energy from the sun. (p. 46)

stamen (STĀ-min)
A stamen is a plant part that makes pollen. (p. 75)

The **pistil** and **stamen** are parts of the flower that help the plant reproduce.

T

thermal energy (THUR-mel EN-er-jē)
Thermal energy is the energy of heat. (p. 20)

transfer (TRANS-fer)
To transfer is to pass from one object to another. (p. 8)

transform (TRANS-form)
To transform is to change. (p. 12)

transmit (trans-MIT)
To transmit means to send. (pp. 60)

tsunami (soo-NAH-mē)
A tsunami is a series of ocean waves caused by an underwater earthquake, volcano, or landslide. (p. 131)

V

volcano (vol-CĀ-nō)
A volcano is formed when molten rock deep inside Earth rises to the surface. (pp. 130)

W

wave (WĀV)
A wave transfers energy from one place or object to another. (p. 52)

wavelength (WĀV-length)
A wavelength is the distance from one crest to the next crest in a wave. (p. 55)

weathering (WE-thur-ing)
Weathering is the breaking apart, wearing away, or dissolving of rock. (pp. 110, 117)

wind energy (WIND EN-er-jē)
Wind energy is energy from wind. (p. 46)

Weathering of rock can produce unusual shapes.

Index

A

Ammonite fossils, 159

Amplifier, 25

Amplitude of waves, 54–55, 58–59

Andes Mountain Range, 146

Animals
- **of Central Plains grasslands,** 106–107
- **changes to Earth's surface,** 122–123
- **communication with dogs,** 94–97
- **eastern diamondback rattlesnake,** 93
- **of Eastern temperate forest,** 108–109
- **elephants,** 78–83
- **fossils of,** 158
- **in Pacific Northwest forest,** 102–103
- **recycling waste of,** 34–35
- **senses of,** 86–89, 92–93
- **as source of fossil fuels,** 44
- **of Southwest desert,** 104–105
- **study of movements of,** 68–69
- **wolves,** 84–85

Animal tracker, 68–69

Argument construction, 76–77, 84–85

Ash, 139

B

Backbone of an elephant, 83

Bacteria, 36

Badlands National Park, 158–159

Balls
- **change of motion,** 8–11
- **speed of,** 4–7
- **transformation of energy of,** 12–13

Bar codes, 66–67

Barrel cactus, 105

Baseball
- **change of motion of,** 8–11
- **sounds of,** 12–13
- **speed of ball,** 4–7

Base isolation, 152–153

Batteries, 26, 42

Batters, 8–9

Biodigester, 34–36

Bison, 106

Bones of an elephant, 82–83

Borneo, use of resources, 32

Borneo orangutan, 70–71

Brachiopods, 164

Brain
- **of an elephant,** 81
- **processing of sensory information,** 86–89

Bridges, construction of, 140–141

British Thermal Units (BTU), 50–51

Buildings
- **earthquake-resistant construction,** 140, 150–157
- **earthquakes' effect on,** 132–133, 134–135

Butter, melting of, 22–23

Buttercup, 76–77

C

Cacti, 105

Cairo, Egypt, 36

California
- **San Andreas Fault,** 148
- **San Francisco-Oakland Bay Bridge,** 140–141
- **Yosemite National Park,** 121

Campfire, 20–21

Carbon dioxide
- **plants' use of,** 75
- **from volcanoes,** 143

Carnival, 2–3, 24–25

Cell phones, 60–63

Central Plains grassland, 101, 106–107

Chad, Africa, 41

Chemical reactions, 16

Chemical weathering, 110

Climate
- **of Central Plains grasslands,** 101, 106–107
- **in Eastern temperate forest,** 101, 108–109
- **in Pacific Northwest forest,** 100, 102–103
- **of Southwest desert,** 100, 104–105

Clothes dryer, 21

Clouded leopard, senses of, 86–89

Coal, 44–45, 49

Code, 64–65

Collision, transfer of energy and change of motion, 8–11

Colorado River, 162–163

Communication
- **cell phones,** 62–63
- **with dogs,** 94–97
- **of elephants,** 78
- **tracking animals,** 68–69

Core-suspended isolation system (CSI), 152–153

Crisis mapper, 166–167

Culhane, Thomas Taha Rassam, 32–38

D

Data, analysis and interpretation, 148–149

Deciduous trees, 108–109

Deep ocean trench, 147

Deer Isle, Maine, 113

Deposition, 113, 114–115

Deserts, 100, 104–105

Digital map, 60–61

Digitized information
- **for cell phones,** 62–63
- **for GPS,** 60–61

Dog psychology, 94–97

Dog Psychology Center, 94

Dog Whisperer, 94–97

Dogwood trees, 102, 109

Douglas fir tree, 103

E

Early warning systems
- **for earthquakes and volcanoes,** 142–143
- **for tsunamis,** 144–145

Ears
- **of an elephant,** 79
- **hearing,** 13, 87

Earthquakes
- **building to minimize damage,** 140–141, 150–157
- **as cause of landslides,** 125
- **as cause of tsunamis,** 126, 136–137, 144, 150–151
- **causes and damage of,** 130–135
- **early warning systems,** 142–143
- **in Haiti,** 166–167
- **locations of,** 148–149

Earth's surface
- **Badlands National Park,** 158–159
- **early warning systems for natural hazards,** 142–145
- **earthquakes,** 126, 130–135, 136–137, 148–149, 150–151, 166–167
- **effects of ice,** 110, 120–121
- **landslides,** 124–125
- **living things' effects on,** 110, 112, 122–123
- **patterns of water and land features,** 146–149
- **reducing the impact of natural hazards,** 140–141
- **tsunamis,** 131, 136–137
- **volcanoes,** 130–131, 138–139
- **weathering, erosion, and deposition on,** 110–119, 126–129, 162–165

Eastern temperate forest, 101, 108–109

East Pacific Rise, 146

Ecosystems
- **desert,** 100, 104–105
- **grassland,** 101, 106–107
- **Pacific Northwest forest,** 100, 102–103
- **temperate forest,** 101, 108–109

Electrical current, 24–25

Electrical energy (electricity)
- **electric circuits,** 26–29
- **electric current,** 24–25
- **generation of,** 30–31, 36, 44–47
- **transfer and transformation of,** 24–25, 42–43

Electric circuits, 26–29

Electric guitar, 25

Elephant
- **bones and muscles of,** 82–83
- **communication of,** 78
- **external structures of,** 78–79
- **internal organs of,** 80–81

El Salvador, 124

Energy
- **of baseball's motion,** 4–11
- **of carnivals,** 2–3, 24–25
- **consumption in the U.S.,** 50–51
- **of earthquakes,** 132
- **electricity,** 24–25, 42–43
- **finding solution to problems,** 32–43
- **generation of electricity,** 30–31, 36, 44–47
- **heat,** 20–23
- **of light,** 16–19, 34, 38–41, 46–49
- **of motion,** 30–31
- **nonrenewable resources,** 44–45, 48–49, 50–51
- **nuclear power,** 45, 49
- **renewable resources,** 46–51
- **resources and the environment,** 48–49
- **sound,** 12–15

speed and, 4–7
sustainable energy, 32, 34
transfer of, 8–9, 14–19, 24–25, 52–53, 56–57
transformation of, 12–13, 16–19, 24–47
in waves, 52–53
from wind, 30–31, 46–49

Environment
effect of energy resources on, 48–49
use of waste products, 34–37

Erosion
definition of, 112
gravity's role in, 124–125
by ice, 120–121
model of, 118–119
observation of and solutions to, 126–129
prevention of, 129
by water, 112, 116–117, 126
by wind, 112, 114–115

Eruption, 138–139

Esophagus of an elephant, 81

Evacuation, 142

Evergreen trees, 102, 103

Evidence, identifying, 162–165

Explorers
Culhane, Thomas Taha Rassam, 32–37

Eyes
of an elephant, 79
sight, 16, 87–91

F

Fault lines, 132–133, 146–147, 160–161

Feet of an elephant, 79, 83

Fertilizer, 36

Fires, 107

Flower
of a buttercup, 76–77
of a wild rose, 72–73, 74–75

Fog, 102

Food of elephants, 78, 80

Force, 6–7, 8–11

Forests
Eastern temperate forest, 101, 108–109
Pacific Northwest forest, 100, 102–103

Fossil fuels
consumption in the U.S., 50–51
effect on the environment, 48–49
as nonrenewable resources, 44–45

Fossils, 158–159, 164

Fruit of a wild rose, 75

G

Game buzzer, 42–43

Gases from volcanoes, 139, 143

Gastropod fossil, 165

Generators, 30–31, 44–45

Geysers, 160

Glaciers, 120–121

Global Positioning System (GPS), 60–61

Gophers, 123

Grand Canyon, 162–163

Grasslands, 101, 106–107

Gravity
causing landslides, 124–125
erosion and deposition by, 112

Great Rift Valley, 149

Great Tōhoku Earthquake, 136–137

H

Hair dryer, 21

Haiti, 166–167

Harris's hawks, 104

Hawai'i, 148

Hearing, 13, 87

Heart of an elephant, 81

Heat
motion transformed into, 12
transfer of energy, 20–23
transformation into motion, 44
transformation of electricity into, 25

Hot springs, 160

Hydroelectric dams, 46, 48–49

I

Ice, changes in Earth's surface, 110, 112, 113, 120–121

Iceland, 160–161

Igneous rock, 160

Iguana, 104

Information technology
cell phones, 62–63
GPS, 60–61
using a code, 64–65

Insects, 75, 107, 123

Intestines of an elephant, 80

Investigations
- **earthquakes,** 134–135
- **electric circuits,** 28–29
- **heat,** 22–23
- **how waves move objects,** 56–57
- **how we see,** 90–91
- **light,** 18–19
- **motion,** 10–11
- **sound,** 14–15
- **speed,** 6–7
- **use a code,** 64–65
- **wavelength and amplitude,** 58–59
- **weathering and erosion,** 118–119

Iris, 103

J

Japan
- **building to minimize damage from natural hazards,** 150–153
- **earthquake in,** 132–133
- **islands of,** 149
- **tsunami in,** 136–137, 150–151
- **volcanoes on,** 98–99, 138–139

Jaws of an elephant, 83

K

Kalahari Desert, South Africa, 122

L

Lakes, formation of, 160–161

Lake Tingvallavatn, Iceland, 161

Land and landforms
- **in Badlands National Park,** 158–159
- **ice's effect on,** 120–121
- **living things' effects on,** 110, 112, 122–123
- **mid-ocean ridges,** 146–147, 160–161
- **mountain ranges,** 146
- **patterns of,** 146–149
- **rift valley,** 160–161
- **weathering, erosion, and deposition,** 110–121

Landslides
- **cause and effect of,** 124–125
- **as cause of tsunamis,** 131, 136, 144

Lava, 138, 160

Leaves
- **of a buttercup,** 76–77
- **of deciduous trees,** 108–109
- **of a wild rose,** 72–73

Legs of an elephant, 79

Light
- **sight and,** 88–91
- **transfer and transformation of,** 16–19, 38–41
- **transformation of electricity into,** 24–25, 26–27

Limestone, 164–165

Liquefaction, 134–135

Liver of an elephant, 80

Living things. ***See*** **Animals; Plants**

Long bones, 82

Loudness, 58

Lungs of an elephant, 80

M

Magma, 138–139, 143

Mali, food preparation in, 39

Mariana Trench, 147

Masaaki Saruta, 150–153

Matter, vibration of particles, 20

Meerkats, 122–123

Meier, Patrick, 166–167

Metamorphic rock, 164

Methane gas production, 34–35, 36

Mid-Atlantic Ridge, 160–161

Mid-ocean ridge, 146–147, 160–161

Migration of animals, 68–69

Millan, Cesar, 94–97

Models
- **of reflection of light,** 91
- **of senses,** 92–93
- **of weathering and erosion,** 118–119, 126–129

Moles, 123

Morse, Samuel, 64–65

Morse code, 64–65

Motion
- **of a baseball,** 4–11
- **change with collisions,** 8–11
- **energy of,** 30–31
- **transformation into electricity,** 44, 46–47
- **transformation into sound and heat,** 12–13
- **transformation of electricity into,** 24–25
- **transformation of energy of to electricity,** 30–31
- **of water waves,** 52–53

Mountain ranges
- **on land,** 146
- **in the ocean,** 146–147, 160–161

Movement
- **bones and muscles enabling,** 82–83
- **earthquakes causing,** 130–135
- **erosion and deposition of sediment,** 112–113
- **landslides,** 124

of rock and sediment, 112–126
of soil and rock by animals, 122–123

Mt. Sakurajima, Japan, 138–139

Muscles of an elephant, 82–83

My Science Notebook, 5, 7, 9, 11, 13, 15, 17, 19, 21, 23, 25, 27, 29, 31, 37, 38, 42, 45, 47, 49, 51, 53, 55, 57, 59, 61, 63, 65, 66–67, 73, 75, 76, 79, 81, 83, 84, 87, 89, 91, 93, 101, 103, 105, 107, 109, 111, 113, 115, 117, 119, 121, 123, 125, 126, 128, 131, 133, 135, 137, 139, 141, 143, 145, 147, 149, 153, 154, 159, 161, 165

N

National Geographic Science Careers
animal tracker, 68–69
crisis mapper, 166–167
dog whisperer, 94–97

Natural gas, 44–45, 49

Natural hazards
early warning systems, 142–145
earthquakes, 130–135
reducing the impact of, 140–141, 150–157
tsunamis, 131, 136–137
volcanoes, 130–131, 138–139

Nonrenewable energy resources
consumption in the U.S., 50–51
effect on the environment, 48–49
types of, 44–45

Northern Saw-whet Owl, 109

Nose, sense of smell, 87

Nuclear power, 45, 49

Nutrients, 75

O

Oceans
deep ocean trench, 146–147
mid-ocean ridge, 146–147, 160–161

Oil, 44–45, 49

Organisms. *See* Animals; Plants

Osaka, Japan, 132–133

Oxygen, animals' use of, 80

P

Pacific Northwest forest, 100, 102–103

Pelvis of an elephant, 82

Petals
of a buttercup, 76–77
of a wild rose, 73

Pistil of a wild rose, 74–75

Pitch, 58

Pitchers, 4–6

Plants
of Central Plains grasslands, 101, 106–107
changes to Earth's surface, 110, 123
of Eastern temperate forest, 101, 108–109
fossils of, 158
in Pacific Northwest forest, 100, 102–103
as source of fossil fuels, 44
of Southwest desert, 100, 104–105

Pollen of a wild rose, 75

Popcorn, 21

Power plant, 24, 26, 44–45

Prairie chicken, 106

Prairies, 101, 106–107

Precipitation
causing landslides, 125
in Central Plains grasslands, 101, 106–107
in Eastern temperate forest, 101, 108
erosion due to, 126–129
in Pacific Northwest forest, 100, 102–103
in Southwest desert, 100, 104–105

Psychology, 94

Purple mat, 105

R

Radio signals
tracking animals with, 68–69
transmitting GPS information, 60–61
transmitting to cell phones, 62–63

Rainfall
in Central Plains grasslands, 101, 106–107
in Eastern temperate forest, 101, 108
erosion due to, 126–129
landslides caused by, 125
in Pacific Northwest forest, 100, 102–103
in Southwest desert, 100, 104–105

Rain forest, 33

Receiver, 60–61

Recycling, 32–37

Redwall limestone, 164

Reflection of light, 88–91

Renewable energy resources
consumption in the U.S., 50–51
effect on the environment, 48–49
types of, 46–49

Reproduction, of plants, 72–73, 75

Ribs of an elephant, 82

Rift, 160–161

Rift valley, 160–161

Rocks
- **effects of ice,** 120–121
- **at fault lines,** 132
- **igneous rock,** 160
- **metamorphic rock,** 164
- **roots' effect on,** 123
- **sedimentary rock,** 158–159, 162–165
- **in volcanic eruption,** 138
- **weathering, erosion, and deposition of,** 112–113, 114–115

Roots
- **of a buttercup,** 76–77
- **changes to Earth's surface,** 123
- **of a wild rose,** 72–73, 75

S

Safety and Security Center, Japan, 152–153

Saguaro cactus, 104

Sakurajima volcano, Japan, 98–99

San Andreas Fault, 148

Sand dunes, 113, 114–115

Sandstone, 118–119, 164–165

San Francisco-Oakland Bay Bridge, 140–141

Satellite, 60–61, 145

Schist, 164

Sediment
- **erosion and deposition of,** 112–119
- **weathering forming,** 110–111

Sedimentary rock
- **formation of,** 158–159
- **in the Grand Canyon,** 162–165

Seeds
- **of desert plants,** 105
- **of a wild rose,** 72–73, 75

Seismographs, 142

Seismometer, 142

Sense receptors, 86–87, 88–89

Senses, 86–93

Shale, 164–165

Shimizu Corporation, 150–153

Sideband snail, 103

Sight, 16, 87–91

Skeletal muscles, 82

Skeleton of an elephant, 82–83

Skin of an elephant, 78

Skull of an elephant, 83

Smartphones, 60–61, 62–63, 66

Smell, 87

Soil erosion, 126–129

Solar CITIES, 33

Solar cooker, 38–41

Solar energy, 46–49

Sonora Desert, 105

SOS, 64

Sound
- **digitization of,** 62–63
- **hearing,** 86–87
- **transfer of energy,** 14–15
- **transformation of electricity into,** 24–25
- **transformation of motion into,** 12–13
- **transmission of elephant sounds,** 78–79
- **as waves,** 58

Southwest desert, 100, 104–105

Sparrows, 68–69

Speed
- **change in,** 9
- **energy and,** 4–7, 10–11

Stamens
- **of a buttercup,** 76–77
- **of a wild rose,** 74–75

Stems
- **of a buttercup,** 76–77
- **of a wild rose,** 72–73, 74–75

Stomach of an elephant, 80

Sun, energy from, 16–19, 34, 75

Surfer, 52–53

Sustainable energy, 32, 34

Switch, 26, 42

T

Tanzania, energy needs in, 33, 34–35

Taste, 87

Taste buds, 87

Technology
- **for crisis mapping,** 166–167
- **for generating energy,** 50–51
- **GPS,** 60–61
- **for reducing damage of natural hazards,** 140–141, 150–157
- **seismometer,** 142
- **tilt meters,** 143
- **tsunami warning buoy,** 144–145

Teeth of an elephant, 81

Telegraph, 64–65

Temperate forests, 108–109

Temperature, monitoring in volcanoes, 143

Thermal energy, 20–23

Thinking Skills
- **analyze,** 59, 75, 81, 119, 135, 147, 149
- **analyze and explain your results,** 40, 42
- **apply,** 9, 21, 25, 61, 91, 119, 133
- **ask,** 11
- **cause and effect,** 75, 103, 115, 123, 125, 133, 161
- **classify,** 115
- **compare,** 5, 15, 63, 65, 76, 79, 84, 109, 119, 121, 139
- **compare your data,** 40
- **conclude,** 29, 57, 113
- **construct an argument,** 76, 84
- **contrast,** 15, 79, 101, 109, 139
- **define,** 21, 25, 89, 101, 109, 123, 131, 133, 159
- **define the problem,** 38, 42, 66, 126, 154
- **describe,** 7, 27, 29, 53, 59, 79, 103, 107, 115, 143, 145, 161
- **discuss your model,** 93
- **draw conclusions,** 159
- **evaluate,** 49, 73, 79, 141
- **explain,** 5, 7, 31, 37, 51, 53, 65, 73, 83, 89, 91, 105, 111, 117, 121, 125, 135, 137, 143, 145, 153, 159
- **find a solution,** 42, 66, 126, 154
- **generalize,** 17, 76, 84, 101, 113, 117
- **give evidence,** 15, 19, 23
- **identify,** 37, 45, 75, 81, 87, 103, 139
- **infer,** 9, 13, 19, 51, 55, 83, 89, 105, 107, 109, 125, 131, 161, 165
- **interpret,** 45, 147, 149
- **interpret diagrams,** 165
- **list,** 45, 47, 73, 76, 84, 131
- **make a model,** 93
- **make judgments,** 49
- **name,** 9, 83
- **observe,** 15, 23
- **predict,** 11, 19, 23, 27, 29, 51
- **present your results or solution,** 40, 42, 128, 156
- **recall,** 13, 17, 25, 31, 55, 105, 111, 113, 137
- **refine or change your solution,** 40, 42, 67, 128, 156
- **relate,** 21, 47, 57, 87, 153
- **research,** 139
- **restate,** 61
- **review,** 141
- **revise,** 135
- **revise your model,** 93
- **sequence,** 63, 81, 165
- **share your model or solution,** 67, 93
- **summarize,** 107, 137, 153
- **test your solution,** 40, 42, 67, 128, 156

Think Like an Engineer
- **Building for the Future case study,** 150–153
- **Compare Multiple Solutions,** 66–67
- **Design, Test, and Refine a Device,** 38–43
- **Finding Solutions to Energy Problems case study,** 32–37
- **Generate and Compare Solutions,** 154–157
- **Make Observations,** 126–129

Think Like a Scientist
- **Analyze and Interpret Data,** 148–149
- **Construct an Argument,** 76–77, 84–85
- **Identify Evidence,** 162–165
- **Obtain and Combine Information,** 50
- **Use a Model,** 92–93

Thorns, 73

Tiltmeters, 143

Toadstool caprocks, 115

Toilet wastes, 36

Touch, 87

Transfer of energy
- **of baseball's motion,** 8–11
- **electricity,** 24–29
- **heat,** 20–23
- **light,** 16–21
- **sound,** 12–15
- **in waves,** 52–53, 56–57

Transformation of energy
- **electricity into other forms,** 24–29
- **by game buzzer,** 43
- **generation of electricity,** 30–31, 44–47
- **by "home-scale" biodigester,** 34–37
- **light from the Sun,** 16–19, 34, 38–41, 46–49
- **of motion into sound and heat,** 12–13
- **recycling waste products,** 32–39
- **by a solar cooker,** 38–41

Transmission of waves, 60–61

Trees
- **in Central Plains grasslands,** 106–107
- **changes to Earth's surface,** 123
- **in Pacific Northwest forest,** 100, 102–103
- **in temperate forests,** 108–109

Trillium, 109

Trilobite fossil, 164

Trunk of an elephant, 79

Tsunamis
- **aid to survivors of,** 166–167
- **building to minimize damage,** 150–153
- **causes and damage of,** 131, 136–137
- **detection technology,** 144–145

Tsunami warning buoy, 144–145

Turbines, 30–31, 44–45, 46, 47

U

United States
- **Badlands National Park,** 158–159
- **Central Plains grassland,** 106–107
- **consumption of resources in,** 50–51
- **Eastern temperate forest,** 101, 108–109
- **Grand Canyon in,** 162–163
- **Pacific Northwest forest,** 100, 102–103
- **Southwest desert,** 100, 104–105
- **Yosemite National Park,** 120–121
- **Zion Canyon, Utah,** 116–117

V

Veins of plant leaves, 75

Vibration Control Engineering Group, 151

Vibrations
- **of particles of matter,** 20
- **of sound,** 12–15, 86
- **technology for reducing damage from,** 140–141, 150–157

Virgin River, 116–117

Volcanoes
- **cause and effect of eruption,** 130–131, 138–139
- **as cause of landslides,** 125
- **as cause of tsunamis,** 126, 136
- **early warning systems,** 142–143
- **of Hawai'i,** 148
- **in Iceland,** 160
- **in Japan,** 98–99, 138–139
- **locations of,** 148–149

W

Waffle iron, 25

Waste products, 34–37

Water
- **animals' use of,** 80
- **erosion and deposition by,** 112–113, 116–117
- **monitoring in volcanoes,** 143
- **patterns of,** 146–149
- **transportation of in plants,** 75
- **weathering by,** 110–111

Water turbines, 30, 46

Water waves, 52–55, 131, 136–137, 144–145

Wavelength, 55, 58–59

Waves
- **how they move objects,** 56–57
- **properties of,** 54–55, 58–59
- **radio signals,** 60–63
- **transfer of energy,** 52–53
- **tsunamis,** 131, 136–137, 144–145

Weathering
- **causes of,** 110–111, 114–115
- **formation of sediment,** 112
- **model of,** 118–119

Western tanager, 103

Whiskers, 87

White-tailed deer, 109

Wikelski, Martin, 68–69

Wild rose
- **external structures of,** 72–73
- **internal structures of,** 74–75

Wind
- **generation of electricity,** 30–31, 46–49
- **weathering, erosion, and deposition by,** 110–115

Wind turbines, 30–31, 47

Wolves, structures of, 84–85

Woodchucks, 123

Worms, 123

Y

Yosemite National Park, 120–121

Z

Zabaleen trash pickers, 36

Zion Canyon, Utah, 116–117

Photographic and Illustrator Credits

Front Matter

Title Page ©Diane Cook & Len Jenshel/National Geographic Creative. **ii–iii** ©Josh Westrich/Bridge/Corbis. **iv–v** ©Frans Lanting/National Geographic Creative. **vi–vii** ©Egmont Strigl/imagebroker/Corbis.

Physical Science: Energy / Waves: Waves and Information

2–3 ©good4nothing/Flicker/Getty Images. **4–5** ©Jim Cummins/Cusp/Corbis. **5** (cl) ©Spectruminfo/Shutterstock.com. (cr) ©Spectruminfo/Shutterstock.com. **6–7** ©Bunsen Bookworm/Solus/Corbis. **6** (cl) ©Michael Goss Photography/National Geographic Learning. (c) ©National Geographic School Publishing. **7** (tl) ©Michael Goss Photography/National Geographic Learning. (tr) ©Michael Goss Photography/National Geographic Learning. **8–9** ©Tim McGuire/Photolibrary/Getty Images. **10–11** ©Rob Friedman/E+/Getty Images. **10** (cl) ©Michael Goss Photography/National Geographic Learning. (c) ©Michael Goss Photography/National Geographic Learning. **11** (tl) ©Michael Goss Photography/National Geographic Learning. (tr) ©Michael Goss Photography/National Geographic Learning. (cl) ©Michael Goss Photography/National Geographic Learning. **12–13** ©Jim Cummins/Flirt/Corbis. **12** (cl) ©Stockbyte/Stockbyte/Getty Images. **13** (tl) ©IE149/Image Source Plus/Alamy. (tr) ©Joseph Sohm/Visions of America/age fotostock. **14–15** ©Grant Faint/The Image Bank/Getty Images. **14** (c) ©National Geographic School Publishing. (cl) ©National Geographic Learning. **15** (tl) ©Michael Goss Photography/National Geographic Learning. (tr) ©Michael Goss Photography/National Geographic Learning. **16–17** ©Courtesy of NASA/SDO and the AIA, EVE, and HMI science teams. **18–19** ©Chris Minerva/Stockbyte/Getty Images. **18** (cl) ©iStockphoto.com/mikemphoto. (c) ©Jeanine Childs/National Geographic Learning. **19** (tl) ©Jeanine Childs/National Geographic Learning. (tr) ©Jeanine Childs/National Geographic Learning. **20–21** ©David Spurdens/Terra/Corbis. **21** (tr) ©Michael Haegele/Corbis. (cr) ©George Doyle/Stockbytee/Getty Images. **22–23** ©Ha Huynh/E+/Getty Images. **22** (bl) ©Michael Goss Photography/National Geographic Learning. (bc) ©Michael Goss Photography/National Geographic Learning. (br) ©Michael Goss Photography/National Geographic Learning. **23** (tl) ©Michael Goss Photography/National Geographic Learning. (cl) ©Michael Goss Photography/National Geographic Learning. **24–25** ©ngkaki/E+/Getty Images. **25** (tr) ©Creativ Studio Heinemann/Westend61 GmbH/Alamy. (cr) ©Thomas Northcute/Photodisc/Getty Images. **26–27** ©George Diebold/Photographer's Choice/Getty Images. **28–29** ©Yuji Kotani/Digital Visione/Getty Images. **28** (cl) ©Jennifer Shaffer/National Geographic School Publishing. (c) ©National Geographic School Publishing. **29** (tl) ©Jennifer Shaffer/National Geographic School Publishing. (cl) ©Jennifer Shaffer/National Geographic School Publishing. **30–31** ©Brian Lawrence/Photographer's Choice/Getty Images. **31** (tr) ©Claus Lunau/Bonnier Publications/Photo Researchers, Inc. **32–33** ©T.H. Culhane. **33** (tr) ©John Livzey. **34–35** ©Hanna Fathy/T.H. Culhane. **34** (tr) ©T.H. Culhane. (cr) ©T.H. Culhane. (br) ©T.H. Culhane. **36–37** ©Myriam Abdel Aziz. **38–39** ©Joerg Boethling/Alamy. **40–41** ©Orjan F. Ellingvag/Dagens Naringsliv/Corbis News Premium/Corbis. **42–43** ©photonic 15/Alamy. **44–45** ©JAMES FORTE/National Geographic. **44** (tr) ©Ocean/Corbis. **45** (tl) ©Lester Lefkowitz/Iconica/Getty Images. (tr) ©Martin Muránsky/Alamy. **46–47** ©Rafa Irusta/Shutterstock.com. **46** (tr) ©Education Images/UIG/Universal Images Group/Getty Images. **47** (tr) ©Brian Lawrence/Photographer's Choice/Getty Images. **48–49** ©Ashley Cooper/Global Warming Images/Alamy. **49** (tc) ©JAMES FORTE/National Geographic. (tc) ©Lester Lefkowitz/Iconica/Getty Images. (c) ©Ocean/Corbis. (c) ©Martin Muránsky/Alamy. (c) ©Rafa Irusta/Shutterstock.com. (bc) ©Brian Lawrence/Photographer's Choice/Getty Images. (bc) ©Education Images/UIG/Universal Images Group/Getty Images. **50–51** © Guy Vanderelst/Photographer's Choice RF/Getty Images. **52–53** ©Andrey Artykov/Vetta/Getty Images. **53** (t) ©EpicStockMedia/Shutterstock.com. **54–55** ©Mustafa Deliormanli/istockphoto/Getty Images. **56–57** ©Terry A Parker/All Canada Photos/Getty Images. **56** (bl) ©Michael Goss Photography/National Geographic Learning. (bc) ©National Geographic Learning. **57** (tl) ©Michael Goss Photography/National Geographic Learning. (tr) ©Michael Goss Photography/National Geographic Learning. **58–59** ©Gerard Lacz Images/SuperStock. **58** (cl) ©Michael Goss Photography/National Geographic Learning. (c) ©National Geographic School Publishing. (bl) ©Michael Goss Photography/National Geographic Learning. (bc) ©Michael Goss Photography/National Geographic Learning. **59** (tl) ©Michael Goss Photography/National Geographic Learning. (tr) ©Michael Goss Photography/National Geographic Learning. **60–61** ©i love images/men's lifestyle/Alamy. **60** (bc) ©National Geographic Learning. (cl) ©National Geographic Learning. **62–63** ©Lane Oatey/Blue Jean Images/Getty Images. **62** (tr) ©Bruce Laurance/Blend/Corbis. **64–65** ©Tetra Images/Getty Images. **64** (cl) ©Jennifer Shaffer/National Geographic School Publishing. **65** (tl) ©Michael Goss Photography/National Geographic Learning. (tr) ©Michael Goss Photography/National Geographic Learning. **66–67** ©Daniel Koebe/Fancy/Age Fotostock. **68–69** ©Caters News/ZUMAPRESS.com/Newscom. **68** (bl) ©Chicago Tribune/McClatchy-Tribune/Getty Images. **69** (tr) ©Roland Knauer/Alamy.

Life Science: Structure, Function, and Information Processing

70–71 ©Rolf Nussbaumer Photography/Alamy. **72–73** ©Klaus Honal/Encyclopedia/Corbis. **74** (bl) ©Odilon Dimier/ZenShui/Corbis. **76–77** ©Kathy Collins/Photographer's Choice/Getty Images. **78–79** ©Joel Sartore/National Geographic Creative. **80–81** ©Joel Sartore/National Geographic Creative. **81** (cr) ©imagebroker/Alamy. **82–83** ©Joel Sartore/National Geographic Creative. **84–85** ©Joel Sartore/National Geographic Creative. **86–87** ©Joel Sartore/National Geographic Creative. **88** ©Animal Imagery/Alamy. **90–91** ©Tetra Images/Getty Images. **90** (cl) ©Jennifer Shaffer/National Geographic Learning. (c) ©Jeanine Childs/National Geographic Learning. (cr) ©Jeanine Childs/National Geographic Learning. **91** (tr) ©Jeanine Childs/National Geographic Learning. (cr) ©Jeanine Childs/National Geographic Learning. **92–93** ©U.S. Fish and Wildlife Service. **93** (bc) ©Paul Sutherland/National Geographic Creative. **94–95** ©Evan Hurd/Alamy. **95** (tr) ©Mark Thiessen/National Geographic Creative. **96–97** © Mark Thiessen/National Geographic Creative.

Earth Science: Earth's Systems: Processes That Shape the Earth

98–99 ©National News/ZUMA Press/Newscom. **100** (cl) ©Greg Vaughn/Alamy. (br) ©Tonda/Age Fotostock. **101** (tr) ©John Elk/Lonely Planet Images/Getty Images. (cr) ©Tim Mainiero/Alamy.

102–103 ©Greg Vaughn/Alamy. **103** (tl) ©Robert C. Paulson/Alamy. (cl) ©Thomas Chamberlin/Alamy. (cr) ©FLPA/Bob Gibbons/FLPA/Age Fotostock. (tr) ©ClassicStock/Alamy. **104–105** ©Tonda/Age Fotostock. **104** (cl) ©Tim Zurowski/Age Fotostock. (bl) ©John Cancalosi/Age Fotostock. **105** (tr) ©George Grall/National Geographic Creative. **105** (cr) ©Bob Gibbons/Alamy. **106–107** ©John Elk/Lonely Planet Images/ Getty Images. **106** (cl) ©Kallista Images/Kallista Images/Getty Images. (bl) ©Joel Sartore/National Geographic Creative. **107** (tr) ©Donald Erickson/E+/Getty Images. (cr) ©Joel Sartore/National Geographic Creative. **108–109** ©Tim Mainiero/Alamy. **109** (tr) ©Raymond Gehman/ National Geographic Creative. (cl) ©Marvin Dembinsky Photo Associates/Alamy. (cr) ©Mathew Levine/Flickr/Getty Images. (bl) ©H. Mark Weidman Photography/Alamy. **110–111** ©Taylor S. Kennedy/ National Geographic Creative. **112–113** ©James Randklev/Corbis. **114–115** ©apdesign/Shutterstock.com. **115** (c) ©Yva Momatiuk & John Eastcott/Minden Pictures. **116–117** ©Michele Falzone/Photographer's Choice RF/Getty Images. **117** (cr) ©Oleksandr Buzko/Alamy. **118–119** ©George F. Mobley/National Geographic Creative. **118** (c) ©National Geographic School Publishing. **119** (cl) ©National Geographic School Publishing. (cr) ©National Geographic School Publishing. **120–121** ©Creatas/Jupiter Images. **120** (tr) ©Carr Clifton/Minden Pictures. **121** (tr) ©Corbis RF/Alamy. (tl) ©Alex Neauville/Shutterstock.com. **122–123** ©Simon King/Nature Picture Library. **123** ©geogphotos/ Alamy. **124–125** ©G. R. Roberts/Natural Sciences Image Library/Visuals Unlimited/Getty Images. **124** ©AP Images/LA PRENSA GRAFICA. **126–127** ©D. P. Burnside/Science Source/Photo Researchers. **128–129** ©David R. Frazier/Photo Researchers. **130–131** ©Ho New/Reuters. **130** (tc) ©TED ALJIBE/AFP/Getty Images. **131** (tr) ©JOANNE DAVIS/ AFP/GETTY IMAGES/Newscom. **132–133** ©Dario Mitidieri/Edit/Getty Images. **134–135** ©Rob Grange/Photographer's Choice/Getty Images. **134** (bl) ©National Geographic School Publishing. **135** (tl) ©National Geographic Learning. (tr) ©National Geographic Learning. (cl) ©Jeanine Childs/National Geographic Learning. (cr) ©Jeanine Childs/ National Geographic Learning. **136–137** ©Miyako City Office/Handout/ Reuters. **136** (cl) ©Mainichi Shimbun/Reuters. **138–139** ©AP Images/ Kagoshima Local Meteorological Observatory. **139** (cr) ©The Asahi Shimbun/Getty Images. **140–141** ©Ezra Shaw/Getty Images. **142–143** ©YAMAPHOTO/Flickr/Getty Images. **142** (cl) ©Inga Spence/Science Source. (cr) ©David Butow/Corbis. **143** ©U. S. Geological Survey. **144–145** ©Jason Edwards/National Geographic Creative. **144** (cl) ©NOAA. **145** (c) ©Precision Graphics. **150–151** ©Dario Mitidieri/Photonica World/ Getty Images. **151** (tr) ©KimimasaMayama/Bloomberg/Getty Images. **152–153** ©KimimasaMayama/Bloomberg/Getty Images. **154–155** ©Roger Ressmeyer/Nomad/Corbis. **156–157** ©Roger Ressmeyer/ Encyclopedia/Corbis. **158–159** ©Tim Fitzharris/Minden Pictures. **159** (tc) ©Layne Kennedy/Corbis. (tr) ©Robert Hynes. (c) ©Layne Kennedy/ Corbis. **160–161** ©Emory Kristof/National Geographic Stock. **162–163** ©Kirk Lougheed/Flickr/Getty Images. **163** (tc) ©Michele Falzone/Alamy. **164** (cl) ©Albert J. Copley/Age Fotostock. (bl) ©NPS/Alamy. (br) ©Peter Essick/National Geographic Creative. **165** ©Tom Bean/Corbis. **166–167** ©Kris Krug. **167** (tr) ©Rebecca Hale/National Geographic Creative.

End Matter

168–169 ©Albert Tan photo/Flickr/Getty Images. **170–171** ©Rick Neves/ Flickr/Getty Images. **172–173** ©blickwinkel/Huetter/Alamy. **174–175** ©Alfons Hauke/Getty Images.

Illustrator Credits

Unless otherwise indicated, all maps were created by Mapping Specialists and all illustrations were created by Precision Graphics.

Text Acknowledgments

94–97 Source: Three Rivers Press (Random House LLC), *Cesar's Way*, 2006.

Content Consultants

Randy L. Bell, Ph.D.
Associate Dean and Professor, Oregon State University

Malcolm B. Butler, Ph.D.
Associate Professor of Science Education, School of Teaching, Learning and Leadership, University of Central Florida

Kathy Cabe Trundle, Ph.D.
Department Head and Professor, STEM Education, North Carolina State University

Judith S. Lederman, Ph.D.
Associate Professor and Director of Teacher Education, Illinois Institute of Technology

Acknowledgments
Grateful acknowledgment is given to the authors, artists, photographers, museums, publishers, and agents for permission to reprint copyrighted material. Every effort has been made to secure the appropriate permission. If any omissions have been made or if corrections are required, please contact the Publisher.

Photographic and Illustrator Credits
Front cover wrap ©Diane Cook & Len Jenshel/National Geographic Creative. **Back cover** (tl) ©John Livzey. (cl) ©Michael Melford/National Geographic Creative.

Acknowledgments and credits continued on page 184.

Visit National Geographic Learning online at NGL.Cengage.com

Visit our corporate website at www.cengage.com

Printed in the USA.
RR Donnelley, Willard, OH

ISBN: 978-12858-46361

14 15 16 17 18 19 20 21 22 23

10 9 8 7 6 5 4 3 2 1

ARTS OF *Diplomacy*

Castle McLaughlin

WITH CONTRIBUTIONS BY
Mike Cross
Pat Courtney Gold
T. Rose Holdcraft
Gaylord Torrence
Anne-Marie Victor-Howe

FOREWORD BY James P. Ronda
PREFACE BY Rubie Watson
PHOTOGRAPHS BY Hillel S. Burger

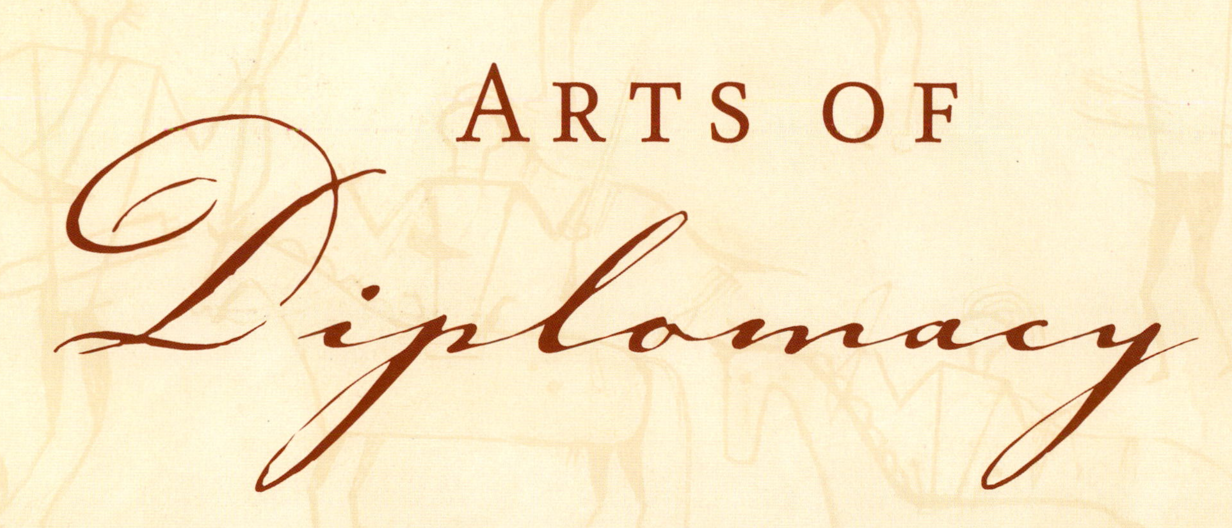

ARTS OF Diplomacy

LEWIS AND CLARK'S INDIAN COLLECTION

Peabody Museum of Archaeology and Ethnology, Harvard University CAMBRIDGE, MASSACHUSETTS
University of Washington Press SEATTLE AND LONDON

University of Washington Press
P.O. Box 50096, Seattle, Washington 98145
www.washington.edu/uwpress

Peabody Museum of Archaeology and Ethnology, Harvard University
11 Divinity Avenue, Cambridge, Massachusetts 02138
www.peabody.harvard.edu/publications

Library of Congress Cataloging-in-Publication Data
McLaughlin, Castle
Arts of diplomacy : Lewis and Clark's Indian collection / Castle McLaughlin with contributions by Mike Cross . . . [et al.] ; photographs by Hillel S. Burger.
p. cm.
Includes bibliographical references and index.
ISBN 0-295-98360-4 (alk. paper)—ISBN 0-295-98361-2 (pbk. : alk. paper)
1. Lewis and Clark Expedition (1804-1806)—Ethnological collections. 2. Peabody Museum of Archaeology and Ethnology. 3. Indians of North America—Commerce—West (U.S.) 4. Indians of North America—Material culture—West (U.S) 5. Indians of North America—Antiquities—Collectors and collecting—West (U.S.) 6. West (U.S.)—Discovery and exploration. 7. West (U.S.)—Antiquities—Collection and preservation. I. Title.

F592.7.M35 2003
978.004'97—dc21 2003053584

♾ The paper used in this publication meets the minimum requirements of the American National Standard for Information Sciences—Permanence of Paper for Printed Library Materials, ANSI Z39.48-1984.

Printed and bound in China by C & C Offset Printing Co., Ltd.

For Sarah Hrdy and Rubie Watson, who made this work possible,

and for all travelers of good heart.

Contents

List of Illustrations viii

Foreword: The Names of the Nations, BY JAMES P. RONDA xiii

Preface, BY RUBIE WATSON xxi

Introduction: Discovering Lewis and Clark's Indian Collection 3

PART ONE The Life History of a Collection

1 The Lewis and Clark Expedition: An American Quest for Commerce and Science 17

The Fabric of Empire 31

2 Up the Missouri: Patterns of Diplomacy and Exchange 33

Glass Beads 38

From Time Immemorial: The Mandan, Hidatsa, and Arikara People, BY MIKE CROSS 48

3 Selections: The Making of a Collection 53

4 Into the Museum: From Gifts to Artifacts 69

PART TWO The Peabody Museum Objects

5 From Warriors and Women Traders: Objects Collected by Meriwether Lewis and William Clark 85

Jo Esther Parshall, Quillwork Artist 91

The Raven Belt Ornaments of Lewis and Clark, BY GAYLORD TORRENCE 106

6 The Army Moves West: The Curious Collection of Lieutenant George C. Hutter 127

Missouri Melodies: Flute Player Keith Bear 141

7 Enigmatic Icons: Objects Probably Collected by Lewis and Clark or by Lieutenant Hutter 147

Butch Thunder Hawk, Painter 159

8 The Language of Pipes 201

Identifying Feathers 220 *Identifying Wood* 225 *Silk Ribbons* 229

9 Grizzly Claws, Garters, and Fashionable Hats: Other Possible Expedition Objects 251

A Wasco Weaver Meets Her Ancestors through Lewis and Clark, BY PAT COURTNEY GOLD 282

Cedar: The Tree of Life 293

Afterword: The Peabody–Monticello Native Arts Project 313

Notes 317 References Cited 325 Acknowledgments 341

Contributors 345 Picture Credits 347 Index 351

Illustrations

FRONTISPIECE Detail of quillwork on otter bag

18–19 MAP: The Corps of Discovery's route

2 Pat Courtney Gold and Anne-Marie Victor-Howe

5 Peale Museum label

7 Daniella Bar-Yosef

10 English brass buttons and Italian glass beads

12 Interior of Northwest Coast whaler's hat

16 Detail of beadwork on quiver

23 Portrait of Thomas Jefferson

25 Portrait of Meriwether Lewis

26 Portrait of William Clark

29 Lewis's woodpecker

31 Trade cloth on a pipe stem

32 Detail of pictographic bison robe

34 Lewis and Clark holding council with Indians

36 Portrait of Black Moccasin

38 Blue glass beads on baby carrier

41 Detail of Indian combat on pictographic bison robe

42 Mandan corn

42 War ax drawing from Lewis and Clark's journals

45 Jefferson Peace and Friendship Medal

46 Brass button

47 Brass button

48 Mike Cross and his daughters

52 Detail of feathers on a calumet

54 Pictograph of a chief on bison robe
55 George Catlin feasting with Four Bears
56 Portrait of Sheheke
60 Detail of biographical war shirt
61 Feathers on a calumet stem
62 Camas bulb, cous roots, and berry cakes
65 Journal drawing of woven hats
68 Self-portrait of Charles Willson Peale
74 Meriwether Lewis in Shoshone costume
76 P. T. Barnum poster
79 Boston Museum advertisement
80 Fire damage at the Boston Museum
81 Northwest Coast cases at the Peabody Museum
84 Detail of raven belt ornament
86 Peale Museum label
87 Otter bag
90 Quillwork on tail of otter bag
91 Jo Esther Parshall and Elaine McLaughlin
92 Jo Esther Parshall, quilling
94 Basketry whaler's hat
95 Basketry whaler's hat
99 Nootka chief wearing whaling hat
101 Top view of whaler's hat
106 Peale Museum label
109 Raven belt ornament
110 Raven belt ornament
111 Raven belt ornament
112 Portrait of Pash-ee-pa-ho
113 Mato Wamniomni wearing raven bustle
123 Autobiographical drawings by Wacochachi
126 Beaded baby carrier
128 Portrait of Lt. George C. Hutter
131 Drawing of the Missouri River
135 Detail of red wool insert on baby carrier
137 Beaded choker
137 Portrait of Mandeh-Pahchu
139 Eagle bone whistle
139 Peale Museum label
140 Wooden flute
141 Flute player Keith Bear
143 Gourd rattle
144 Knife sheath
146 Pictographic bison robe
148 Battle scene on bison robe
151 Drawing of Awatixa Hidatsa village
155 Two mounted warriors on bison robe
157 Shaman on bison robe
159 Butch Thunder Hawk with shield
159 Thunder Hawk's earth pigments
161 Peale Museum label
162 Peale Museum label
165 Quillwork, beads, and horsehair on war shirt
166 Human figure on war shirt
167 Biographical war shirt
169 Reverse side of war shirt
171 Detail of "Sioux-type" side-fold dress
172 Peale Museum label
174 Side-fold dress pattern
176 Painted side-fold dress
179 Detail of painted dress
180 Portrait of Chan-Chä-Uiá-Teüin
181 "Sioux-type" side-fold dress
183 Portrait of Tchon-su-mons-ka
185 Detail of tinklers at dress hemline
187 Cowrie shells on dress yoke
189 Quilled bison robe
190 Hood and straps on quilled bison robe
194 Quill ornaments on bison robe
198 Pendant hooves on bison robe
200 Eagle-feather tips on fan calumet

205 Lakota feast for traders and U.S. officials
209 Portrait of Grizzly Bear
213 Peale silhouette of Tahawarra
214 Drawing of Native American pipe forms
216 Calumet fragments
217 Calumet
218 *Omaha Sacred Pipes of Fellowship*
219 Mallard pelt on a calumet stem
220 Carla Dove
222 Calumet stem
223 Calumet stem
224 Calumet stem
225 Bruce Hoadley
225 Decorative woodworking on pipe stem
227 Calumet stem
227 Calumet stem
227 Calumet stem
227 Calumet stem
227 Detail of decoration on calumet stem
228 Quillwork thunderbird design on pipe stem
229 Silk ribbons on calumet stem
231 Pipe stem
231 Bladder bag
232 Pipe bag
233 Pipe stem
233 Pipe stem
235 Pipe stem
235 Garnet beads on pipe stem
236 Pipe stem
236 Quillwork and beaded fabric on pipe stem
236 Pipe stem
237 Decoration on pipe stem
238 Pipe stem
238 Pipe stem
240 Tube pipe
241 Pipe bowl
241 Pipe bowl
241 Pipe bowl
242 Pipe bowl
242 Pipe bowl
243 Pipe bowl
243 Pipe bowl
244 Pipe bowl
244 Pipe bowl
245 Pipe bowl
246 Pipe bowl
247 Pipe bowl
248 Pipe bowl
248 Colonial eagle motif on pipe bowl
249 Pipe bowl
249 Pipe bowl
250 Quiver with arrows
253 Arrow points
256 Elk antler bow
257 Sinew detail on bow
258 Bow case and quiver with bow and arrows
259 Portrait of Péhriska-Rúhpa
260 Grizzly bear claw ornaments
260 Peale Museum label
261 Daguerreotype of No-che-ninga
264 Fragmentary box-and-border robe
265 Painted and quilled robe
266 Painted box-and-border robe
267 Quilled belt
267 Woven wool sash
268 Beaded garter
269 Roach feather
271 Strings of wampum
274 Root-gathering basket ("sally bag")
278 Plateau Indian woman digging roots

278 Journal drawing of root-digging tool
280 X-ray faces on sally bag
282 Pat Gold with commemorative sally bag
283 Petroglyph of Tsagaglalal
286 Harvesting tule in eastern Oregon
287 Gold basket, *From Generation to Generation*
288 Detail of fiber skirt
289 Woman's fiber skirt
293 Ada Markishtum preparing cedar bark
293 Cedar bark bundle
294 Bark shredder
294 Whalebone bark beater
296 Black-rimmed hat
297 Top view of black-rimmed hat
299 Basketry top hat
300 Basketry sailor's cap
301 *Winter Quarters*
307 Wooden cradle
308 Wooden cradle
310 Journal drawing of cranial modification
312 Recreation of Jefferson's Indian Hall
314 Dennis Fox with the Peabody's pictographic robe
316 Moonrise over the Missouri River
324 Quilled bison robe interior
340 The Peabody Museum of Archaeology and Ethnology
342 Gaylord Torrence
343 Peabody Museum staff

Foreword

The Names of the Nations

James P. Ronda
H. G. BARNARD PROFESSOR OF WESTERN AMERICAN HISTORY
UNIVERSITY OF TULSA

The story of the Lewis and Clark expedition has become an American commonplace, part of the theater of memory we call our national history. For all sorts of reasons, other explorers of the American West—most notably Zebulon Montgomery Pike, John Charles Frémont, and John Wesley Powell—have slipped away into near-obscurity. The Lewis and Clark journey now seems our emblematic road story, and in all its racial and cultural diversity, the expedition has become the representative American community. Jefferson's travelers fill the stage, crowding others off to the wings and beyond. Told by many different voices, the story of Lewis and Clark is almost an American master narrative, prefiguring the shape of things to come. In it we find all the Western stories—moments of high drama, unbridled arrogance, spirited struggle, and sometimes terrible disappointment. If Wallace Stegner was right when he asserted that "the West is America *only more so,*" then the stories that swirl around Lewis and Clark's way west seem the quintessentially American tale.

Because the expedition's story has grown so large and taken on such powerful meanings for so many Americans, we ought to ask hard questions about the journey. What were its origins? How did it fit within an age of Empire and Enlightenment? Who were the actors in a drama that swept across a continental stage? Where should we put this set of events and cast of characters in our widening understanding of the North American past? And, finally, how can we start to recover the many meanings in those objects that linked natives to newcomers? Beginning to answer these questions, we re-chart our own sense of the American cultural landscape.

The traditional answers to those questions can be summed up in a kind of Lewis and Clark catechism. *Question:* What were the origins of the expedition? *Answer:* The enterprise sprang from the mind and imagination of Thomas Jefferson. *Question:* What were the purposes of the expedition? *Answer:* Jefferson sent his explorers to find the elusive Northwest Passage and add to the store of scientific knowledge about the West. *Question:* What were the long-term consequences of the Lewis and Clark venture? *Answer:* The voyage of discovery marked the beginning of Jefferson's "empire for liberty," as well as an expansion of the empire of the mind.

Such textbook wisdom—now firmly embedded in so many popular treatments of the expedition—is brilliantly challenged in Castle McLaughlin's path-breaking book *Arts of Diplomacy*. Taking as her jumping-off point a comprehensive reconsideration of the Native American objects related to Lewis and Clark held by the Peabody Museum, McLaughlin expands the story and enriches our understanding of it. Pipes, ornaments, peace medals, bags, and hats become visible manifestations of mutual discovery, transforming Lewis and Clark and Native people into adventurers, each exploring the other. Objects once seen as simply exotic specimens from a distant past now take their places in a more complex set of cultural encounters. Built on the best historical and anthropological sources, and informed by current critical theory, *Arts of Diplomacy* gives voice to seemingly mute objects and lets readers hear Native voices in the expedition conversation.

Because we often seek to collapse the distance between the past and the present, Jefferson's explorers have been increasingly portrayed as pioneer scientists busy studying Western peoples and landscapes in a ceaseless quest for objective truth. Never mind that the word "scientist" did not enter the English and American vocabularies until the 1830s and that Lewis and Clark were quite innocent of anything we might recognize as the scientific method. Jefferson had his own wry opinion of those we would now call "scientists" in Western exploration. Writing to a French correspondent, the president explained that "These expeditions are so laborious, and hazardous, that men of science, used to the temperature and inactivity of their closet, cannot be induced to undertake them." Perhaps later, "when the routes shall be once open and known," some "men of science" might venture into the West. It is easy to forget that for all his skills as a naturalist, Lewis was not comfortable with Linnaean binomial nomenclature and had to consult reference books carried on the journey. And the objects the explorers "collected"—everything from plant specimens to Indian pipes—are frequently described by modern scholars as part of a growing catalogue of the material culture of the American West. Enlightenment exploration, so many recent writers would have us believe, was just an earlier version of present-day scientific ventures. But Jefferson knew better, and McLaughlin carefully brings us back to a fuller, more Jeffersonian understanding of the expedition and its complex relations with Native peoples.

President Jefferson gave initial expression to the expedition's Indian missions in June 1803, when he completed a draft of instructions for Meriwether Lewis. While the instructions ranged over everything from astronomy and botany to mineralogy and zoology, Jefferson devoted much of the document to relations with Native peoples and inquiries about Indian cultures. His explorers were to record "the names of the nations," a phrase that meant far more than simply noting tribal locations and village names. In a list that would have been familiar to eighteenth-century explorers like Captain James Cook or Admiral Alejandro Malaspina, Jefferson directed his travelers to gather knowledge of Native "traditions, monuments [and] their ordinary occupations in agriculture, fishing, hunting, war, arts & the implements for these." The Indian objects brought back by Lewis and Clark have often been interpreted as the results of the captains' dutiful efforts to fulfill Jefferson's instructions. Lewis and Clark were, so we have been told, pioneering ethnographers, and what they "collected" was meant to serve the interests of disinterested science.

Castle McLaughlin offers us a very different reading of the expedition, its goals, and the objects it carried back to the president. Although some writers continue to describe the expedition in modern, scientific terms, McLaughlin rightly resituates it within the late eighteenth-century worlds of commerce, diplomacy, and imperial expansion. In those intricate and interrelated spheres no one was more important than Sir Joseph Banks. Banks was president of England's Royal Society, advisor to the royal government on matters of empire, and exploration patron without equal. One English adventurer described Banks as "the common Center of we discoverers." Banks and Jefferson had remarkably parallel careers: both served as presidents of important scientific organizations, both carried on wide correspondence with scholars in many fields, and both linked exploration to imperial expansion. While Banks and Jefferson never corresponded on matters of geographical exploration, the Lewis and Clark expedition took place in an intellectual and imperial context shaped by Banks. Like Banks, Jefferson believed that knowledge had to be measured by the yardstick of utility. Even more important, useful knowledge gained by voyages of discovery was always placed in the service of the state. As McLaughlin makes plain, Lewis and Clark were part of an exploration tradition that connected the Enlightenment enterprise to imperial expansion. The eighteenth-century British statesman and political theorist Edmund Burke put it best when he wrote, "now the Great Map of Mankind is unrolld at once." The life of the mind was never to be separate from the growth of the empire. When Lewis and Clark acted as ethnographers, whether recording Native American languages or describing domestic architecture, they worked within the confines of useful knowledge and imperial ambition. McLaughlin's Lewis and Clark are diplomats in buckskin engaged in complex exchanges with Native diplomats adept at the arts of rhetoric and negotiation.

Having fixed Lewis and Clark more fully within the context of late eighteenth-century thought in general and conceptions of knowledge in particular, McLaughlin skillfully moves readers yet another step toward a fuller appreciation of the expedition's Indian objects and objectives. When Jefferson prepared instructions for Lewis in the spring of 1803, negotiations surrounding the Louisiana Purchase—negotiations the president knew little about—were just beginning in Paris. Initial exploration instructions for Lewis and Clark contained nothing about American sovereignty and the requirement that Native peoples honor a new Great Father. Although Levi Lincoln, Jefferson's attorney general and skilled political advisor, recognized that the expedition had "national consequence," direct diplomacy with the tribes up the Missouri and across the mountains was not on the official exploration agenda. Instead, there was considerable talk about "commercial intercourse" between the United States and Native nations and "mutual emporiums, and the articles of most desirable interchange for them and us." Before the Louisiana Purchase the point of contact between Native peoples and Jefferson's explorers was business. While eighteenth-century explorers and their patrons never fully separated commerce and imperial expansion—trade and the flag often marched in company—Jefferson's language before the purchase was almost exclusively commercial.

That changed dramatically after the ratification of the purchase treaty by Congress in October 1803. Now commerce and empire were openly linked, with an increased emphasis on diplomacy. In an often-overlooked letter to Lewis dated January 22, 1804, Jefferson essentially redefined the expedition's key Indian objectives: "Being now become sovereigns of the country," American explorers were to make their way west announcing a new imperial order. Jefferson's letter to Lewis is so fundamental to a proper understanding of expedition–Indian diplomacy that it deserves to be more fully quoted. "It will now be proper you should inform those through whose country you will pass, or whom you may meet, that their late fathers the Spaniards have agreed to withdraw all their troops from all the waters & country of the Missisipi & Missouri, that they have surrendered to us all their subjects Spanish & French settled there, and all their posts & lands: that henceforward we become their fathers and friends, and that we shall endeavor that they shall have no cause to lament the change." Getting the "names of the nations" now meant adding them to the list of tribes within the American domain and part of an expanding territorial and commercial empire. Lewis and Clark were still bound up the wide Missouri to find the elusive Northwest Passage. But as William Clark observed when leaving St. Louis in May 1804, the expedition's "road across the continent" would take it through a "multitude of Indians." And meeting with those Native peoples would be all about diplomacy and the new names of the nations.

Reframing the expedition in terms of useful knowledge rather than objective science, the politics of empire rather than the passions of the mind, and the mutual exchanges of diplomacy rather than one-sided announcements of a new Great Father, McLaughlin then places the surviving Lewis and Clark Indian objects within an appropriate temporal and cultural context. The emphasis has traditionally been on Lewis and Clark as active collectors, as if every pipe or ornament had been selected and obtained by the explorers themselves. But as McLaughlin ably demonstrates, such a view reduces Indians to mere objects for scientific study and drains the life from the objects that passed between natives and newcomers. Paying close attention to both the objects and the written expedition record, McLaughlin tells a different story.

Nothing illustrates that new narrative better than her penetrating discussion of the pipes associated with the expedition. As McLaughlin explains, "Lewis and Clark acquired, or kept, more pipes than any other category of object that they received from Native people." The emphasis here is on the active role of Indians as negotiators, presenting the American visitors with pipes as part of a larger diplomatic and cultural encounter. In a superb chapter titled "The Language of Pipes," McLaughlin describes the central role of pipes and smoking rituals in Native American diplomacy. Lewis and Clark did not "collect" these pipes; the explorers were given them as part of an artfully choreographed diplomatic dance. Just as Jefferson's travelers handed out peace medals and printed certificates, Indian speakers offered pipes as a means to represent and seal diplomatic arrangements. But between the pipes and the medals there were profoundly different worlds of meaning.

For Lewis and Clark the medals signified American power and the presence of a new, sovereign Great Father who demanded prompt obedience and undivided loyalty. The pipes, however, symbolized another, more fluid understanding of diplomacy and power relations. Smoked and passed within a circle, the pipes spoke the language of mutual respect and reciprocal responsibilities. Presenting medals was all about client chiefs and junior partners in trading deals. Just as European diplomats gathered in council chambers to sign carefully worded treaties, Lewis and Clark expected medals and flags to seal binding agreements between the American republic and Native nations. But as McLaughlin so effectively demonstrates, the gift of pipes to Lewis and Clark symbolized a different, although no less formal, way to construct diplomatic arrangements. Located within a world shaped by kinship and mutual obligation, pipes and the rituals surrounding them linked personal responsibility and national honor to the shared benefit of all.

No one paging through *Arts of Diplomacy* should think that this is just another lavishly illustrated coffee-table book. The Lewis and Clark Bicentennial has already generated more than its share of such volumes. Castle McLaughlin and her colleagues have taken the Peabody Museum's magnificent Lewis and Clark Indian collection as the beginning point for a fundamental reassessment of the expedition's Native American missions. The contributions made by this book are many, and they all merit close attention. The objects themselves have been treated to unparalleled curatorial scrutiny, giving us a much clearer sense of their origin and crafting. Gaylord Torrence's essay on the raven ornaments is a model for that kind of critical study. Anne-Marie Victor-Howe contributes substantially to our growing appreciation for the Northwest Coast objects in the Peabody collection. Contemporary Native American voices are here in memorable essays by or sidebars about Mike Cross, Jo Esther Parshall, Keith Bear, Butch Thunder Hawk, and Pat Courtney Gold. All of this is bound together by McLaughlin's compelling narrative and astute analysis. By relocating the Lewis and Clark Indian collection within the larger stories of commercial and imperial expansion, McLaughlin has restored the expedition to its own meaning in time. And even more, by paying attention to diplomacy and the objects that passed from hand to hand in those exchanges, McLaughlin places Native people at the heart of the expedition's life on the road. The voices in the objects now join those already present in the expedition's written record. Taken together, they make the expedition's story an American conversation. In this telling of the Lewis and Clark story, Native people are not passive objects to be acted upon by explorers. Rather, this is a narrative of mutual encounters as told by compelling objects. That those objects were so often chosen by Native people as gifts to Lewis and Clark is yet another reminder of the central role of Indians in our first national road story. Castle McLaughlin's *Arts of Diplomacy* provides a map for that road, a way to catch a glimpse of all those who made the journey and made the journey possible.

Preface

Rubie Watson

WILLIAM AND MURIEL SEABURY HOWELLS DIRECTOR
PEABODY MUSEUM OF ARCHAEOLOGY AND ETHNOLOGY

In 1899, Harvard University's Peabody Museum received an extraordinary collection of artifacts. Known as "the Boston Museum donation," or more simply by its accession number 99-12, the collection was an enormous assemblage of nearly 1,500 rare items of material culture from North America, Africa, and the Pacific Islands. And hidden within 99-12 were what we now recognize to be probably the only surviving Native American artifacts acquired by Lewis and Clark during the epic journey of the Corps of Discovery in the first years of the nineteenth century.

Before the Smithsonian Institution was established as the U.S. national museum in 1846, Charles Willson Peale's Philadelphia Museum served as the unofficial repository of the young nation's "curiosities." In the early pages of this volume, Castle McLaughlin describes the circuitous route by which the ethnographic collections of the Peale Museum made their way in the 1840s to the now defunct Boston Museum, and from there across the Charles

River to Cambridge and the Peabody, where they became the twelfth accession of 1899 (hence, 99-12). At the time, no one appeared to have noticed the significance of the transfer of one of the earliest collections of Plains and Northwest Coast Indian artifacts to America's first ethnographic museum. Indeed, little attention has been paid to 99-12 as a collection, and not until recently have the Lewis and Clark items within the collection been a subject of serious research.

It had long been thought, however, that the Peabody owned the only collection of cultural materials acquired by Lewis and Clark, and as interest in the expedition grew with its approaching bicentennial, it seemed important to study the collection more closely. In 1997, I asked Castle McLaughlin, then a Hrdy postdoctoral fellow and now the Peabody's associate curator of North American ethnography, to research the set of artifacts that had been ascribed to the captains and their journey. She began in the very best curatorial tradition with a two-pronged effort focused on the artifacts themselves and their provenance. Eventually, she broadened her investigation to encompass Lewis and Clark scholarship, the history of U.S.–Indian relations, the history of museums, and detailed materials analysis. Many specialists were consulted during the course of McLaughlin's research. These included Native American artists who are descendants of the tribal members Lewis and Clark encountered during their journey, materials scientists from a number of academic and research institutions, curators at Monticello and the Missouri Historical Society, art historians, anthropologists, and historians of the United States. All who participated in making this volume possible have shown remarkable scholarly integrity and generosity. But the Peabody Museum owes a special debt of gratitude to Castle McLaughlin, whose tenacity, incisive scholarship, and creative vision have directed the research upon which this volume is based.

As expedition member John Ordway wrote in his journal, using a phrase based on a Native American saying, Lewis and Clark "made the road open" to the West. This volume asks us to look at that opening in the terms of those who participated in it. Members of the Corps of Discovery entered an unfamiliar (but by no means unknown) world when they crossed the Missouri River in 1804. The people they met during their journey were variously interested, indifferent, or hostile, depending on their own unique perspectives and histories, as well as the specific circumstances of the encounters. We have Lewis and Clark's journals to guide our understanding of their historic meetings with Indian leaders and the men and women they led. Now, thanks to Castle McLaughlin's work, the historical record has been broadened to include the objects—the "arts of diplomacy"—that the explorers were given or traded during the two-year expedition. Because of the meticulous research of McLaughlin and her colleagues, the painted hides, smoking pipes, whaling hats, and carrying baskets that Lewis and Clark (and other early travelers) sent east can now be added to the record of that fateful set of early contacts.

It is a record that is by no means simple or uncontested. The Peabody Museum's Lewis and Clark project asked a number of questions: What can the objects themselves, and their documentation, tell us about Lewis and Clark? What do they say about the explorers' relations with Indian peoples? What can they reveal about trade and exchange relationships in the early days of the nineteenth century more generally? This book and the extensive research project upon which it is based are testimonials to the power of artifacts—of the tangible past—to open up new historical perspectives. *Arts of Diplomacy* presents an object-driven anthropological history that allows us to explore submerged and often neglected social relationships. Here the focus is not on Lewis and Clark, the expedition leaders, but on the gifts and trade items they received as they traveled from the Missouri River to the Pacific Ocean and back. This focus challenges us to think in new ways about the objects of exchange themselves and, more importantly, it highlights the interactions between Indians and members of the Corps of Discovery that those exchanges made possible.

Arts of Diplomacy is part history, part anthropology, part art history. It is also a detective story that investigates one of the first great national collections of Native American art and artifacts. The analysis of these artifacts not only adds to our understanding of the Lewis and Clark expedition, but also—and in my view more importantly—contributes significantly to the emerging history of nation-to-nation diplomacy that formed an integral part of the early movement of Euro-Americans west of the Missouri River. *Arts of Diplomacy* gives us the history of a collection as well as the history of a set of remarkable encounters between the representatives of a new nation and those who hosted, assisted, and sometimes obstructed their journey. It is a compelling set of stories, as rich and intricate as the objects described and illustrated in the pages that follow.

ARTS OF
Diplomacy

Introduction

DISCOVERING LEWIS AND CLARK'S INDIAN COLLECTION

THIS BOOK TELLS THE STORY OF ONE OF THE UNITED STATES' unique and most treasured collections, the objects that Native American people gave to Meriwether Lewis and William Clark during their epic exploration of North America from 1804 to 1806. The so-called Lewis and Clark collection at Harvard University's Peabody Museum of Archaeology and Ethnology has been known for many years. Its objects have been exhibited at a world's fair,[1] illustrated in countless publications, and replicated for venerable institutions such as Thomas Jefferson's home, Monticello. But how many actual Lewis and Clark objects are there, and how well are they documented? Who made them, and where and how did Lewis and Clark acquire them? Surprisingly, no serious, sustained effort had been made to answer these questions until Rubie Watson, director of the Peabody Museum, commissioned an internal review of the collection in 1997.

Among the specialists who researched the Peabody's Lewis and Clark items were Peabody curatorial associate Anne-Marie Victor-Howe (left) and Wasco fiber artist Pat Courtney Gold (right), seen here examining baskets in the museum's collection.

What began as a part-time investigation soon became a complex, multifaceted project involving virtually every department in the museum and enlisting many outside consultants and advisors. The North American Indian pipes, garments, hats, and robes making up the Lewis and Clark collection had been given to the Peabody by the Boston Museum in 1899; our primary task was to determine which of those objects were indeed associated with the Lewis and Clark expedition. We knew that many of the Peabody's Boston Museum objects had originated in the Peale Museum, Charles Willson Peale's exhibition hall in Philadelphia, where Jefferson, Lewis, and Clark had deposited the original expedition materials. But it was unclear how many expedition objects had survived to be transferred to the Peabody.

Methods of Discovery

In 1905, Peabody Museum assistant curator Charles C. Willoughby published an article in the journal *American Anthropologist* announcing that the museum had acquired at least eleven Lewis and Clark objects from the Boston Museum, including a painted robe, two dresses, three raven ornaments, and a quilled otter bag. Willoughby was convinced that the Boston Museum accession contained additional Lewis and Clark objects for which documentation had been lost, but he never systematically searched for them. Throughout most of the twentieth century, Peabody staff and outside scholars relied on Willoughby's initial definition of the collection. Then, in 1980, historian Charles Coleman Sellers published *Mr. Peale's Museum: Charles Willson Peale and the First Popular Museum of Natural Science and Art.* In it he claimed that fourteen objects at the Peabody could be attributed to Lewis and Clark. Researchers quickly accepted Sellers's account, and his fourteen objects appeared widely in books and exhibits relating to Lewis and Clark's "Corps of Discovery" and to Native American art generally. Sellers, however, had simply made an educated guess based on fragmentary evidence, limited research, and his own biases. The history, origins, and identity of the collection remained poorly understood.

Many early museum collections lack solid documentation. How, then, do curators and communities of scholars go about identifying and "authenticating" objects? Does it really matter whether we can associate objects with specific people? In this case, authentication matters a great deal to many stakeholders in the Lewis and Clark story, especially in light of the bicentennial of their expedition, which this book commemorates. Throughout the bicentennial years, 2003 through 2006, Indian people, museums, government agencies, scholars, communities along the Lewis and Clark trail, and everyday citizens recognize a signal event in American history and reckon with the contemporary legacies of Thomas Jefferson's national vision. Jefferson's purchase of the Louisiana Territory had profound implications for the American nation and for the Native nations who thus became part of the nascent United States.

The Peabody's Lewis and Clark objects are uniquely positioned in this commemorative landscape, for they are the best-documented Native American objects symbolizing the first encounters between Indian nations of western North America and representatives of the United States government.[2] Though often unappreciated in today's museum world (but not by Indian people), the power of original objects to engage meaningfully with the past and the present is incomparable. We at the Peabody Museum felt an obligation to research these objects as thoroughly as possible and to make the results available to all interested parties.

To try to solve the puzzle of which Peabody objects were associated with Lewis and Clark, we pursued two standard museum strategies: documentary research and formal analysis of the objects themselves. Guided by fading ink on object labels, archival manuscripts, and institutional records and armed with several published versions of the Lewis and Clark journals, we attempted to identify which kinds of objects the explorers received from Indian people and under what circumstances. We then tried to trace the fate of those objects after the expedition ended.

In 1809, Charles Willson Peale wrote a list titled "Memorandum of Specimens and Artifacts" in his museum ledger book, describing the more than fifty expedition objects he had received from Thomas Jefferson, Meriwether Lewis, and William Clark. The Peale memorandum proved invaluable, because we assumed that only those Lewis and Clark objects that went first to the Peale Museum and then to the Boston Museum could ultimately have arrived at the Peabody (for more on this history, see chapter 4). Unfortunately, Peale described the Lewis and Clark objects in only the most general terms, rather than elaborating on their appearance in ways that might help us identify them today. A typical entry reads, "A Tobacco Pouch, from the Ioway's." We therefore also examined closely the entire Peale Museum catalogue, which listed donations throughout the period 1804 to 1842.

Indian Necklace, made of the claws of the Grizly Bear— Presented by Capt Lewis and Clark.
13.

Peale Museum labels such as this one were part of the documentation examined by researchers in their efforts to authenticate Lewis and Clark objects in the Peabody collection.

Did Peale, for instance, have other, non–Lewis and Clark tobacco bags that might now be at the Peabody?

Each object at the Peabody that might potentially be linked to Lewis and Clark was evaluated against a set of well-defined criteria. Did the object match a type of item mentioned or described by Lewis and Clark? Was it of the right vintage to have been acquired by them? Did Peale describe receiving such an object? If so, could the potentially matching object at the Peabody have been acquired from the tribe identified by Lewis and Clark as the group of origin? Did it come to the Peabody with a Peale label identifying it as a Lewis and Clark object? Was there other related documentation suggesting a potential link? Was there evidence that would counter a Lewis and Clark identification? If so, what was it?

This evaluation process was complicated by the evident presence at the Peabody of a set of objects collected along the upper Missouri River and donated to the Peale Museum in 1828 by the family of Lieutenant George C. Hutter. Hutter was married to William Clark's niece and was closely associated with Clark in St. Louis. Because previous researchers had postulated that Hutter acquired his objects from Jefferson or from Clark, we decided to include Hutter's objects in our project while further investigating these links.

Each object in the Peabody's Boston Museum accession that matched a category of object listed by Peale (e.g., tobacco bag) was then evaluated on the basis of its formal, material properties. We pursued three interrelated approaches for determining the ages and origins of the candidate objects. First, we consulted widely with colleagues who had experience and expertise in assessing early historic North American objects. Peabody Museum curatorial associate Anne-Marie Victor-Howe was called upon to research the Northwest Coast and Columbia River materials. Art historian Gaylord Torrence, our primary consultant, visited the Peabody twice to evaluate the collection, and we sent photographs of objects to many other specialists. We also solicited opinions about objects from Native American experts including elders, artists, and professional scholars. Second, we engaged scientific specialists and consultants to identify the constituent materials of many objects. We avoided destructive analysis out of sensitivity to Native American concerns, but we were able to identify many organic materials: wood, hair, fabric, pigment, plant fibers, feathers, and hides. We also attempted to identify trade materials such as brass buttons and glass beads on the objects, which we hoped would help to establish their earliest dates of manufacture. Finally, we compared the Peabody objects with objects of similar age and type that are housed in other museum collections around the world.

Daniella Bar-Yosef, Peabody Museum research associate, analyzing shell beads on a calumet.

These efforts, although rewarding, underscored just how challenging it is to identify eighteenth- and early nineteenth-century ethnographic materials from North America. Few such objects exist in world collections, and those that do are often poorly documented. It is doubtful, for example, that anyone really knows what the material culture of the Mandan people looked like in 1800. We know even less about the garments, arms, and accoutrements of Indian peoples who lived on the lower Missouri River, in the trans-Mississippi region, and across the "old Northwest"—peoples such as the Ioways, the Mesquakies (Foxes), and the Kaskaskias. Because intertribal trade was so widespread, it is impossible to assume that any individual object was made by the same group that gave it to Lewis and Clark (we know, in fact, that many were not).

Complicating the picture, many Indian peoples were undergoing both geographical and cultural transitions at the time of the expedition. Some of the groups Lewis and Clark encountered were moving from east to west and were shifting from a Woodlands to a Plains cultural orientation (see map, pp. 18–19). The stylistic and aesthetic properties of many of the objects described in this volume reflect that transition, as well as vestiges of poorly understood eighteenth-century art traditions. Regional styles are more clearly understood than are tribal styles for this transitional period.

Lewis and Clark returned from their journey with objects given to them primarily by Native people living near the Mississippi, Missouri, and Columbia Rivers (unfortunately, no materials from the Shoshones or Nez Perce appear to have survived at the

Peabody). We looked carefully at objects collected in those areas by other early explorers and travelers in the American West and, when they were available, at paintings and drawings chronicling the travelers' encounters with indigenous people. Most of those materials are now in European museums. For example, both Captain James Cook of England and Captain Alejandro Malaspina of Spain visited the Pacific Northwest before 1800, and they acquired woven hats and garments quite similar to examples now at the Peabody. Their materials are held by various institutions, including London's British Museum, the Florence Museum of Anthropology and Ethnology, and the Museo de América in Madrid.

In contrast to collecting in the Northwest, large-scale collecting in the Plains and Prairie regions did not get under way until a generation after Lewis and Clark's expedition, so most comparable materials from those regions were made later. Nevertheless, the objects collected by Maximilian, prince of Wied-Neuwied (in the Linden-Museum, Stuttgart, and the Ethnological Museum Berlin), L. A. Schoch (Berne Historical Museum), and Friedrich Paul Wilhelm, duke of Württemberg (Linden-Museum, Stuttgart; Ethnological Museum Berlin) were useful for our analysis. Also helpful were some of the early objects assembled later by the Arthur Speyer family (Canadian National Museum, Ottawa). Comparable early Plains items were also acquired by the artist George Catlin (Smithsonian Institution) and by Nathan S. Jarvis (Brooklyn Museum). We also benefited from examining collections at the Smithsonian's Museum of Natural History and at the Musée de l'Homme in Paris.

Comparison is a standard approach to identifying museum objects, but it is problematical with materials of such early manufacture. Even some of the best-known exemplars of early tribal styles lack well-documented attributions. Often the cultural identification of an object—say, as a "Mandan" shirt—derives from an earlier curator's best guess, which might have been based on a comparison with another example of similarly uncertain attribution. Once inscribed in museum records and published, such attributions take on an aura of authority. The result is that all like objects, such as so-called Mandan shirts, become mutually referential. The Peabody's Lewis and Clark objects are themselves implicated in such referential networks and have been used as the basis for dating other, similar objects and for attributing their manufacture to specific tribal groups.

Because these objects serve as important signposts in the global network of early Native American art history and material culture, my colleagues and I wanted to share the results of our new research in a way that acknowledged the limits of what we know.

The configuration of the Peabody's Lewis and Clark collection that is presented here differs from past understandings, including those of both Willoughby and Sellers. Our presentation of the collection also deviates from conventional strategies. We decided to present the objects in tiered categories, based on our confidence that a given object was associated with the Corps of Discovery's expedition. By discussing and illustrating aspects of the research process, we have tried to provide readers with a sense of how museums go about documenting their materials.

The cultural identities of the people who made many of the objects that we present remain uncertain. Naming and names are important parts of identity, and we are sensitive to the fact that many Indian people prefer their own self-designations (such as Mesquakie) to older tribal names coined for them by Euro-Americans (such as Fox). But most of our references to tribal groups are drawn from historical documents, and for this project we needed to preserve the trail of provenance that the objects had generated. Using contemporary tribal names to refer to historical groups can be complicated, and in some cases we can identify those groups only very generally, as in the many historical references to objects and people as "Sioux." To avoid confusion and to make the text accessible to readers, we decided to use terms that were consistent with earlier attributions, clarifying them where possible with newer designations. Gaylord Torrence's essay in chapter 5 is an exception; he uses the term "Mesquakie" to refer to the people historically known as Fox.

Assigning dates to the objects was also challenging. To do so, we had to reconcile two aspects of each object: its formal, stylistic features and its known or probable collection history and provenance. For some objects, such as the raven ornaments illustrated in chapter 5 and the Hutter materials in later chapters, firm collection dates informed this process. For many other objects, we can be certain only that they were acquired during the nineteenth century by the Peale Museum, the Boston Museum, or both. All of the objects presented in this book should be of early enough vintage to have been collected by Lewis and Clark. But some of them, such as many of the pipe stems, are formally consistent with comparable objects that were collected throughout the first half of the nineteenth century. Believing that the integrity of the object outweighed an uncertain provenance, we assigned each object a range of dates consistent with its condition and formal stylistic characteristics, factoring in provenance data whenever possible.

Interpretations

Understanding historic objects is an interpretive process that requires recovering many contexts of meaning. As social texts that encode meaningful human activity, objects recall past worlds in a rich and direct way. Many of the objects presented here reflect not only the complex belief systems and societies of the Indian people who made and used them but also the personal lives of individual Native Americans who interacted with Lewis and Clark. They register how such personal interactions were couched within the larger processes of engagement that we now call globalization. The profusion of "exotic" materials found on the pipes and garments in this collection, such as French silk, English wool, and Chinese pigments, evokes the world systematics and experiential textures of Lewis and Clark's moment in history more vividly than can the written word.

Listening to the stories these objects tell sheds new light on the expedition itself. In Part 1, "The Life History of a Collection," I examine how and why Lewis and Clark acquired pipes, robes, hats, and garments from the Indian people they encountered as their small party crossed the continent. It is apparent that the exchange of goods and services between Lewis and Clark and Native Americans played a critical role in the Corps of Discovery's success. This is not surprising. Exchange relations facilitated early contact between Western explorers and Native peoples throughout the world, including North America. By carefully reconstructing the nature and scope of exchange events during the Lewis and Clark expedition, we can come to see these objects not merely as museum artifacts but as animate social symbols, objects that both created history and became a part of it.

English brass buttons and Italian glass beads adorn the yoke of a side-fold dress. Tokens of emergent globalization, exotic materials from Asia and Europe were available to Native artists along the Missouri River by the early nineteenth century. (Detail of 99-12-10/53047; see p. 181.)

In chapter 1, "The Lewis and Clark Expedition: An American Quest for Commerce and Science," I begin to argue that the expedition is best understood when framed in terms of larger social histories. Foregrounding the political and economic objectives of Jefferson's "darling project," I emphasize that national diplomatic agendas, not scientific studies, motivated Lewis and Clark's relationships with Indian peoples. Chapter 2, "Up the Missouri: Patterns of Diplomacy and Exchange," describes how the Lewis and Clark party and leaders of the Native peoples they met created those necessary diplomatic relations through the reciprocal exchange of diverse goods and services. Exchange was a customary ritual in the intercultural politics of the frontier, and established protocols guided Lewis and Clark's encounters. By exchanging diplomatic gifts with Indian leaders, Lewis and Clark created both short- and long-term alliances and forged the first official relationships between the United States government and Indian nations west of the Missouri River. Establishing those ties enabled the two men to barter for the horses, food, clothing, and information they needed to sustain and guide their party across the continent. In his essay "From Time Immemorial," Mike Cross describes life among the Mandans, Hidatsas, and Arikaras following their contact with Lewis and Clark from a contemporary Mandan-Hidatsa perspective.

But intercultural diplomacy proved more complex and nuanced than Lewis and Clark anticipated. Depending upon the independent political agendas of tribes and tribal leaders, Indian people responded differently to the captains' overtures. Both customary cultural practices and the political strategies of tribal leaders influenced the nature of exchange relations. In chapter 3, "Selections: The Making of a Collection," I examine the specific objects the explorers brought back and try to reconstruct how and why they became objects of exchange. Most of them were conventional diplomatic gifts, given to create and affirm alliances, and I argue that Lewis and Clark chose to convey them to the U.S. president because they understood the gifts' intended meaning. The explorers created a collection from the individual objects they received, but the objects themselves were selected by Indian leaders who used gift giving to comment upon and further their own objectives with the exploring party.

Few of the objects that Lewis and Clark brought back survive today. In chapter 4, "Into the Museum: From Gifts to Artifacts," I briefly recount what is known about what happened to them after the expedition. Unfortunately, expedition objects kept by Thomas Jefferson and William Clark disappeared after their deaths. But Jefferson, Lewis, and Clark donated a great number of their Indian gifts to Peale's Philadelphia Museum, where they were transformed into museum specimens and artifacts. After Peale's museum failed, the items were purchased by P. T. Barnum and by the Boston Museum before being transferred, as institutional gifts, to the Peabody in 1899. Throughout the nineteenth century, remarkably little effort was made to preserve or interpret the objects as a collection, and their historic value has been recognized only recently.

In Part 2, "The Peabody Museum Objects," Anne-Marie Victor-Howe and I discuss the individual items that remain today at Harvard's Peabody Museum of Archaeology and Ethnology. First, in chapter 5, "From Warriors and Women Traders," we present the six objects that we believe were acquired during the expedition by Meriwether Lewis and William Clark. Among this core group are an otter bag and two basketry whaling hats from the Northwest Coast, the latter a type that Lewis and Clark described at length in their journals. There also are three men's raven ornaments, probably worn as bustles by members of a military society. Little has been written about the history and significance of raven ornaments, and Gaylord Torrence adds considerably to our knowledge in "The Raven Belt Ornaments of Lewis and Clark," the essay he contributed to this book.

Next, in chapter 6, "The Army Moves West," I introduce six objects that were donated to the Peale Museum by the family of Lieutenant George Hutter in 1828 and examine the potential relationship between Hutter's collection and that of William Clark. Both the Lewis and Clark party and Hutter acquired many of the same kinds of things. Chapter 7 presents "Enigmatic Icons," five remarkable robes and other garments that could have been collected either during the 1804–6 expedition or by Hutter in 1825–26. In chapter 8, "The Language of Pipes," I describe the integral role that pipe ceremonialism played during the Corps of Discovery's journey and illustrate the spectacular pipes and calumets that the Peabody received from the Boston Museum, some of which may have been given to Lewis and Clark.

Finally, in chapter 9, "Grizzly Claws, Garters, and Fashionable Hats," Anne-Marie Victor-Howe and I present additional Peabody objects that are of the right age and type to have been collected by Lewis and Clark but which lack direct evidence linking them to the expedition. All of them are early-nineteenth-century objects that correspond to categories of things Peale received from the expedition. Some of these objects have recently been identified as possible Lewis and Clark objects. Others, such as a Wasco-Wishram "sally bag," have well-established identities as expedition artifacts. In a compelling and personal essay titled "A Wasco Weaver Meets Her Ancestors through Lewis and Clark," Pat Courtney Gold recounts how that bag inspired her own efforts to revive Columbia River basketry traditions.

The interior of this Northwest Coast whaler's hat shows the catalogue number that marks it as a Peabody Museum artifact. (Detail of 99-12-10/53080; see p. 95.)

Throughout the book, we have added commentary on the objects by contemporary Indian artists. Since 2001, Butch Thunder Hawk, Jo Esther Parshall, Dennis Fox, and other artists have been collaborating with the Peabody Museum and the historic site of

Monticello, Jefferson's home and plantation, to produce new objects for Jefferson's famous "Indian Hall," which has been recreated for the Lewis and Clark Bicentennial. Through endeavors such as the Peabody-Monticello Native Arts Project, the bicentennial commemorations have provided opportunities for engaging historical legacies in a meaningful way and for creating new relationships between museums and Native Americans.

In the coming years, it is likely that many people, both Indians and non-Indians, will draw on the objects presented here to tell both old and new stories, to grapple with history, and to forge new understandings. This interpretive engagement has already begun to take many forms (e.g., Gilman 2003). We offer our book, then, not as an end to the story of Lewis and Clark and their contacts with the Native peoples of North America but as a first chapter in a reengagement with the rich history of cultural exchange that this museum collection represents.

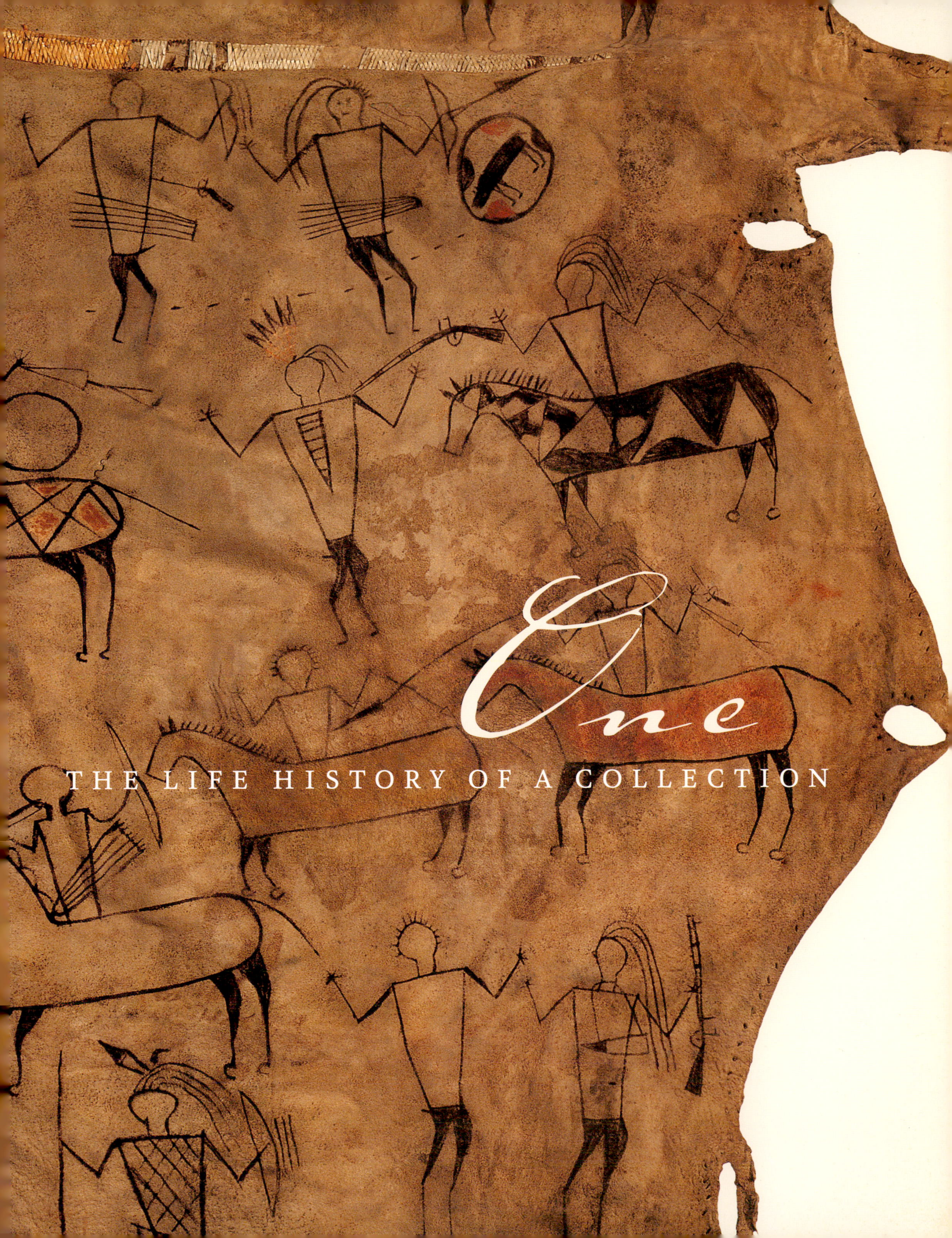
One
THE LIFE HISTORY OF A COLLECTION

Chapter 1

The Lewis and Clark Expedition

An American Quest for Commerce and Science

The story of the Lewis and Clark expedition is a powerful national origin myth. Each generation reinterprets the expedition's significance to the nation's past in terms of contemporary values, issues, and concerns. Through this process of reinterpretation, we discover fresh significance in the simple yet complex story of individual personalities, empire building, legacies and first contact, nationalism and multiculturalism. The surviving artifacts of the expedition can be understood only in the context of the meanings we assign to the journey as both an act in history and a collection of meaningful narratives.

My interpretation of the expedition, briefly presented here, departs from the prevailing view. Many scholars currently regard Lewis and Clark's trek as a unilateral project of the Western Enlightenment and of Thomas Jefferson's commitment to science. Seen in the context of Western expansion and colonialism, the expedition helps us understand how

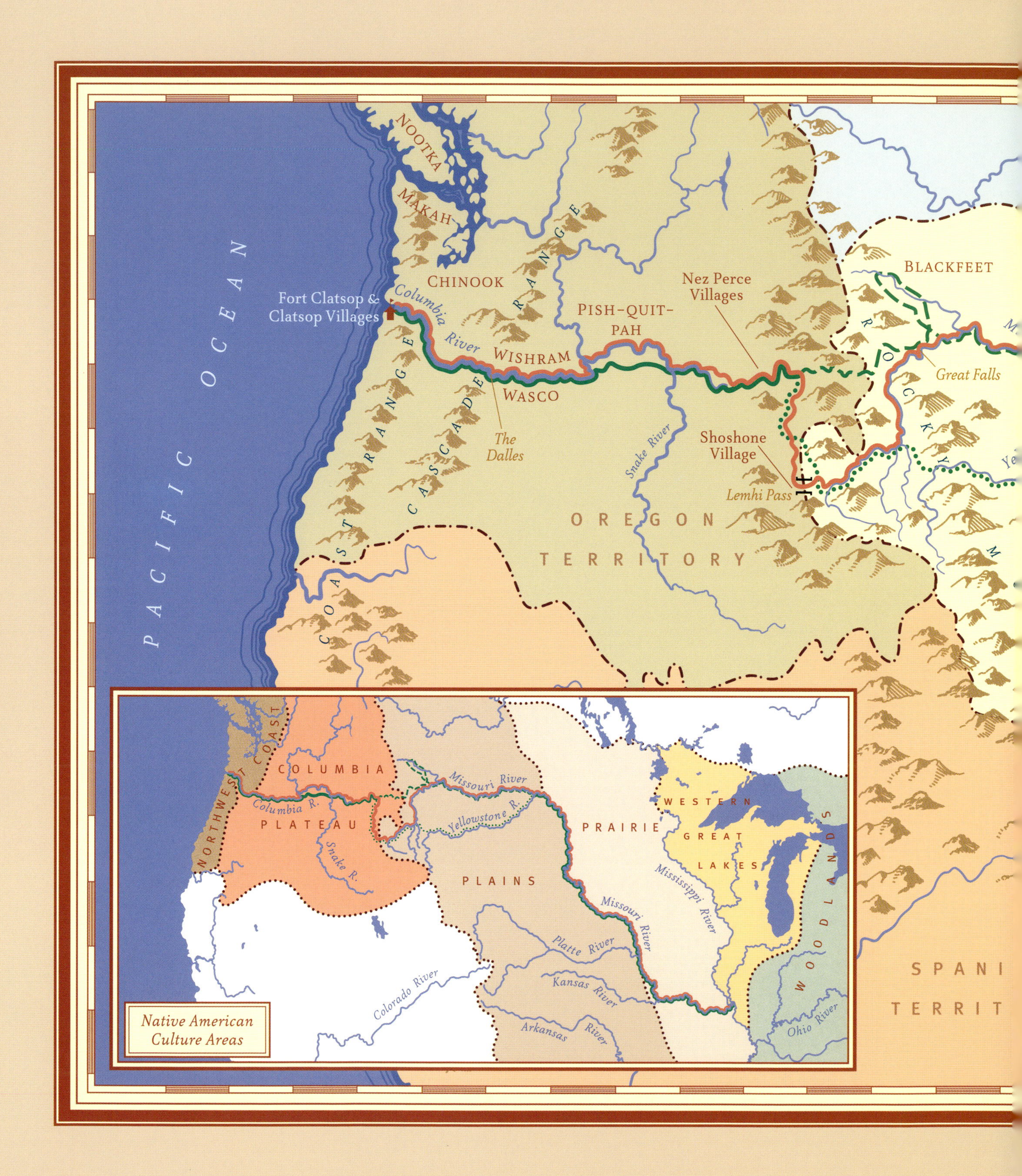

NOOTKA
MAKAH
CHINOOK
PACIFIC OCEAN
Fort Clatsop & Clatsop Villages
Columbia River
CASCADE RANGE
COAST RANGE
WISHRAM
WASCO
The Dalles
PISH-QUIT-PAH
Nez Perce Villages
Snake River
Shoshone Village
Lemhi Pass
OREGON TERRITORY
BLACKFEET
Great Falls
SPANI
TERRIT
Native American Culture Areas
NORTHWEST COAST
COLUMBIA
PLATEAU
Columbia R.
Snake R.
Missouri River
Yellowstone R.
PLAINS
PRAIRIE
WESTERN GREAT LAKES
WOODLANDS
Mississippi River
Missouri River
Platte River
Kansas River
Arkansas River
Colorado River
Ohio River

The Corps of Discovery's route across the continent, 1804–6, showing international political territories, Native American culture areas, and selected tribal locations.

issues of power infuse seemingly neutral scientific agendas and meanings. Jefferson, the architect and mastermind of the Corps of Discovery, certainly envisioned that the mission would contribute to scientific knowledge. But reducing the enterprise to a modern scientific expedition perpetuates a type of historiography that mutes the intentions and understandings of the original participants, whose experiences were forged in the political and cultural climate of the eighteenth century. The epistemology of pre-Darwinian "science" cannot easily be conflated with contemporary understandings; it requires its own historiographic attention.

Moreover, I believe the goal of expanding the empires of the mind was secondary to Jefferson's overt goal of nation building. The Corps of Discovery was commissioned primarily to stabilize a fledgling nation through political consolidation and economic growth. Lewis and Clark's relations with Indian peoples were shaped by the ways in which these objectives intersected with those of Native nations, and their consequent encounters provide the context for understanding the objects Lewis and Clark collected.

It is also important to recognize that although the expedition's mythic status in American history is singular, neither the U.S. government's interests in Indian peoples nor its methods of conducting Indian diplomacy were unique. The experiences Lewis and Clark had with Indian people and the objects they acquired from them were shaped largely by existing patterns of interaction that had been established by their British, French, and Spanish competitors for empire, not by their specific endeavor and personal perspectives.

Exploration and Empire

Exploration, politics, economics, diplomacy, and science were all intertwined in the Lewis and Clark expedition. It can be claimed as a scientific quest, a military expedition, a commercial enterprise, or, as Lewis described it, a "tour"—all of which it was. As James Ronda (1998a), Gary Moulton (1998), and others have shown, the historiography of the expedition reflects a corresponding breadth of interpretive emphases. Nicholas Biddle (1814) and Elliot Coues (1893), the first two editors of Lewis and Clark's journals, omitted portions of the explorers' natural history and "ethnographic" data, in part because, by the time the journals were published, their achievements in these areas had been eclipsed by those of other explorers and disciplinary pioneers and had lost their novelty (Ronda 1998a). Early- to mid-twentieth-century historians, while acknowledging the many dimensions of the expedition, viewed it primarily against the larger backdrop of international political rivalries and national expansion (e.g., DeVoto 1952; Jackson 1981). Historian Arthur Woodward (1970:33) summed up the early nineteenth century as "a historic period rife with intrigue—when the United States was attempting desperately to establish her sovereignty on frontiers where Great Britain, France, and Spain had trod for two centuries."

Since the 1960s, scholars have overwhelmingly emphasized the scientific and ideological dimensions of the expedition over the political and economic, either by analyzing Lewis and Clark's scientific practices (such as classification and naming) as de facto colonizing strategies or by downplaying material factors altogether. So dominant has this interpretive framework become that John L. Allen was recently moved to write that for the first time, popular and scholarly perspectives were merging into a single "coherent view of the expedition as the 'transcendental achievement' of exploration in the Age of the Enlightenment" (Allen 1998:277). Some scholars have naturalized the "scientizing" of the expedition as simply the recovery of its inherent meaning, long overlooked because of the delayed publication of Lewis and Clark's journals. However, larger trends in social science theory and in mainstream culture since the 1960s (such as the popularization of natural history and critical analyses of science) have no doubt nurtured and shaped this developing—and retroactively constructed—narrative of the expedition. Framing the expedition as an ostensibly detached, scientific discovery of the American West resituates the Lewis and Clark epic in terms of current popular interest and veils the interrelated political and economic underpinnings of the enterprise.

Contributing to the instability of meaning assigned to the expedition is the fact that it occurred at a political, economic, and ideological crossroads that invites interpretive travel in many directions. Claiming meaning in any one direction runs the risk of confusing outcomes for motives and threatens to obscure the essentially transitional and inchoate quality of the era in which the expedition took place. In retrospect, science, nation building, conquest, collecting, and the displacement of Native peoples seem to be foregone conclusions, inherently linked in a chain of logical and mutually reinforcing progression. Lewis and Clark are then easily reduced to twin "structural reflexes," essentially acting out these larger processes and agendas.

But in the late eighteenth and early nineteenth centuries, American nationalism, society, and science were not yet the normative institutions and shared structures that they now appear to be—they were embryonic and contested (e.g., Appleby 2000). Religious revivals vied with rationalism, and the nation was poorly defined in both a geographic and an imaginative sense. Centuries of international competition for North American land and resources loomed large in the minds of American politicians, and, as subsequently evidenced by the War of 1812, U.S.–British relations were particularly tenuous. International relations and boundaries remained unresolved in the vast hinterlands and at the margins of empire, where no single nation had definitively captured the fur trade.

Science was certainly an important subtext of the expedition; Jefferson, a leading figure in the "American Enlightenment," was president not only of the nation but also of the American Philosophical Society. But the expedition was intended as much more—and at the same time somewhat less—than a scientific venture. Conceived in a climate of political uncertainty and international rivalry for the North American continent, national expansion and commercial growth were Jefferson's primary and overtly expressed motives for commissioning the Corps of Discovery. In proposing the first U.S. government–sponsored exploration of the West to Congress in January 1803, he stated its purpose as "extending the external commerce of the U.S." by locating the Northwest Passage to the Pacific Ocean. Noting that because it would "incidentally advance the geographical knowledge of our own continent," Jefferson observed that it might be cast as "a literary pursuit," so that "the nation claiming the territory [Spain] would not be disposed to view it with jealousy" (Jackson 1978, 1:13). Indeed, the idea of the expedition as a "literary pursuit" served as a sort of cover story to which Jefferson, Lewis, and Clark appealed when interacting with representatives of rival powers, such as traders whom the explorers encountered. In council with Indians, however, they explained their mission as a commercial reconnaissance, stressing the future benefits to tribes.

Jefferson took pains to keep the mission secret, for fear that internal political enemies (primarily Federalists) who opposed territorial expansion might undermine the project. He also was prone to representing the mission differently to different audiences and at various points in his life. Ronda (1998b) has suggested that Jefferson, prompted by his attorney general, Levi Lincoln, might initially have added the scientific agenda of the expedition so that some contribution might be claimed if the party failed to reach its authorized objectives. In the years following the expedition, during which Jefferson retired from public life and devoted himself to the private pursuit of knowledge and farming, he reflected on the undertaking as a more organic whole, often foregrounding the value of the geographic and natural history knowledge that had been gained.

Although he had long incubated the idea of the expedition, Jefferson's secret plea to Congress in 1803 was galvanized by the publication in 1801 of Alexander Mackenzie's book *Voyages from Montreal.* In that work, Mackenzie described his recent continental crossing from interior Canada to the Pacific Ocean and urged the British to secure for themselves the fur trade of the Northwest. France, Spain, and England had long anticipated the existence of a northwest passage that might link fur production in the Great Lakes and upper Missouri River drainage to the Pacific and beyond. The late-eighteenth-century voyages of Malaspina, Vancouver, Cook, and mariners from the Boston area to the sea otter grounds along the rugged north Pacific Coast had shifted international competition for North American resources into that arena. Jefferson instructed Lewis that the object of his mission was "to explore the Missouri river, & such principal streams of it, as, by its course and communication with the waters of the Pacific ocean . . . may offer the most direct & practical water communication across this continent" (Jackson 1978, 1:61).

Jefferson's instructions to Lewis couch both the quest for and the application of knowledge in instrumental terms: the country soon to be called Louisiana Territory was to be mapped and measured to assert U.S. claims over those of rival nations and to facilitate future settlement and economic development. The Indian trade was both an incentive to that goal and, as demonstrated by the competition waged by England, France, and the United States for Indian allies, a strategy for achieving it. Indian policy in colonial America was dominated by the regulation of trade, and Jefferson, like most other political figures of the day, viewed mercantile relations as the first step toward integrating Native peoples into the national society. Of the Native inhabitants in the vast terra incognita of Louisiana, Jefferson wrote to Lewis: "The commerce which may be carried out with the people inhabiting the line you will pursue, renders a knolege [*sic*] of those people important" (Jackson 1978, 1:62). Samuel Latham Mitchill, a naturalist, congressman, and distinguished associate of Jefferson's, entertained Lewis in Washington after the expedition. He later reported that "Lewis thought the signal advantage to be ultimately derived from their geological and zoological discoveries would be the establishment of a trading post at the mouth of the Columbia River in order to expedite the commerce in trading furs to China" (Mitchill 1826:29).

Portrait of Thomas Jefferson, from life, by Charles Willson Peale, 1791–92.

In essential respects, Jefferson's instructions to Lewis differed little from those written by James Mackay, who, acting in the interests of the Spanish Crown, in 1796 gave John Evans directions "for crossing the continent in order to discover a passage from the sources of the Missouri to the Pacific Ocean." Both Mackay and Jefferson emphasized the importance of taking accurate geographical measurements, recording distances and climatic data, noting the topography, plants, animals, and minerals encountered, collecting unknown or unusual materials, and treating Indian people with all possible civility (Nasatir 1952, 2:410–14). Similarly, Lewis and Clark's most formal presentation of information about Native peoples, their "Estimate of the Eastern Indians," distilled information about the tribes located between St. Louis and Fort Mandan into categories with overt political and economic utility. Arranged in tabular form and sent to the U.S. secretary of war, the document summarized data regarding each tribe's location, village and population size, military strength, trading activities (types and quantities of furs produced, estimated annual returns, and so forth), and allies and enemies (Moulton 1987, 3:386–450). Also indicated was "the place at which it would be mutually advantageous to form the principal establishment in order to Supply the Several nations with Merchindize" (Moulton 1987, 3:389).

Such strategic knowledge was also eagerly sought by France, England, and Spain, and comparable documents can be found in their respective colonial administrative records. The Louisiana governor Kerlérec, for example, compiled a similar record in 1758 when he listed tribes allied with France in Louisiana and Illinois (Nasatir 1952, 1:51–55). Another example may be found in Trudeau's 1796 account of the Indian nations of the upper Missouri (Nasatir 1952, 2:383–85).

Science and Wonder in the Early American Republic

Jefferson's pragmatic nationalism was essentially inseparable from his prodigious intellectual curiosity and his commitment to science, learning, the arts, and philosophy. For members of the American Enlightenment, little meaningful distinction existed between science, industry, the arts, patriotism, and faith in a divine order (e.g., May 1976). Scientific discoveries were essentially inductive apprehensions of a marvelously complex and bountiful creation that, once illuminated by rationality, would aid human progress. All science had instrumental applications, and the resulting commercial growth would legitimate the philosophical project of advancing humanity through the creation of a democratic civil society in the new United States.

Jefferson's famously eclectic interests and mastery of wide-ranging domains of inquiry are both testaments to individual talent and symptomatic of the formative state of the sciences in early America. Only twenty years later, the government-sponsored expedition of Major Stephen H. Long would select participants on the basis of their training in emergent professional domains such as natural history, but Jefferson apparently believed that Lewis alone could master sufficient scientific knowledge and method to accomplish the requisite tasks. To prepare him for the expedition, Jefferson arranged for Lewis to be tutored briefly by some of the leading scientists of the day, Philadelphia men who were pioneering the creation of distinct disciplines such as botany, medicine, and astronomy. As James Ronda points out in the Foreword to this volume, the term "scientist" did not enter the American lexicon until the 1830s. Writing to Benjamin Smith Barton, a physician and naturalist who helped prepare Lewis for the expedition, Jefferson explained that he had chosen Lewis not for his scientific training, which was minimal, but because he had "firmness of constitution & character, prudence, habits adapted to the woods, & a familiarity with the Indian manners and character" (Jackson 1978, 1:17).

Portrait of Meriwether Lewis, from life, by Charles Willson Peale, 1807.

Although Jefferson's passion for learning and his encouragement of science were so great that his epoch in American science has been called "the Age of Jefferson" (Greene 1984), it is important to contextualize the exploratory scientific practice of the day and its role in the expedition. Even as he and the expedition he launched struck a blow for science, Jefferson stood on a fault line of epistemic transition. Shortly after pitching the expedition to Congress, he speculated excitedly that it might reveal living mammoths and giant ground sloths. It would be many years before paleontology and other natural sciences emerged from this conjectural stage.

Lewis and Clark, both professional soldiers, were at best amateur scientists even by the measure of their time. Natural history was then essentially a descriptive enterprise; Linnaeus had provided a rational classificatory framework for organizing observed relationships among living species, but the historical explanations of Darwin lay decades in the future. Jefferson was a deist, committed to the principle of the "great chain of being," which assumed that divine creation had occurred as a single and

Portrait of William Clark, from life, by Charles Willson Peale, 1807–8.

complete act—hence his belief that mammoths remained a living species. He was dismissive of attempts by his contemporaries to construct evolutionary theory in the earth sciences, regarding them as irrelevant (Boorstin 1993 [1948]:30–32; Greene 1984:31–32). Jefferson, then, rejected the historicizing models that later would provide a single, progressive narrative framework for understanding the development and diversity of life forms and so give rise to anthropology under the rubric of the natural sciences.

Lewis and Clark proved themselves to be careful observers of the natural world, reasonably skilled handlers of the sextant and compass, and accurate cartographers. They traveled with reference books such as Linnaeus's two-volume work on botany and systematically collected a broad suite of observational data and an array of natural history specimens. Their legacy as scientists, however, derives from their having been the first to describe and collect previously unrecorded plant and animal species (e.g., Cutright 1969), not from conceptual insights. For those, they intended to rely on members of the Philadelphia intelligentsia, who would analyze and write up the natural history data for the planned account of their western "tour." [1]

In short, although Lewis and Clark's practices of observation, description, and collecting helped constitute an emergent set of beliefs and practices known as science, they were hardly exemplars of a normative, scientific worldview. Even among the most learned segments of society, incipient scientific frameworks had yet to vanquish speculation, folk belief, and vernacular taxonomies (e.g., Appleby 2000; Ritvo 1997). Some of Lewis and Clark's ideas about medical practices, animal behavior, and natural processes reveal them to have been men formed as much by eighteenth-century folk culture as by emergent rationalism.

While Jefferson, Lewis, and Clark lived at the dawn of American science, they also dwelt in the twilight of the "age of the marvelous" (e.g., Kenseth 1991), a time that has been characterized as an interregnum between religious explanations and those provided by scientific rationalization (Pomian 1990). Novel, unusual, and rare elements of both nature and society were generally thought of as "curiosities," a term highlighting their wondrous qualities and the absence of established classificatory meaning. In the early nineteenth century, the exotic, singular qualities of native armaments, garments, and other "artificial [man-made] curiosities" excited fascination and speculation as much as they invited the objectification of scientific scrutiny. Despite the self-consciousness with which Lewis and Clark observed and recorded natural and cultural phenomena, they also highlighted the novelty of those phenomena by calling them curiosities (e.g., Moulton 1993, 8:164). And although the explorers were aware of the interest that learned men were beginning to cultivate in antiquities and non-Western material culture, there was as yet no discipline of anthropology to organize and instruct such curiosity. Not until after 1850 would early American anthropologists and scientific institutions begin to provide generalizing and comparative frameworks for the classification and interpretation of such objects. The notion of "ethnographic" materials had not yet been formulated.

Jefferson hoped that the knowledge Lewis and Clark would gain about Indian peoples would both contribute to scholarship and result in practical and diplomatic applications. Although he was a keen "natural philosopher" with a lifelong interest in Indian peoples and their history, the president's initial instructions to Lewis may have included few questions about Indians. He solicited recommendations from colleagues about how to guide the explorers' inquiries among Native Americans (Ronda 1984:2, 2000:33–36), directing them to learn as much as possible about the languages, economies, beliefs, marriage customs, kinship systems, martial practices, and recreations of the people they encountered (Jackson 1978, 1:157–61). How did they plant crops, define crimes, and receive guests? What, Jefferson wondered, were their homes, garments, and tools made of? A student of historical linguistics, Jefferson also asked Lewis and Clark to record tribal vocabularies (none survives), in the hope that they would shed light on the origins of and relationships between Native peoples. But no well-formulated notions existed regarding the relationship between culture and objects, and Jefferson gave the Corps no specific directives to acquire such things.

Lewis and Clark as Collectors

In contrast to their purposeful collecting of natural history items, Lewis and Clark received Indian-made objects primarily in the course of trade and diplomatic pursuits. In this respect, it is somewhat misleading to consider them collectors of these kinds of things at all, given that "collecting" has come to imply the purposeful acquisition of selected items that are generally understood to be interrelated (e.g., Belk et al. 1991; Pearce 1995). Unsystematic in their accumulation of objects of material culture, Lewis and Clark left few records regarding their receipt of the Indian garments, pipes, hats, and bows and arrows that later made their way into museum collections. While carefully recording the locations of observed or collected plants, rocks, animals, and topographic landmarks, as well as distances and astronomical readings, Lewis and Clark so seldom mentioned in their journals having received Indian-made objects that only a few of the surviving objects can be traced to encounters with specific people.

Our knowledge of which Indian objects Lewis and Clark kept and presented to President Jefferson and to the Peale Museum is based on a few surviving documents from the expedition and its aftermath. The first of these is a letter with an enclosed packing list that Lewis sent to Jefferson from Fort Mandan in April 1805. Having spent its first winter among the Mandans, the Corps was preparing to embark into land uncharted by Euro-Americans. Part of the crew was sent back to St. Louis in the keelboat, into which was loaded the first shipment of natural history specimens and Native American items destined for presidential consideration. Animal skins, the horns of mule deer, elk, and mountain sheep, minerals, insects, plant specimens, Indian robes and garments, and even living magpies, a prairie chicken, and a prairie dog were carefully packed into boxes and trunks. Written material was also dispatched, including some journals, a map, muster rolls, and Clark's "Estimate of Eastern Indians." This would be the largest and last shipment that Lewis and Clark sent while the expedition was in progress. The Fort Mandan letter and packing list reveal that whereas Lewis and Clark scrupulously observed, gathered, and collected insects, birds, animals, and plants, they regarded Indian materials in a different light.

The Fort Mandan packing list is one of the few descriptions that Lewis and Clark left regarding the Indian items they received. As reproduced in Gary Moulton's edition of the Lewis and Clark journals (Moulton 1987, 3:329–31), Lewis summarized the Native American items sent to Jefferson in the spring of 1805 as follows—with Lewis's original spelling and capitalization retained:[2]

Box No. 1: a Mandan bow an quiver of arrows—with some Ricara's [Arikara] tobacco seed.

Box No. 2: contains 4 Buffalow *Robes,* and a ear of Mandan Corn.

Box No. 4: 1 Earthen pot Such as the Mandans Manufacture and use for culinary purposes.

In a Large Trunk: 1 Minitarra Buffalow robe Containing Some articles of Indian dress. 1 Mandan robe Containing a dressed Lousirva Skin, and 2 Cased Skins of the Burrowing Squirel of the Praries. 1 Buffalow robe painted by a mandan man representing a battle fought 8 years Since by the Sioux & Ricaras against the mandans, *menitarras & Ah wah har ways (Mandans &c. on horseback[)].*[3]

In his accompanying letter to Jefferson, Lewis drew particular attention to the mineral specimens he was sending, noting that they had been numbered and correlated with a map of the Missouri River that he was forwarding to the secretary of war. "These have been forwarded," he continued, "with a view of their being presented to the Philosophical society of Philadelphia, in order that they may under their direction be examined or analyzed." The "other articles," he added, "are intended particularly for yourself, to be retained, or disposed off [*sic*] as you may think proper." He then turned his attention to apprising the president of the geography of the country through which they had passed, the status of the journey, and their future plans (Jackson 1978, 1:231–34).

The "other articles" to which Lewis referred were the natural history specimens and the objects of Indian manufacture. The invoice and later correspondence make it apparent that the botanical samples and the mammal skins, like the mineral specimens, were systematically numbered (Jackson 1978, 1:250). Through their collection, preservation, and labeling, they had been transformed into "specimens," a term Lewis himself used in describing them in the packing list.

The ethnographic objects were not so transformed; Lewis lacked the conceptual framework to see them as "artifacts" or "specimens." Instead, and in contrast to the natural history objects, he enumerated the ethnographic materials in cursory fashion, describing the appearance and explaining the significance of only one item, a painted buffalo robe.

The Corps of Discovery did, however, come into possession of a great variety of such materials. We can form some understanding of how and where they acquired these items by considering the explorers' relationships with Indian people, by closely reading

Lewis's woodpecker (*Melanerpes lewis*) at Harvard's Museum of Natural History is the holotype, or type specimen, for the species. Lewis and Clark probably collected this specimen in what is now Idaho. It is believed to be the only surviving natural history object from the expedition.

their journals and other surviving documents, and by considering the nature of the objects themselves. It is apparent that members of the expedition engaged in a lively, multifaceted exchange of goods and services with the people they met as they traveled. This exchange provided a crucial framework for social interaction and played a vital role in the expedition's success.

THE FABRIC OF EMPIRE

Fabrics in profusion dress the pipe stems in this collection: satin ribbons, plain and fancy tapes, lengths of woven and dyed wool cloth, and Native-woven wool tapes. European fabrics were valued trade items and diplomatic gifts, and Lewis and Clark took stores of such goods with them when they crossed North America. Cloth is emblematic of life on the early-nineteenth-century American frontier, where peoples, ideas, and products from distant parts of the globe met and became transformed.

By the time of the Lewis and Clark expedition, the Americas and Europe had been engaged in energetic commerce for nearly two hundred years. Cloth was one of the most important commodities in the emerging world economy, and England, France, the Netherlands, and other centers of loomed-cloth production competed fiercely for the growing export market. In North America, Indian alliances and trade were integral to the contests for control of the continent between France, England, Spain, and, after 1776, the new United States. As early as 1650 cloth had become a significant component of trade and diplomatic relations with Indian peoples.

European countries competed for North American resources and power through the Indian trade. French and English manufacturers catered to Native people's preference for wool cloth showing an undyed white selvage, such as the trade cloth on this pipe stem. (Detail of 99-12-10/53105.2; see p. 233.)

As Native groups developed preferences for certain types and colors of cloth, particularly woven and dyed wools, European cloth producers and traders in North America responded to their demands. English blankets and broadcloths such as strouds and duffels, particularly those dyed blue or scarlet (like those worn by the French and English militaries), were widely favored. Lewis and Clark distributed red and blue wool blankets, leggings, breechclouts, and coats. Red and blue have cosmological significance in cultures throughout Native North America, and it is reasonable to assume that color symbolism played a role in the preference for these colored fabrics.

French and English manufacturers also responded to the desires of Indian consumers for cloth of certain weights, widths, and patterns. Many Indians, for example, seem to have sought out lengths of wool fabric that retained an undyed, white selvage (fabric edge), and this preference may have influenced European dying techniques. Most of the strips of red wool cloth bound to the calumets in this collection share this feature.

Also present on the Peabody pipe stems are woven tapes or gartering made of wool, produced either on factory looms or on small, home-style wooden tape looms. Gartering and tapes were used to construct garments and served as decorative trim on clothing and upholstery. Lewis and Clark gave a variety of woven tapes to Indian women, including "Dutch tape" and "noneso-pretty." By the time of the expedition, Indian women of the Woodlands and Great Lakes were already using small wooden hand looms (called "heddles") and worsted wool trade yarns to weave similar trims and lengths. They produced woven wool tapes, sashes, and garters, sometimes incorporating glass beads into their weaving. Six of the pipes in this collection are wrapped with tapes that could be of either Native or European manufacture.

T. Rose Holdcraft and Castle McLaughlin

Chapter 2

Up the Missouri

Patterns of Diplomacy and Exchange

As they led their small Corps of Discovery through unfamiliar terrain, Lewis and Clark faced their greatest challenge in negotiating with representatives of independent Indian tribes, men who were savvy traders, formidable warriors, and experienced political negotiators with their own agendas and goals. Diverse languages, cultures, and economic interests separated the American explorers from the more than forty different nations they encountered as they made their way up the Missouri River and west to the Pacific Coast.

Both Indian people and members of the Corps of Discovery used the meaningful exchange of goods and services as their primary strategy for communicating with each other. Both relied on established rituals of "forest diplomacy" and trading protocols that had become conventionalized by French, British, and Spanish patterns of interaction with Indian groups. Most of these protocols, such as ritual gift giving, derived from indigenous

Captains Lewis & Clark holding a Council with the Indians, an engraving from an 1810 account of the expedition by Sergeant Patrick Gass (see Moulton 1996, vol. 10). The artist depicted Indian men from the familiar Eastern Woodlands rather than from the West.

diplomatic practices. Everywhere they traveled, Lewis and Clark signaled their peaceable intent, attempted to establish alliances, and provisioned their party by participating in formal and informal gift giving and bartering. These modes of exchange played a crucial role in the expedition's success.

Exchange, however, was neither a straightforward transfer of material goods nor a unilateral process. Although conventional protocols served as templates for Indian-white interaction, Lewis and Clark learned that in practice, exchange was a complex symphony, attuned to the nuances of social life. Exchange events registered the mutual discovery, appropriation, and strategizing reflective of both the immediate and the long-term interests of the participants. For their part, Lewis and Clark hoped to unify the interests of Indian peoples and join them under the American flag. This task proved more difficult than they imagined when they set off up the Missouri River with their small flotilla in the spring of 1804.

"Making the Road Open": The Politics of First Encounters

Lewis and Clark's primary charge with regard to Indian peoples was a diplomatic one. As ambassadors of the fledgling United States government, they were to initiate political and economic relationships with Native nations, gather information, and comport themselves in a manner that would foster the future growth of those ties. Jefferson clearly articulated these intertwined diplomatic and economic objectives in the instructions that he addressed to Meriwether Lewis in June 1803:

> In all your intercourse with the natives, treat them in the most friendly & consiliatory manner which their own conduct will admit; allay all jealousies as to the object of your journey, satisfy them of it's innocence, make them acquainted with the position, extent, character, peaceable & commercial dispositions of the U.S. of our wish to be neighborly, friendly & useful to them, & of our dispositions to a commercial intercourse with them; confer with them on the points most convenient as mutual emporiums, and the articles of most desireable interchange for them & us. (Jackson 1978, 1:64)[1]

Except in a few instances, Lewis and Clark were not the first Euro-Americans to encounter members of the Indian tribes west of the Missouri River, but they were the first official representatives of the new U.S. government to do so. In explaining their purpose to the Indians they met, Lewis and Clark stated that they were not traders but had come "to make the road open for the [American] traders to come" (Ordway in Moulton 1995, 9:49). "Making the road open" required enacting customary rituals and etiquette that had guided cross-cultural diplomacy throughout North America for several centuries. These practices included mutual gift exchanges and feasting, addressing one another by kinship terms such as "brother," "father," and "child," and displaying and exchanging key diplomatic symbols including pipes, wampum, robes, and peace medals. By extending these indigenous diplomatic protocols into their relations with outsiders, Indian peoples drew Europeans—initially on the Native Americans' own terms—into a web of interpersonal ties and mutual obligations.

Chief among these protocols was reciprocal gift giving, the sine qua non of social intercourse among Indian peoples and a prerequisite to forming new relationships. In Native societies, virtually all forms of exchange and diplomacy proceeded through the idiom of reciprocal "gifting." Consequently, the presentation of gifts became a ritual of intercultural diplomacy during the earliest phases of the European exploration and settlement of North America—a necessary prelude to more pragmatic trade and an integral aspect of diplomatic negotiations (e.g., Jacobs 1950). On an individual level, gift giving might take the form of presenting tobacco and sundries to a desired trading partner. For colonial governments and mercantile agents seeking to create or affirm alliances with tribes or bands, gifting involved the repeated expenditure of large

Black Moccasin, a Hidatsa chief, by George Catlin, 1832. Black Moccasin remembered Lewis and Clark to the artist, who rendered this portrait nearly thirty years after the expedition.

quantities of trade goods. In return, Native people engaged in trade, received visitors with hospitality, and formed military and political alliances.

As a form of exchange, gifting foregrounds social relationships and is idealized as an expression of friendship and mutual reciprocity. In Native societies, generosity was highly valued, and liberal gifting demonstrated the giver's status, sincerity, and social fluency. But gifting entails a subtle politics. Gifts seldom, if ever, represent isolated transactions; they create, acknowledge, and fulfill social debts and obligations, strengthening social ties and implying the expectation of equivalent returns in the form of goods and services (Davis 2000; Mauss 1954; Thomas 1991). Both Euro-Americans and Native Americans used gifting as a strategy for initiating trade relationships and cultivating well-placed allies. With equal fluency, both put objects into motion in order to pledge and affirm alliances, demonstrate status and power, leverage negotiations, secure resources, and forgive transgressions (e.g., Peers 1987). Lewis and Clark would learn about and participate in all of these practices during their two-year journey across the continent.

Bartering also played an important role in early Indian-white relations. The willingness to trade communicates peaceable intent, recognizes and creates reciprocal interests, and establishes a basis for extending social relationships. Most Europeans and Americans with whom Indians interacted were traders, and the tribes who encountered Lewis and Clark expected the explorers to exchange some of their goods as an indication of sincerity and goodwill. Gifting and bartering were the means by which the Lewis and Clark party provisioned themselves and ensured their safe passage across the continent. Only through exchange could they gain entry into Indian communities, deliver Jefferson's message about future intercourse, and hope to understand Indian responses.

But opening the road for a new "Great Father" would not be easy among peoples who already had established political and economic relationships with competing

European powers as well as with many other tribes. The Corps of Discovery's route up the Missouri and across the Rocky Mountains to the Pacific took them straight through two of the most dynamic indigenous trade systems in North America. One, the middle Missouri system, was centered at the semipermanent Mandan, Hidatsa, and Arikara villages; the second, the Pacific Coast–Columbia Plateau system, was concentrated along the Columbia River (Ewers 1968; Wood 1980). Both trade systems consisted of extensive intertribal networks of kinship, exchange, affiliation, and enmity. French, Spanish, and British traders had inserted themselves into these exchange fields, creating a transcultural system that provided most tribes with direct or indirect access to European trade goods. Lewis and Clark observed the extent of these networks firsthand, finding Spanish mules, horses, and horse gear in Shoshone camps near what is now the Idaho-Montana border (Moulton 1988, 5:92). On their return east, they presented the Peale Museum with material evidence of these far-flung interaction spheres: a "Spanish Dollar obtained by Captn. M. Lewis, from the Pallotepallers—a nation inhabiting Lewis's River within the Plains of the Columbia, who had never previously seen white-men."

Aware that he would have to communicate the "peaceable & commercial dispositions of the U.S. " by presenting gifts and trading, Lewis purchased and sorted quantities of trade goods and presents before leaving the East Coast. Wool cloth, Euro-American garments, peace medals, pigments, glass beads, tobacco, and a variety of metal objects—fishhooks, scissors, awls, tomahawks, knives, musket balls, and needles—were packed into fourteen wrapped and numbered bales and a box (Moulton 1987, 3:492–505). At Jefferson's request, Lewis included several corn mills, destined for the horticultural Mandans. Some of the bales were designated for known tribal groups, whereas others were prepared for the "foreign nations" dwelling upriver from the Mandans. The contents of the bales were organized according to the social status of the intended recipients, including chiefs and their wives, elders, men and women "of consideration," other adult men and women, and young people. Silver peace medals in four or five graduated sizes and several series were included in the bales according to the perceived importance of tribal groups and their leaders (Prucha 1971:17). Only three of the largest Jefferson series medals were included, designated for the Omahas, Arikaras, and Mandans.

In selecting his purchases, Lewis drew on several centuries of eastern forest diplomacy and on known Indian preferences and expectations regarding the quality and types of materials to be offered (e.g., Ray 1980). Metal tools, colored glass beads, varieties of cloth, and tobacco were standard gifts and trade currency; tobacco, peace

GLASS BEADS

Columbus first introduced colored glass beads to the Americas, and they became so thoroughly integrated into Native life that today they are considered emblematic of Indian identity. Before contact with Europeans, Native Americans created beads from shell, bone, stone, and other natural materials, and Indian women embellished garments with dyed porcupine and bird quills and plant fibers. Although some Native American women continued to use only traditional materials in their work, many others appreciated glass beads for their color, luminosity, and ease of use. Women used them to make jewelry as well as to decorate ritual objects, garments, and even tools.

At the time of the Lewis and Clark expedition, glass beads were among the most sought-after Euro-American trade items. Despite local preferences for certain colors and sizes, Lewis and Clark (and other travelers) found that blue beads were valued most highly, followed by white. Clark noted that among the Indians along the Columbia River, "blue beads occupy the place which gold has with us. White beads may be considered as our silver" (Jackson 1978, 2:529).

Of the great variety of glass beads traded into North America during the several centuries of the fur trade, most, though not all, were manufactured in Italy, a center of glass production since the Middle Ages. French, English, Russian, and Spanish traders carried the beads to Indians living west of the Missouri River during the eighteenth century. In preparation for the first U.S. expedition into that country, Lewis purchased stores of glass beads to be distributed as "Indian presents," including sky blue, yellow, red, and white seed beads and faceted "mock garnets."

The glass trade beads sewn onto objects in this collection are of two sizes: small seed beads and the slightly larger "pound" or "pony" beads. The beads most commonly seen on objects in the Peabody's Lewis and Clark and Hutter collections are drawn (hollow cane) translucent beads of a distinctive blue-green. These were used on the beaded baby carrier (p. 126), one of

Before the expedition, Lewis purchased glass beads in a variety of colors, only to discover that most Indian people would trade only for blue or white ones. Blue glass beads similar to those on this baby carrier appear on several objects in the Peabody collection. (Detail of 99-12-10/53016; see p. 126.)

the side-fold dresses (p. 181), and the bow case and quiver (p. 259). They were made from a tube of blown glass, then shaped through re-heating and tumbling. Bill Billeck, of the Smithsonian's National Museum of Natural History, noted that some 100,000 beads of this type were excavated at the Leavenworth site, a former Arikara village visited by Lewis and Clark on their way up the Missouri in 1804.

The white, black, and dark blue pony and seed beads on objects in this collection, such as the pony beads on the beaded choker (p. 137), are types known to have been used during the first quarter of the nineteenth century. Others, such as the blue-gray beads on the painted side-fold dress (p. 176) are more unusual. Small green, red, and lavender seed beads occur as isolates or in small numbers on some Peabody objects, but the origins and dates of manufacture of these rare beads are poorly understood.

medals, flags, and wampum were essential for conducting diplomatic negotiations. Lewis chose his gifts well. While wintering at Fort Mandan, Clark wrote, "The nations in every quarter I am told are fond of Blue Beeds, red Paint, Knives, axes, Guns & ammunition." Other prized commodities exchanged among the Missouri River tribes, Clark added, were horses, the feathers of the "Calumet [golden] eagle," arrow points, pipes, and hides from the rare white buffalo (Moulton 1987, 3:484). Lewis later advised Jefferson that the most valued trade items were blue beads, brass buttons, knives, battle axes and tomahawks, awls, glover's needles, iron combs, and brass kettles. He said that were he to repeat the expedition, blue beads would make up at least half of his supplies (Jackson 1978, 1:374–75). Clearly, Indian people were drawn not only to the overt utility of European-made weapons and tools but also to objects imbued with less tangible kinds of power and efficacy.

Lewis and Clark used these stores as basic currency for negotiating their way across the country, "opening the road" to friendship through the cross-cultural idiom of exchange and alliance building. The contexts of the Corps' relationships with Indian peoples were many and their character varied, evoking several sometimes overlapping modes of exchange. These included the formal exchange of diplomatic gifts by leaders, generally directed toward long-term goals; more pragmatic barter for immediate necessities such as horses and food; the "purchase" of sexual favors and information; and the informal exchange of gifts between individuals. The nature of these transactions often blurred the distinction between gifts and barter, and not all exchanges were material. The sharing of both cooked meals and raw foodstuffs, often critical to the Corps' survival, was also a fundamentally important symbolic bonding mechanism. And members of the Corps and their Indian hosts provided each other with a wide array of services and favors expressive of social allegiance. Expedition members offered medical treatment, metal-working skills, and even military support, whereas Indians furnished horse care, loaned canoes, shared information, and provided sexual companionship. When their trade goods became depleted, the Corps improvised, substituting pierced coins for peace medals and presenting Indians with the hides of animals they had shot, bear claws, their own personal possessions (including garments and armaments), and objects received from other tribes.

Lewis and Clark's relationships with Indian groups and associated transactions expressed a wide range of individual short-term agendas and interpersonal dynamics as well as the long-range views of each nation. Several important structural factors shaped the responses of tribal peoples to the Corps' diplomatic overtures. Most fundamentally, relations between Lewis and Clark and the Indians they met were conditioned by the ways in which tribal groups were differentially situated in existing political and economic networks.

The Corps' ambassadorial mission was least successful among tribes that were already well positioned in regional political economies. The Lakotas (Teton Sioux), for example, had no incentive to forge relations with a distant, unknown United States government—quite the contrary. Well supplied with arms and other goods by British traders on the Mississippi River and by their eastern relatives, the Lakotas were a dominant force in the trafficking of European commodities along the Missouri River. The Arikaras were closely bound to the Sioux in the trading network, supplying the Lakotas with food staples, dressed hides, garments, horses, and other goods that Lakota people then exchanged for guns, ammunition, and steel tools. The Arikaras also supported Lakota military actions against the Mandans and Hidatsas, important intermediaries in the regional trade with British fur companies in Canada (Jablow 1994 [1951]).

To thwart the development of trading alliances between French and Spanish traders in St. Louis and the upper Missouri village tribes, powerful and autonomous Lakota bands intimidated Euro-Americans traveling up the Missouri, seizing their boats and goods. This strategy proved so successful that white trading and exploratory parties often conveyed an extra canoe loaded with trade goods, hoping to buy their passage upriver (e.g., Nasatir 1952, 1:88, 97, 2:267–76; Thwaites 1904a:103). Lewis and Clark's refusal to capitulate to Lakota demands and their declared purpose of opening direct trade with all Missouri River tribes nearly resulted in physical violence with the Lakotas when the Corps ascended the Missouri in August 1804 (Ronda 1984:27–41). Sioux people continued to threaten the Corps during their winter with the Mandans, and when Lewis and Clark prevailed upon the Arikaras to make overtures of peace to the Mandans, Arikara emissaries were beaten by their Sioux allies (Ronda 1984:107–10).

By the time they wintered with the Mandans, Lewis and Clark had begun to appreciate the extent of intertribal relationships and the ways in which European trade goods influenced their dynamics. In council with Mandan leaders, Clark argued that the Arikaras persisted in their alliance with the Sioux not because they were inherently "bad men" but because "you know that the Ricarees, are Dependant on the Sceaux for their guns, powder & Ball, and it was policy in them to keep on as good terms as possible with the Siaux until they had Some other means of getting those articles" (Moulton 1987, 3:246).

Lewis and Clark's stated claims to American political authority, demands for intertribal peace, and intention to regulate commerce threatened the status quo and were received differently in different camps. Groups who were advantaged by existing balances of power were alarmed by the Americans' declared aim of opening trade with all tribes, which would offer their enemies arms and other trade goods.[2] Resistance was expressed in many forms. The only violent interaction between the Corps and Native people, an occasion when a small party under Lewis's command killed two Piegan Blackfeet men, unfolded when young warriors tried to take the Americans' guns and horses. As James Ronda (1984:239–42) has noted, this incident may have been precipitated by Lewis's announcement that he and Clark had unified the enemies of the Blackfeet and would be providing them with traders.

Combat between Sioux warriors and earthlodge villagers is depicted on this Peabody bison robe in a pictographic representation of political economy along the upper Missouri River. (Detail of 99-12-10/53121; see p. 146.)

Lewis and Clark were also at a disadvantage among tribes accustomed to European traders, and they often found it difficult to communicate an abstract message of exploration and future advantages. "We want one thing for our nation very much," the Yankton leader Arcawecharchi (Half Man) told them: "we have no trader, and [are] often in want of good[s]" (Moulton 1987, 3:30). On occasion the Corps' relatively meager stores of goods threatened to undermine its diplomatic success. Even the Mandans, with whom the Corps developed especially close ties, initially expressed disappointment at the paucity of their presents, and some chiefs refused to accept their gifts altogether, thereby opting out of a potential relationship.

The Corps compensated for its material shortages by providing other goods and services during the months that the men camped near the Mandan villages. Expedition blacksmiths proved critical in this regard, spending a busy winter making war axes and repairing metal tools, for which the party received corn and other supplies. Some of the corn trade at Fort Mandan was conducted by women, the gardeners in Mandan and Hidatsa society. The trader François A. Larocque, who wintered with the neighboring Hidatsa tribe during Lewis and Clark's stay, wrote in January 1805 that the American party "have a very Expert smith, who is always employed in making dift. things & working for the Indians, who are grown very fond of them although they disliked them at first" (Wood and Thiessen 1985:149–50). The Hidatsas, who were allied with British North West Company traders, remained aloof from the Americans, and some of their chiefs declared that "had these Whites come amongst us with charitable

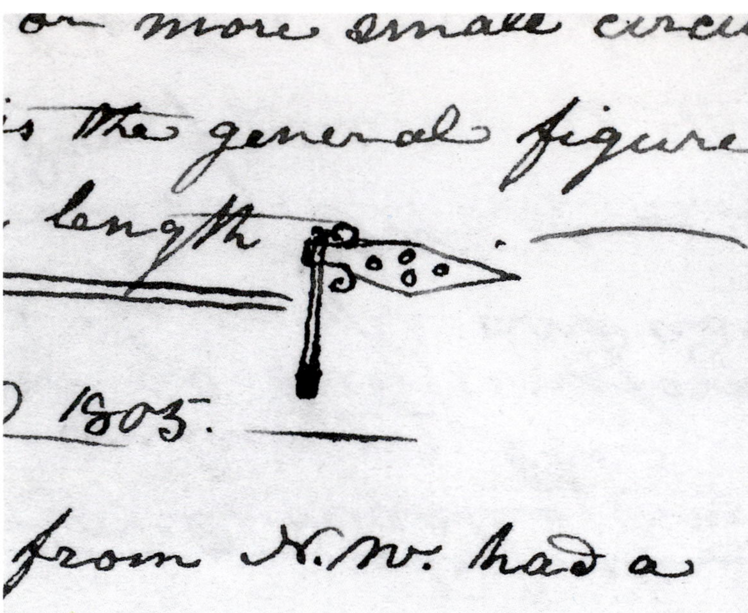

Top: Mandan corn. Archaeologist and nurseryman George Will presented these corn specimens to the Peabody Museum in 1915–17.

Bottom: Drawing of a spontoon-type "battle ax" or tomahawk from the Lewis and Clark journals, February 5, 1805. Lewis and Clark's expedition members survived the winter of 1804–5 at Fort Mandan by trading metal goods and metal scraps for Indian corn and other garden staples.

views they would have loaded their Great Boat with necessities" (McKenzie in Wood and Thiessen 1985:232). On occasions when exchange could not be conducted through the idiom of social relationships, Indians also attempted to "leverage" the Corps' goods through pilage, "theft," and intimidation.[3] (The Lewis and Clark party itself illicitly took a canoe from some Clatsops when preparing to leave the Pacific Coast, ostensibly in retaliation for a Clatsop theft of the Americans' elk meat; see Ronda 1984:210–11.)

At the other extreme from the Sioux, Native peoples living in the trade bottleneck of the Rocky Mountains, such as the Lemhi Shoshones, had much to gain by forming ties with a nation that promised to deliver goods and services. Surrounded by well-armed enemies such as the Blackfeet, Atsinas, and Hidatsas, the Shoshone band led by Cameahwait, the brother of Sakakawea, the well-known Lemhi Shoshone woman who traveled with Lewis and Clark from the Mandan villages to the Pacific Ocean,[4] had almost no guns and few powerful allies. Clark described them as "excessive pore [poor]" (Moulton 1988, 5:139). "They told me, Lewis wrote, "that to avoid their enemies who were eternally harassing them that they were obliged to remain in the interior of these mountains . . . where they suffered . . . great hardships for the want of food[,] sometimes living for weeks without meat and only a little fish roots and berries." But this, Cameahwait observed, "would not be the case if we had guns, we could then live in the country of buffaloe and eat as our enemies do and not be compelled to hide ourselves in these mountains and live on roots and berries as the bears do" (Moulton 1988, 5:91). Unlike residents of the indigenous trading emporiums along the Missouri and Columbia Rivers, Cameahwait's people had seen few Europeans, if any. Lewis observed that "every article about us appeared to excite astonishment in the[i]r minds; the appearance of the men, their arms, the canoes, our manner of working them, the black man [Y]ork and the segacity of my dog were equally objects of admiration" (Moulton 1988, 5:112).

Exchange Events

Virtually every encounter between the Corps and Native people began with the display and exchange of objects. To broadcast a desire for peaceful interaction, Lewis and Clark often displayed or sent gifts in advance of the party. Upon sighting a lone Shoshone horseman, for example, Lewis hastily indicated his intention to trade through a succession of signs, including waving and then spreading a blanket, tying goods to a pole erected over his fire, and even attaching some "trinkets" around the necks of Indian-owned dogs that approached while their masters hesitated. These stratagems, Lewis believed, would serve as tokens that "we were friendly and white persons" (Moulton 1988, 5:70). Similarly, upon first seeing members of the party, Indians often reacted by presenting personal possessions or customary diplomatic gifts—sometimes as acts of supplication. When the expedition surprised a small Shoshone camp, for instance, the frightened occupants "offered us every thing they had, which was but little; they offered us collars of elks tusks which their children woar[,] Salmon ber[r]ies &c." (Moulton 1988, 5:145). The captains returned these offerings, adding to them various small trade goods as a gesture of reassurance.

Formal greeting ceremonies and meetings occasioned larger-scale, ritualized exchanges conducted by representatives of each group. The parties traded speeches, shared food, and reciprocally presented objects symbolizing allegiance and friendship. Gifts were often exchanged as a prelude to the proceedings, as expressions of hospitality. Several Missouri River groups carried Lewis and Clark into their camps on buffalo robes. During Lewis's first few moments with Cameahwait's Shoshones, he smoked with them, presented beads and vermilion paint, and "gave [Cameahwait] the [U.S.] flag which I informed him was an emblem of peace . . . and now that it had been received by him it was to be respected as the bond of union between us" (Moulton 1988, 5:79–80). The two groups then proceeded to Cameahwait's camp, where, seated on green boughs and dressed antelope hides under the canopy of a lodge, they smoked and parlayed more earnestly, creating a ritual space that Lewis poetically described as "this little magic circle" (Moulton 1988, 5:80). Clark recorded that when he joined them the following day, the Shoshone chiefs immediately "tied to my hair Six

Small pieces of Shells resembling *perl* which is highly valued by those people and is pr[o]cured from the nations residing near the *Sea Coast* (Moulton 1988, 5:114).

Formal diplomatic councils, which Lewis and Clark conducted with representatives of Indian peoples east of the Rocky Mountains, unfolded in accordance with a widely shared protocol: they were convened with ritual pipe smoking, which was followed by speech making and an exchange of objects invested with political symbolism. Feasting and a solemnizing round of the pipe might follow, while more generalized gift giving and trading often developed after formalities had concluded. On these occasions, Lewis observed, "points of etiquet are quite as much attended to by the Indians as among scivilized nations" (Moulton 1988, 5:112). A council between a Yankton group and the Corps of Discovery is recounted in detail in chapter 8, "The Language of Pipes."

Among the diplomatic conventions and rituals developed by Lewis and Clark's French, Spanish, and English precursors were recognizing chiefs in three status categories (principal, second, and third chiefs), "dressing" them in Euro-American military regalia to acknowledge their standing vis-à-vis colonial officials, and presenting them with conventional symbols of their office (Ewers 1997; Nasatir 1952). These presentation gifts, which Lewis carefully assembled, included European-style shirts, jackets, and hats (some plumed), silver gorgets and armbands, red and blue wool trade cloth, wampum, shell "hair pipes," peace medals, flags, and written "commissions" testifying to the authority of individual chiefs (Ewers 1997; Prucha 1971; Woodward 1970).

The dressing of chiefs, presented as an act of homage and respect, may have originated as an intertribal diplomatic gesture. It was a standard practice among both European officials in the Northeast and interior fur traders by the time of Lewis and Clark. A Yankton warrior named Tar-ro-mo-nee commented approvingly on Lewis and Clark's fidelity to this custom. After watching the captains present ceremonial regalia to another man, Shake Hand, Tar-ro-mo-nee remarked that he was "verry glad you have made this man our great Chief, the British & Spaniards have acknowledged him before but never Cloathed him" (Moulton 1987, 3:29–30).

Silver peace medals, flags, and commissions were particularly significant symbols of state power and mutual allegiance. European nations had distributed these to their Indian allies for generations; the new American nation simply issued new versions bearing their own national symbolism and imagery. John Evans and Jean Baptiste Truteau, who worked for the Spanish Missouri Company, had presented Spanish flags and medals to leaders of Missouri River tribes, including the Mandans, in 1795–96.

Truteau observed that the Arikaras venerated these objects, keeping them in medicine bundles and smoking them with sweetgrass when they were unwrapped (Ewers 1997:105).

Lewis and Clark presented medals to make a direct statement about the transition of power in North America, and the historian Francis Paul Prucha (1971:24) noted that they did so "with a certain abandon," in contrast to conservative European conventions. Lewis and Clark were eager to encourage Indian peoples to disavow their existing allegiances to other countries, particularly England. While at Fort Mandan, they warned the North West Company trader Larocque against distributing British medals and flags (Wood and Thiessen 1985:138).

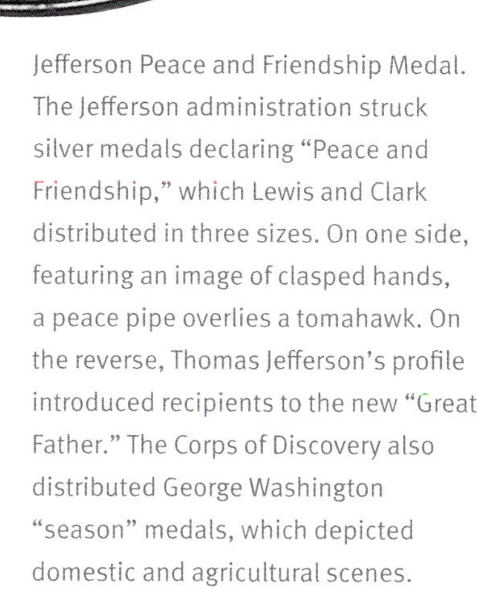

Jefferson Peace and Friendship Medal. The Jefferson administration struck silver medals declaring "Peace and Friendship," which Lewis and Clark distributed in three sizes. On one side, featuring an image of clasped hands, a peace pipe overlies a tomahawk. On the reverse, Thomas Jefferson's profile introduced recipients to the new "Great Father." The Corps of Discovery also distributed George Washington "season" medals, which depicted domestic and agricultural scenes.

Members of the Corps of Discovery also bartered with Indians everywhere they traveled. Once formal diplomacy had concluded, Native people themselves initiated much of the trade. Sergeant Patrick Gass, who kept a journal of the expedition, wrote at Fort Mandan that "[a] great number of the natives came with corn, beans and mockasins to trade, for which they would take anything—old shirts, buttons, awls, knives, and the like articles" (Moulton 1996, 10:67). But although the Indians were interested in all manner of Euro-American goods, the Corps relied heavily on the widespread demand for three commodities—metal, beads, and tobacco—to barter for food and other necessary goods and services. At Fort Mandan, the party exchanged metal tools and their repair for Mandan corn and garden produce; on the Pacific Coast, where the expedition spent its second winter in proximity to Clatsop groups, the men exchanged metal fishhooks for local roots, berries, and other foodstuffs. Two days before their Pacific Coast Christmas, for example, Clark wrote, "2 Canoes of Indians Came up to day. I purchased 3 mats verry neetly made, 2 bags made with Flags verry neetly made, those the *Clatsops* Carry ther fish in. [A]lso a Panthor Skin and Some Lickorish roots, for which I gave a worn out file, 6 fish hooks &Some Pounded fish" (Moulton 1990, 6:135).

Barter dominated interactions between the Corps of Discovery and the peoples who lived along the Columbia River and near the shores of the Pacific Ocean, a trading hub where indigenous groups conducted lively commerce with other tribes as well as with the crews of European and American trading ships (e.g., Malloy 2000). There the explorers encountered Indian tribes who were already armed with muskets, wearing European-style sailor's outfits, and boiling water in brass tea kettles. Along the Columbia River, the explorers often met Indian trading parties traveling by canoe. Writing about the Chinooks, whom Lewis famously described as being "great higlers in trade," Clark declared, "It is a bad practice to receive a present from those Indians as they are never Satisfied for what they re[ce]ive in return if ten times the value of the articles they gave" (Moulton 1990, 6:61). When Cuscalah, a Clatsop chief, visited their camp and made a present of mats and roots to both Lewis and Clark, he specified that he expected metal files in return. Rather than trade, the captains chose to return the mats, "which displeased Cuscalah a little" (Moulton 1990, 6:136).

Lewis and Clark negotiated most of the expedition's strategic and utilitarian trading activities and managed the supply of trade goods. Lewis, for example, received three good horses from Shoshones for "a uniform coat, a pair of legings, a few handkerchiefs, three knives and some other small articles" (Moulton 1988, 5:117). Crew members were permitted to exchange gifts and to barter for personal mementos and for sexual favors from Native women (see Moulton 1990, 6:74, 1991, 7:247). Indians were interested in all manner of objects associated with Europeans, and both the captains and enlisted men made use of their own possessions to effect exchanges. On the Pacific Coast, Lewis traded one of his old coats and a vest for a sea otter skin (Moulton 1990, 6:346); a dressed sea otter skin was among the personal effects sent to William Clark when Lewis died en route to Washington in the fall of 1809 (Jackson 1978, 2:471).

Brass buttons. Lewis reported to Jefferson that Indians traded eagerly for brass buttons like these from the yoke of a Peabody side-fold dress. The "back marks" (seen here and on facing page) were unregulated marketing slogans. (Detail of 99-12-10/53047; see p. 181.)

In late May 1806, as they prepared to recross the Rockies on their journey home, Lewis and Clark distributed their dwindling stock of trade items to members of their party, "with a view that each should purchase [from the Nez Perce] therewith a parsel of roots and bread from the natives as his stores for the rocky mts" (Moulton 1991, 7:275). "Each man's stock in trade," Lewis recorded, "amounts to no more than one awl, one Kniting pin, a half an ounce of vermillion, two needles, a few scanes of thread and about a yard of ribbon; a slender stock indeed." The enlisted men augmented this stock by offering the brass buttons from their clothing. When weather delayed their departure for a month, Lewis chronicled their declining trading position:

> McNeal and [Y]ork were sent on a trading voyage over the river this morning, having exhausted all our merchandize we are obliged to have recourse to every subterfuge in order to prepare . . . to meet that wretched portion of our

journy, the Rocky Mountain. . . . Our traders McNeal and York were furnished with the buttons which Capt. C and myself cut off our coats, some eye water and Bsilicon which we made for that purpose and some Phials and small tin boxes which I had brought out with Phosphorous. [I]n the evening they returned with about 3 bushels of roots and some bread having made a successful voyage, not much less pleasing to us tha[n] the return of a good cargo to an East Indian Merchent. (Moulton 1991, 7:325).

In all of these transactions, values were flexible and negotiated. As they traveled, Lewis and Clark developed a sense of the appropriate and expected measure of goods and services necessary to initiate social contact, conduct successful diplomatic encounters, procure supplies, calm misunderstandings, and demonstrate appreciation. But because the expectations and desires of both parties were complex and situational, each transaction was also in effect a calculated negotiation in which "value" was largely a product of the exchange context. Sometimes no agreement was reached. The expedition journals contain numerous descriptions of failed attempts at both gift exchange and barter, especially near the confluence of the Columbia River and the Pacific Ocean, where Lewis and Clark often found themselves unable to meet the prices that prosperous Native people asked for food and trade goods. "Those people ask double & tribble the value of everry thing . . . and never take less than the full value of any thing," the explorers reported. "[T]hey prise only Blue & white beeds, files[,] fish hooks and Tobacco—Tobacco and Blue beeds principally" (Moulton 1990, 6:133–34).

Relations between the Corps of Discovery and Indian people were forged, mediated, and enacted through the exchange of goods and services. These relationships, however tenuous, were critical to the immediate survival of the Lewis and Clark party and to the success of their diplomatic mission. And some of the materials they received would become part of one of the earliest assembled collections of Native American objects.

FROM TIME IMMEMORIAL

The Mandan, Hidatsa, and Arikara People

Mike Cross

In the days of Thomas Jefferson, the peoples called Mandans, Hidatsas, and Arikaras—my ancestors—were living on the banks of the Missouri River. They watched as the boats of Lewis and Clark arrived at their homes.

The strength was ours, then, to offer friendship or withhold it. We asked little for what we gave, and what we gave was often beyond price. Let it not be thought that we held in low esteem the earthlodge villages lining the shores of the river, our planted fields of corn, beans, squashes, sunflowers, pumpkins, melons, and tobacco, or the timbered forests. We shared what was dearest to us and never questioned our wisdom—at least not in those first, early days. The time would come when our eyes would sicken with longing for what we could no longer see.

Need we recite what was in our sharing? Lodging, guides, information. Missouri, Yellowstone, Klamath, Columbia. Knives, horses, robes, shelter. Food. Millions of buffalo and other wild game. Water and green hills. Hardwood leaves flashing in the sun. River waters startled by waterfowl rising at dawn. Big-eyed deer at the edge of a glade. Antelope on the open buffalo grass prairies.

Ours was not an unused land. Our families had their agricultural plantings here. We knew the uses of forest and rivers and streams. Our commerce knew the farthest traces eastward and westward, north and south. We mined and quarried. We left our monuments in many livable river valleys. Ours was not a land untouched but a land as yet unspoiled.

Mike Cross and his daughters, Kara and Amanda, cultivate a traditional garden at their home in New Town, North Dakota.

Our greeting, as we gather now for the bicentennial of the Corps of Discovery, is spoken in part out of sadness. Our ancestors were vital to the success of Lewis and Clark's expedition, and it is said that the explorers would not have succeeded without their help. When my father, in his first term as chairman of the Mandans, Hidatsas, and Arikaras—the Three Affiliated Tribes—met with Franklin Delano Roosevelt in 1944, he told the American president, "We have proof as long ago as 1804 . . . when the Lewis and Clark Expedition came up the river to explore the Louisiana Purchase[;] we furnished a tribal guide, a woman named Sacagawea, to show the party the way to the Pacific Coast. . . . [O]ur [t]ribe played an important part in claiming the territory of the Northwest of the United States." Since that time, our history has been less happy.

In memorial

I wish all Americans could have seen our beautiful valley. In the early fall days of September and October, the colors of our large forests of golden cottonwoods, yellow ash, red cedar, the river willows, the gray-green buffalo-berry trees with their bright, shiny red fruit, the dusty-looking pink plums and purple grapes hanging from their vines along the elbow of the sacred river, A-wah-tee, where it turned southeast toward Nishu from Lucky Mound, Independence, and Shell Creek—these will always be fixed in our tribal memories. The sacred waters that came from the western mountains, the earth smells, the muted beauty of the calls of mourning doves echoed in the early mornings of autumn. Indian women would place cornballs made from their vegetable gardens and wild fruit they harvested along the river's edge as offerings to the river spirit for the gift of his protection. These things were sacred to our people, part of a traditional culture that was complex and deeply felt.

In the early nineteenth century, when the Lewis and Clark expedition arrived, ours was still a "traditional" Indian culture, even though we had already experienced extensive contact with whites. Indeed, the traditions and ancient ways had begun to change with the arrival of the first European explorers and traders in the 1700s. These men brought with them not only their dreams of seeing new lands and finding fortune but also the most destructive agent Indians have ever known: smallpox. It is possible to say that smallpox has been the driving force behind all major changes that have shaped the lives of members of the Three Affiliated Tribes since then. From the first epidemic in 1781, when the Mandans, Hidatsas, and Arikaras lost over half their populations, to the last catastrophic outbreak in 1837, the tribes were devastated by disease, weakened by attacks from the enemy Sioux, forced to move from their earlier homes, and scattered along the upper Missouri River (Cash and Wolff 1974:41). In June 1837 the Mandans had numbered between 1,600 and 2,000 individuals. Just four months later, in October, only 138 Mandans were left alive (Thornton 1987:96).

Eventually, the three tribes set up their winter camp together at a bend in the Missouri. They called the spot "Like-a-Fishhook Village" (Cash and Wolff 1974:44). The forty-some years spent there have been called "the twilight of traditional native lifeways" (Gilman and Schneider 1987:337). The women began planting traditional gardens again, and although they had lost many of the ceremonies and songs, which had died along with their owners, the people joined together in creating new ones.

But there was no hope of returning completely to the traditional way of life. The great herds of buffalo were gone, and grain farming and cattle ranching had replaced hunting as the primary source of subsistence. Religion, too, was forever changed: for each man and woman lost to disease, the tribes also lost sacred bundles, medicine, and ceremonies. And then the missionaries arrived on the reservation, working to convert to Christianity as many Mandans, Hidatsas, and Arikaras as possible.

When Poor Wolf, second chief of the Hidatsas, at last agreed to give up his sacred bundle articles, which included a bear's arm, a crane's head, an owl's head, a buffalo skull, and some braided sweetgrass, he spoke of the "agony" he felt: "Must I give up the old Indian songs, which are a part of the life of our people? Must I give up the charms that I have carried on my body for years and which I believe have defended me from demons? My body is tattooed to show my allegiance to various spirits. How can I cut these out of my flesh?" (Meyer 1977: 128–29). Even more catastrophic than the physical effects of smallpox was the emotional and spiritual suffering of the people. Within the short span of sixty years, the Mandans, Hidatsas, and Arikaras had lost their homelands, families, religion, and culture.

The people were now vulnerable, too, to the Indian agents who came to Like-a-Fishhook Village to persuade them to abandon traditional ways. The agents' objective was to "civilize" the people, a task they believed they could accomplish by turning the Indians into farmers. They sent the children away to boarding schools, where they were not allowed to speak their native language, had their hair cut short, and were taught to be ashamed of their Indian heritage. After the passage of the Dawes Allotment Act in 1887, which gave every Native male head-of-household 160 acres of land to farm, Like-a-Fishhook Village was completely abandoned, its former residents scattered throughout the reservation on small "allotted" plots of land (Spotted Wolf 1972).

The three tribes managed to maintain some of their former social connections in the small communities they set up throughout the reservation—Nishu, Shell Creek, Independence, Elbowoods, Red Butte–Little Missouri, Charging Eagle, Beaver Creek, and Lucky Mound. In 1893, the Fort Berthold Indian Agency was moved to Elbowoods, where it remained for the next sixty-odd years. Once more the people began piecing their lives back together, little by little. Again they planted traditional gardens, and their children, though educated in the white man's schools, were taught what their native cultures were like before the great epidemics.

And then disaster struck again, this time in the form of Garrison Dam, a project that would provide irrigation for white farmers in North Dakota but would flood the land on which all eight of the existing Indian communities were built. Construction of the dam began in October 1947 and was completed in 1954. The people were given just six months to move their homes, their belongings, and their families before their land was completely covered in water. The Indians lost everything—the natural resources of the area, the floodplain timber that provided logs for their houses, the wild fruits and berries that were an important part of their diet, and the springs and creeks that provided a natural water supply (Meyer 1977:220). The Mandans, Hidatsas, and Arikaras would never again live on the rich bottomlands of the Missouri River—the land guaranteed to them "in perpetuity" only one hundred years before.

Today the Mandan, Hidatsa, and Arikara nations reside on the Fort Berthold Indian Reservation in North Dakota, around the Missouri River. Most of the younger generation is now of mixed blood—either Mandan-Hidatsa or Arikara mixed with another tribe, such as Crow or Sioux.

My grandfather, Old Dog, married Many Dances in the 1890s. Their parents were Iron Eyes and Yellow Corn and Many Bears and Sweet Grass, all Mandan and Hidatsa. Many Bears and Sweet Grass, Iron Eyes and Yellow Corn had survived the smallpox epidemic of 1837. They lived first in the two traditional villages along the Knife River that the people had created after the epidemic of 1781, then in Like-a-Fishhook Village. Old Dog and Many Dances lived at Like-a-Fishhook Village until the missionaries and Indian agents moved them to land allotments in different communities around the reservation. Finally they lived at Old Dog's Flat near the Fort Berthold Agency, which became known as the village of Elbowoods. My father, Yellow Eagle (Thomas Martin Old Dog Cross), lived there, too. As tribal chairman he devoted years to fighting the Garrison Dam project. I spent my childhood on my father's farming and ranching operation near Elbowoods. I was twelve when the flooding began and my family, my aunties, and my other relatives were forced to move to different white towns off the reservation.

After the loss of our valley we often spoke of the big gardens almost every family had grown. The Missouri River, streams, springs, and wells were utilized to water the gardens during the hot summer months. Sometimes motorized irrigation pumps would draw water from the river to the upper banks. We grew a large variety of vegetables and preserved them to last throughout the year: potatoes, beans, corn, squashes, peas, onions, carrots, Indian corn, cabbage, turnips, radishes, pumpkins, cucumbers, watermelons. Some of the seeds, passed along from harvest to harvest, were unique to Fort Berthold: for example, Arikara and Hidatsa squash, Hidatsa and Mandan beans, sunflowers, and unique varieties of Indian corn from all three tribes.

Despite the forced changes of the last two hundred years, the Three Affiliated Tribes are still here, our culture and traditions still alive. As I wrote to my daughters at college, "We have developed methods for coping with change. . . . We do not stubbornly hang onto our old ways, nor do we accept the new without question. Instead, the success of the Indian people today on the Fort Berthold Indian Reservation is owed to our ability to continually renew our culture inside the American culture. The Mandan, Hidatsa, and Arikara families today still use herbal and ceremonial medicines, tribal rites of passage, music, religion, stories, and other tribal customs to meet their needs right along with their automobiles, microwave ovens, color television, and K-Mart, Target, and Wal-Mart merchandise."

I have never lived in an earthlodge. But today we are reestablishing the small-scale agricultural pursuit of family gardens, even if they are no longer planted in the Knife River villagers' traditional way, with acres and acres of red and black corn and white and yellow squash growing along the banks of the Missouri. We cannot speak the beautiful old languages of long ago, and we do not know the old songs and prayers that were ours when we were the Nueta, Hidatsa, and Sahnish, before the white man came to change our way of life forever. We do, however, still have a distinct and unique culture we are proud of. And there are continual signs of its renewal—in the gardens we plant, in the new herds of buffalo and elk we manage, in the celebrations of life we practice together as the indigenous people of this Earth we call our mother.

We can still sit on the banks of our sacred river. We still visit with our grandparents, our aunties and uncles, today's elders who hold stories of the past. The events of the past, the good and the bad, the changes and the preservation of tradition make us who we are today, the grandchildren of the ancient ones who lived in those wonderful villages along the Knife River—not so many lifetimes ago—when Lewis and Clark achieved the first major goal of their carefully planned expedition. They traveled into an unknown land, reached the homeland of the Mandan and Hidatsa people, learned from them, enjoyed their hospitality for the winter, and received their generous, humanistic, and compassionate assistance in making the rest of the journey the successful and remarkable event that we Americans recognize it as today.

The Mandans, Hidatsas, and Arikaras still live on what remains of our traditional homelands. We are recognized as a sovereign nation by the United States government. And we have never ceased honoring the land. As our fathers did before us, we honor it by living in beauty and sharing with the stranger. May you never honor America less.

> Through the deeds of our Grandfathers we will protect
> the value and principles they have told us.
> Our Grandfathers' relatives on this Earth
> will stand forever.
> I cherish the Grandfathers who have gone on before us;
> because of these virtues I humble myself.
> Our Grandfathers' people on this Earth
> will exist and stand forever.
>
> —*Plains Indian Flag Song*

Chapter 3

Selections

The Making of a Collection

Only a few of the gifts and trade goods that Indian people presented to members of the Corps of Discovery were destined to become historical artifacts. Members of the exploring party wore—and wore out—many of the moccasins, hats, and garments they acquired as they traveled. They traded some things a second time, and still other items were damaged or lost. Lewis and Clark and their men also kept some of the objects Indians gave them as personal mementos of their experiences or passed them along to their loved ones as gifts. The captains were not self-conscious collectors of Indian materials, yet after the expedition they did choose to give certain objects to the president and to Charles Willson Peale's Philadelphia Museum, effectively creating a collection of interrelated objects.

Lewis and Clark did not actively select the individual objects that made up their collection. Instead, Indian people did—and this is what makes the collection uniquely valuable.

Pictograph of a chief, identified by his pipe and feather bonnet, on the Peabody's elaborately decorated battle robe. Trading and exchanging gifts were important diplomatic skills for Native American leaders. (Detail of 99-12-10/53121; see p. 146.)

By looking closely at the specific objects the explorers are known to have brought back from their journey and reconstructing the contexts in which the objects were exchanged, we can gain insights into why Indian people chose to present those particular kinds of things and why the captains in turn decided to present them to their young country and its leader. The logic of the collection as a whole becomes apparent only by viewing Lewis and Clark and the Native people they met as active agents whose selections were informed by individual moments of mutual understanding.

We know through written documents—correspondence between Clark, Lewis, and Jefferson; the Fort Mandan packing list; Jefferson's records that mention expedition objects at Monticello; and the Peale Museum accession ledger—which objects Lewis and Clark sent or brought back in an official capacity. The Peale memorandum is especially important, because it identifies the large portion of expedition materials that Lewis, Clark, and Jefferson selected to form a museum collection. It also records the genesis of the objects that are at the Peabody Museum today. Although not all of the objects listed in the memorandum have survived, the document allows us to reconstruct the overall character of the original collection, a process necessary for understanding the individual objects that remain.

According to the Peale memorandum (see chapter 4), Lewis and Clark presented more than seventy Indian objects to the museum. Because Peale did not list every item individually (arrows, for example, are listed in lots), it is impossible to get an exact count of the objects, but the memorandum does reveal the kinds of things Peale received. Ceremonial smoking pipes (14) and tobacco bags (5) together make up the largest category of material. The second largest category consists of eight garments and robes, a number that swells to thirteen if two belts and three basketry hats are included. Six types of plant materials represent the third largest category. Other objects listed include peace medals, wampum, a Spanish dollar, and several woven bags from the Columbia River.

One can begin to recover the history and meanings of these objects of exchange by tracing their occasional appearances in the expedition journals. In addition, accounts

George Catlin's drawing of himself being hosted in the earthlodge of the venerable Mandan chief Four Bears in the early 1830s. Feasting and pipe smoking were integral aspects of Indian hospitality.

of other early Indian-white encounters document the significance of exchange and reveal patterns in the kinds of things that were customarily conveyed in various social contexts. These accounts from early explorers, government officials, and traders suggest that the objects Lewis and Clark acquired were typical of the kinds of things Indians customarily gave or traded to European and American visitors. There is no evidence indicating that Lewis and Clark actively sought to acquire these particular objects or followed a set of criteria about what to collect. Instead, the explorers themselves seem to have recognized that what they received from Indians were not simply inanimate objects or artifacts but social symbols, emblematic of Indian hospitality and of the establishment of ties between Indian nations and the United States.

As a group, the objects that Lewis and Clark brought back may reflect the agency of the givers in a way that is rare in museum collections. Lewis and Clark seem to have presented to Jefferson and the Peale Museum the very categories of objects that the Indians they met considered appropriate gift and trade items. Most of them are what the anthropologist Arjun Appadurai (1986:16) has called "commodities by destination," that is, objects intended principally for exchange. Others may be considered "commodities by metamorphosis," things made for another purpose but voluntarily exchanged because they were appropriate to give in certain contexts. Calumet pipes, robes, and hide garments are classic examples of such commodities from the Plains region. Indeed, the meaning of many of these objects was generated in part through the exchange process itself.

Most of the objects that Lewis and Clark presented to Jefferson and Peale seem to have been diplomatic gifts, a broad category that I divide into "gifts of state" and "chiefly gifts" and discuss in more detail later. Other objects were acquired through trade and barter, and the explorers may have regarded these as curiosities or mementos. In addition, the captains commissioned Indian women to make native garments and hats for them as they traveled. The ethnobotanical specimens in the collection were probably gathered expressly for Jefferson, who was keenly interested in horticulture and the future of Indian agriculture. Finally, the explorers acquired a few objects by means that we would now regard as illegitimate.

By selectively conveying diplomatic gifts to Jefferson, Lewis and Clark in effect served as liaisons between the president and the leaders of distant nations, bridging the miles between them with objects. Jefferson kept some of the expedition objects and exhibited them in his Indian hall at Monticello. He clearly recognized them as social symbols, not scientific artifacts. In a letter to Lewis inviting him to visit Monticello with the Mandan chief Sheheke (Big White), Jefferson expressed the hope that Sheheke would see "in what manner I have arranged the tokens of friendship I have recieved from his country particularly as well as from other Indian friends: that I am in fact preparing a kind of Indian hall" (Jackson 1978, 1:351). The expression "tokens of friendship" was not simply a felicitous turn of phrase but almost certainly a reference to the "covenant chain" or "chain of friendship," the dominant metaphor of Indian-white diplomatic discourse in the Northeast (e.g., National Museums of Canada 1980). The chain of friendship was predicated on reciprocal gift giving, with the idea that the "chain" of alliance between groups would rust unless lubricated by periodic exchange. Leaders of Indian peoples presented Jefferson with diplomatic gifts throughout his political career. Some of these he kept at Monticello, but many others he gave to the Peale Museum, as did the explorers themselves.

Portrait of Sheheke, a Mandan leader, by C. B. J. Fevret de Saint-Memin, 1807. After the expedition, Sheheke accompanied Lewis back to Washington, D.C., to meet President Jefferson.

Objects of Honor: Gifts of State and Chiefly Gifts

Broadly speaking, diplomatic gifts might include anything given to initiate or further a political relationship, from explicit national symbols to more informal, personal items and privileges. Here I draw attention to two important modes of diplomatic gifting that can be discriminated in the expedition's written record: the giving of "gifts of state" and the giving of "chiefly gifts." Both are evidenced by the kinds of objects Lewis and Clark presented to Jefferson and the Peale Museum, but these modes represent points on a continuum rather than distinctive forms of exchange. They are often interdependent: chiefs generally presented both kinds of diplomatic gifts, and it is sometimes unclear whether such items were directed to Lewis and Clark as individuals or as U.S. representatives. Overlapping sets of objects were considered appropriate gifts in either context. Gifts of state and chiefly gifts were keyed, however, to relatively distinct social overtures, so it is useful to distinguish their respective places in diplomatic protocol.

Gifts of State

Formal diplomatic encounters between Lewis and Clark and tribal representatives were nearly always consummated by the mutual exchange of objects symbolizing group identity and alliance. Lewis and Clark followed established protocols and made their mission manifest by presenting Indian leaders with objects such as medals engraved with the expression "Peace and Friendship" that symbolized U.S. national identity, leadership, and allegiance. In turn, Indian leaders generally reciprocated with the kinds of gifts customarily given to forge alliances and open exchange relations between indigenous groups. Some of the objects given during these formal diplomatic occasions assume the character of gifts of state, that is, things given by one government or people to another by representatives of those larger entities.

Sometimes the "nation-to-nation" connotation of goods exchanged during the expedition was explicitly stated. For example, following a portentous council during which Nez Perce tribal leaders resolved to endorse American policy, Nez Perce leaders presented Lewis and Clark with two especially fine horses, "at the insistence of the [Nez Perce] nation" (Moulton 1991, 7:247). In other cases, gifts of state can be identified from the context of the presentation and the character of the objects given.

Calumet pipes, for instance, were emblems of and vehicles for intertribal and cross-cultural diplomacy throughout much of interior North America (see chapter 8).

They were integral for forging relationships and symbolic of group identity, leadership, and political interaction. Chiefs and leading men customarily presented calumets in contexts of intertribal adoption and diplomacy, and they extended this protocol into their external political relations with European powers. Chiefs and leaders have historically given council pipes to U.S. presidents, military leaders, and important visitors on behalf of their communities. The fourteen pipes and calumets donated to the Peale Museum, some of them certainly smoked by Lewis and Clark and Native leaders, were given to the captains by representatives of Indian peoples to pledge their sincerity in entering into formal relations with representatives of the U.S. government. By presenting such powerful symbols of their nations and their own status, Indian leaders hoped to elicit a similar degree of fidelity from the United States. The Peale memorandum identifies the tribal origins of all of the pipes and attributes many of them to particular Indian leaders.

Indian leaders also gave Lewis and Clark British peace medals (minted during the reign of King George III) and wampum, two classic examples of gifts of state that the explorers in turn deposited in the Peale Museum. By relinquishing their British peace medals to U.S. representatives, Sauk and Fox leaders acknowledged the change in dominion over the North American continent. While this may have been only a diplomatic gesture, it was an act of courtesy and recognition directed toward the new nation. During his post-expedition career as superintendent of Indian affairs in St. Louis, William Clark received Spanish, English, and American peace medals from Indians with whom he conducted political negotiations (Ewers 1967:58). Peale also recorded that "Different nations" presented Lewis and Clark with wampum, an overt political symbol that was often exchanged to initiate relationships and to communicate a desire for peace.

Chiefly Gifts

Like gifts of state, chiefly gifts figured prominently in cross-cultural diplomatic encounters during the nineteenth century. Chiefs and leading men were expected to practice the cardinal virtue of generosity toward members of their own communities, guests, and even enemies. By gifting extravagantly, great men displayed their selflessness and concern for others, redistributed valuable resources, and demonstrated confidence that they could marshal more. Giving also cultivated allies. Since gifting implied reciprocity, lavish presents could neutralize potential enemies and compel declared friends to honor their commitments. Gifting, then, was an important political skill, and chiefs and leaders commemorated their acts of generosity as well as their feats in war. On the Plains, they often painted their robes and garments with iconic figures or tallies representing the number of valuable presents they had made (Brownstone 2001:78–83; Thwaites 1906a:264).

By giving away his most prized and symbolically charged possessions, such as war regalia, a chief demonstrated that he was a person of transcendent consequence and honor. Chiefs and leading warriors often gave their personal garments and military regalia to, or exchanged them with, their counterparts in other tribes to express their largesse (McKenzie in Wood and Thiessen 1985:281) and to create fictive kinship ties (Thwaites 1906a:320). Chiefs extended this practice to Euro-American military men and trading partners in order to create alliances and to reciprocate the Euro-American "dressing" of Indian leaders.

Friedrich Paul Wilhelm, the duke of Württemberg, for example, received chiefly gifts when he traveled through the trans-Mississippi West in 1822–24. Describing his visit to the lodge of the Oto chief Isch-nan-uanky, he wrote, "The chief brought leather cushions to sit on, said to be a distinction, and gave me some presents as a counter compliment for those that I had given him. These presents consisted of all sorts of ornaments which the Indians, especially the warriors, value most highly" (Württemberg 1973:381). Following a feast and council, George Catlin, the American painter of western landscapes and Indians, observed the Lakota chief Ha-wan-je-tah (One Horn) giving Major Sanford, the Indian agent, his own eagle feather headdress, war shirt, leggings, grizzly claw necklace, and moccasins, along with a pipe (Catlin 1973 [1844], 1:229). Chiefs who visited Washington, D.C., as tribal delegates often presented their personal regalia, including war shirts and feather bonnets, to the president and other government officials (Viola 1981:108–9).

Tunnachemootoolt (Broken Arm), a Nez Perce chief, gave his shirt to Captain Clark and his leggings to Captain Lewis (Moulton 1991, 7:249). Lewis and Clark also exchanged clothing with Mandan and Shoshone leaders during the expedition. The Shoshone chief Cameahwait dressed Lewis in an entire ceremonial outfit, including a beaver, ermine, and shell "tippet," or cape (see p. 74 and chapter 7). Lewis kept and presented this ensemble to the Peale Museum, where it was probably registered as "The Dress worn by Captn. Lewis &c."

Lewis and Clark transferred to the Peale Museum a number of other chiefly gifts, among them "Feathers which were at various times presented to Captn. Lewis and Clarke by the principal Chiefs of the nations inhabiting the Plains of Columbia, whose

Biographical war shirt. Several Indian chiefs gave Lewis and Clark their own shirts, sometimes in exchange for those worn by the expedition leaders. (Detail of 99-12-10/53041; see p. 167.)

custom it is to express the sincerity of their friendship by cutting feathers from the crowns of their War Caps and bestowing them on each as they esteem" (Jackson 1978, 2:476–77). Others included the "Tail feathers of the Eagle" in the "pattern for a war-cap," "2 ornaments, worn around the neck, by the nations of the Plains of the Columbia," Sioux leggings "ornamented with the hair &scalps taken by the Indian who wore it," and "Legings of the Pallatepallers" [Nez Perce] (Jackson 1978, 2:476–77).

In addition, Lewis and Clark presented the Peale Museum with a scrap of hide from a white bison, an enormously valued commodity and ceremonial item among the nations of the Missouri River. It is unlikely that the explorers bartered, or even could have bartered, for an item so valuable. The trader Charles McKenzie declared that "when a man has a white Buffalo hide his fortune is made" (Wood and Thiessen 1985:282). William Clark also received from the Mandans a pair of "Chiefs Mockersons with white Buffalow Skin tops," which he sent home from Fort Mandan to his brother Jonathan. Clark noted in his accompanying letter that "a white buffalow Skin Sels in this Countrey for about fifteen horses" (Holmberg 2002:84). These white bison hide items must have been chiefly gifts, perhaps presented by either Black Cat or Sheheke, two Mandan chiefs who were particularly hospitable to the captains. A bow that Lewis and Clark presented to Peale might also have been given to Lewis by Black Cat.

Feathers on a calumet stem. Indian leaders also gave Lewis and Clark feathers that symbolized their war honors and leadership status. (Detail of 99-12-10/53099.2; see p. 222.)

Unfortunately, most of the chiefly gifts transferred to Peale and Jefferson have since been lost. The Peabody Museum does, however, have several raven belt ornaments, commonly called "crow belts," of the type worn by members of soldier societies in the plains and prairies along the Missouri River. These are listed in the Peale memorandum as "A Dress made of Crow or Raven Skins, worn by the Police Officiers of the Saux, nation." Indian leaders are known to have presented raven ornaments and other military regalia to U.S. officials on diplomatic occasions. Like feather bonnets, such objects epitomize the chiefly gift, given by a great man to honor the recipient and, in doing so, to reflect honorably upon himself.

Perhaps the most extravagant chiefly gifts presented to Lewis and Clark were offered by Nez Perce leaders, who, although they did not eat horses, gave the party horses to butcher when the Americans were desperate for meat. Lewis wrote with uncommon feeling of one such incident, which occurred upon the expedition's return from the Pacific. Having left the Pacific Coast early in the spring of 1806, the Corps of Discovery found the Bitterroot Mountains still too snow-packed to climb when they reentered Nez Perce country. The group arrived at the village of Chief Tunnachemootoolt (Broken Arm) along "Commearp" Creek (*qémyexp,* now known as Lawyer Creek) and found the chief flying the flag that Lewis and Clark had given him the previous year. Lewis wrote that upon greeting the villagers,

> we collected the Cheifs and men of consideration[,] smoked with them and stated our situation with rispect to provision. [T]he Cheif spoke to his people and they produced us about 2 bushels of the Quawmas ro[o]ts dryed, four cakes of the bread of cows [cous root] and a dryed salmon trout. We thanked them for this store of provisions but informed them that our men not being accustomed to live on roots alone we feared it would make them sick, to obviate which we proposed exchangeing a good horse in reather low order for a young horse in tolerable order with a view to kill. [T]he hospitality of the cheif revolted at the aydea of an exchange, he told us that his young men had a great abundance of young horses and if we wished to eat them we should by furnished with as many as we wanted. . . . This is a much greater act of hospitality than we have witnessed from any nation or tribe since we have passed the Rocky mountains. [I]n short be it spoken to their immortal honor it is the only act which deserves the appellation of hospitallity which we have witnessed in this quarter. (Moulton 1991, 7:237–38)

Among the items that Lewis and Clark gave to the Peale Museum upon their return were samples of many of the plant foods that served as dietary staples in the Columbia Plateau region, including roots and root-based "breads." The Peale memorandum suggests that several such specimens were provided to illustrate the Corps' diet as it made its arduous crossings of the Rockies, certainly the most difficult parts of the trip. Like the specimens of Arikara tobacco and Mandan corn that Lewis and Clark sent to Jefferson, these samples might have been collected and preserved in part because of the president's deep interest in horticulture.

Camas bulb, cous roots, and berry cakes. Indians traded native foods such as these to the Corps of Discovery, sustaining the exploring party while it crossed the Rocky Mountains.

One plant sample, though, can be considered a kind of diplomatic gift. The Peale ledger records the accession of "Roots, presented to Captn. Lewis on the 24th of June 1806 by Neeshneparkeooh [Neeshneparkkeook], the great chief of the Pottotepallers [Nez Perce] as an emblem of the poverty of his nation which he described in a very pathetic manner."[1] In fact, Neeshneparkkeook probably made use of a widespread Native American rhetorical trope by asking the United States to "take pity" on his people, reinforcing his speech by presenting roots as "an emblem of [their] poverty." Prayers to spiritual powers were (and are) often framed as appeals for pity, and Indian people used the same supplicatory discourse in negotiating relationships with traders and U.S. officials.

Consistent with Native speech events recorded by many other traders and government officials, the speeches delivered to Lewis and Clark by tribal representatives often included requests that the United States "take pity" on their people. For instance, the Yankton chief Shake Hand, whom Lewis and Clark "clothed" on their way up the Missouri River, told them: "We are verry glad you would take pitty on [us] this Day, we are pore and have no powder and ball. My Father—We are very Sorry our women are naked and all our children, no petiecoats or cloathes" (Moulton 1987, 3:28). Likewise, the invocation of pity plays a prominent role in the recorded speech of an intertribal delegation to Jefferson and his secretary of war on January 4, 1806 (Jackson 1978, 1:284–89).

In her study of fur trade narratives, Elizabeth Vibert (1997:149–56) pointed out that whereas Euro-Americans often interpreted such petitions as begging, for Indian peoples the idiom of pity couched a request in terms of respect and humility. In presenting the roots, Neeshneparkkeook was acknowledging the power of the United States and graphically asking that the national leaders remember and aid his people by sending traders to them.

Offering and sharing food are universal gestures of hospitality and important accompaniments to social intercourse and diplomacy. Two items that Lewis and Clark brought back from the Columbia River area, "A Bag prepared of grass by the Pishquilpahs" and "A water-cup of the natives who resides in the Plains of Columbia, made of the same grass," may have been either given to Lewis and Clark as hospitality gifts or traded as part of a more pragmatic exchange. The "bag," which may be the Peabody's Sahaptin root-gathering basket (see chapter 9), could have contained edible roots when it was presented to Lewis and Clark. Columbia River tribes were savvy and experienced traders, and Lewis and Clark's party was in particular need of food when it interacted with the Native peoples of that region, which lies between the Rocky Mountains and the Pacific Coast.

Barter, Commissions, and Curiosities

Members of the Corps of Discovery bartered with Indian people everywhere they traveled, both as a gesture of goodwill and to gain provisions. From the Plains, bison robes and dressed hide garments exemplify what Appadurai (1986:15) called "quintessential commodities"—things customarily used for exchange. They were such staples of intertribal trade and diplomatic gifting on the Plains that today it is often impossible to assign tribal attributions to nineteenth-century examples in museum collections. Although only two buffalo robes are listed in the Peale memorandum, Lewis and Clark are known to have sent seven robes to Jefferson from Fort Mandan in April 1805; four more, undoubtedly acquired during the return portion of their trip, may have been sent by Lewis to Jefferson in 1806–7.[2] The expedition's journals record the receipt of many more robes, especially during the Mandan winter. Most of the robes sent from Fort Mandan and by Clark after the expedition were described as Mandan, whereas the two robes delivered to the Peale Museum were recorded as "Menetarre" (Hidatsa) and "Scioux, or Soue" (Sioux). We know that both Sioux and Mandan chiefs gave decorated robes to the explorers during diplomatic councils and that some chiefs presented Lewis and Clark with garments. It is certain that the explorers acquired other robes and items of clothing through barter.

Plains Indian people often presented robes and garments to treaty delegates, traders, and U.S. military officials in contexts of contact and intercultural diplomacy. When, for example, the Canadian trader François A. Larocque "cloathed" and presented gifts to Crow chiefs to open trade with them in 1805, they reciprocated with robes, a mountain lion skin, four shirts, two women's dresses, two elk skins, three saddles, and thirteen pairs of leggings (Wood and Thiessen 1985:171).

Whereas Native people regarded robes and garments as appropriate objects of exchange, such items were initially of little or no interest to commercial Anglo-European traders, who were focused on acquiring the pelts of small fur-bearing mammals. Jean Baptiste Truteau, who traveled up the Missouri in 1794–95 to assert Spanish trade rights, charged that fellow trader Jacques d'Eglise had misrepresented Spanish interests because he "neglected to encourage the hunting of beaver and other good furs, being too much bent upon trading for robes, leggings, moccasins, women's garments, dry cow hides, etc. which sufficed [only] to meet the needs of the Indians." The Indians, Truteau wrote, "find it extraordinary that I refuse to trade for this sort of article" (Nasatir 1952, 1:298). As late as the 1830s, the German Prince Maximilian of Wied-Neuwied, while exploring the upper Missouri River, wrote that "the Indians . . . have frequently nothing to offer for barter but their dresses, and painted buffalo robes," noting that traders regularly exchanged for goods of such marginal economic interest only in order "not to create any dissatisfaction" (Thwaites 1905d:381).

-quently in the water also renders those articles of dress i-
-convenient. they wear a hat of a conic figure without a
brim confined on the head by means of a string which
passes under the chin and is attached to the two opposi-
sides of a secondary rim within the hat. the hat at top
terminates in a pointed knob of a connic form also or i[n]
this shape. these hats are made of the bark of ceda[r]
and beargrass wrought with the fingers so closely that
it casts the rain most effectually in the sh
give them for their own use or that just d
hats they work various figures of different
most commonly only black and white are
figures are faint representations of the whale
the harpooneers striking them. sometimes
triangles &c. The form of knife which seem
by these people is a double edged and doubl
-gar; the handle being in the middle, and th
unequal lengths, the longest usually from 9 to ten in
long and the shorter one from four to five. these kn
they carry with them habitually and most usually i[n]
the hand, sometimes exposed but most usually parti-
larly when in company with strangers, under their robe
with this knife they cut and clense their fish make
their arrows &c. this is somewhat the form of the knife
A
A is a small loop of a strong twine through wh
they sometimes insert the thumb in order to prevent its be
wrested from their hand.

es in a poin
pe. th
grass wr
the rain mo
n for their

Portion of a journal page from January 30, 1806. Rainproof woven hats were among the items of clothing that Lewis and Clark commissioned Native American women to make for their use during the expedition. While they wintered near the Pacific Ocean at Fort Clatsop, the two explorers described and made small sketches of these onion-dome hats in their journals.

Members of the Corps of Discovery traded frequently for moccasins in addition to robes, and Lewis and Clark later gave two pairs to Peale, one identified as "Otoe" and the other "Soue."

On a number of occasions, Lewis and Clark commissioned Native women to make utilitarian items of apparel for themselves and their crew. While spending a bitterly cold winter at Fort Mandan, Clark had a cap made from the skin of a lynx or bobcat

(Moulton 1987, 3:256), and while on the Pacific Coast the following December, Lewis purchased seven lynx and bobcat robes to make a coat (Moulton 1990, 6:124, 208).[3] Lewis may have commissioned the garment as a gift for Jefferson, who had apparently expressed interest in receiving a "Leopard or tyger skin, such as the covers of our saddles were cut of" (Jackson 1981:159, n. 33).

The captains also observed and extolled the virtues of the woven fiber hats made by the Natives of the moist Pacific rainforest west of the Rocky Mountains. They paid Indian women with trade goods to weave some of the rainproof hats for themselves and members of their party, and they later donated several of the hats to Peale (see chapters 5 and 9). Lewis and Clark may have regarded the hats as curiosities or unusual cultural mementos specific to certain groups. In that category might also be included the "great number of arrows from different tribes of Saux" that they presented to Peale (see chapter 9). Another curiosity, an Indian pot, was probably given to the explorers by a local person near St. Louis who found the piece when digging a well near "St. Gennevieve."

Illegitimate Acquisitions

A group of objects listed in the Peale memorandum as "Stone, Spear points, from the natives inhabiting the Rocky Mountains" was probably acquired somewhat by chance and under unfortunate circumstances. On August 22, 1805, Lewis chronicled an adventure that had befallen George Drouillard ("Drewyer") while he was hunting near the Lemhi River in present-day Idaho. After visiting with a small party of Shoshones whom he encountered, Drouillard forgot to pick up his gun when he rose to catch his grazing horse. A young Shoshone man took it and fled, with Drouillard in pursuit. After regaining the gun, Drouillard collected some of the objects the Shoshones had left in their haste to abandon their campsite. Drouillard returned to his own camp bearing dressed and undressed skins, silk-grass bags full of berries and berry cakes, parfleche containers loaded with roots, a bone or antler flint-knapping tool, and pieces of flint and obsidian. Lewis recognized the illicit nature of Drouillard's appropriation; in his journal, he described this group of objects as "plunder" (Moulton 1988, 5:141–43).

The following year, Lewis himself appropriated several objects by even more dubious means (albeit one practiced by warriors the world over) when he took the amulets from the shields of two young Piegan Blackfeet men he and his party had killed during an altercation over horses and guns. These amulets, which can be regarded only as war trophies, were later presented to the Peale Museum along with a brief account of their acquisition. When *Poulson's Daily Advertiser,* a Philadelphia newspaper, announced the museum's acquisition of materials from the expedition, the paper tellingly described the amulets as having been "presented to the party by themselves."[4]

The great majority of the Indian materials that Lewis and Clark gave to the Peale Museum, however, can be accounted for as customary and symbolically potent diplomatic objects. The resulting collection, still partly intact, encodes the agency of the Indian peoples who accepted a group of strangers into their communities and counciled with them about their common future. Lewis and Clark collected natural history specimens in the service of science, and they knew that Jefferson would find Indian objects of interest. But they seem to have recognized the meaning of the gifts they received and to have taken care to convey them to the president and ensure their future preservation. There is no evidence that they purposefully sought particular kinds of things or encouraged Indians to give or alienate certain objects. Such efforts might well have undermined their objective to "open the road" for future friendship and commerce.

Chapter 4

Into the Museum

From Gifts to Artifacts

Lewis and Clark's Indian objects have journeyed great distances and circulated through many spheres of meaning and exchange. After the expedition, they traveled through the storerooms and exhibit galleries of some of America's most storied museums: the Peale Museum in Philadelphia, P. T. Barnum's American Museum, the Boston Museum, and finally Harvard's Peabody Museum of Archaeology and Ethnology. Once in a museum setting, they were catalogued and transformed into artifacts, objects suspended in place and preserved for study and public viewing. Their perceived value, however, has varied widely over the past two hundred years. Here I provide a brief account of their post-expedition history and explain how some of the objects came to rest at the Peabody Museum.

Charles Willson Peale's famous self-portrait, *The Artist in His Museum,* 1822. Lewis, Clark, and President Jefferson all presented Peale with expedition artifacts for his Philadelphia Museum.

Peale's Philadelphia Museum

During the three years immediately following the expedition, Jefferson, Lewis, and Clark deposited more than seventy Indian objects in Charles Willson Peale's museum, also known as the Philadelphia Museum. Peale was an artist, an ardent patriot, a dedicated amateur scientist, and an active member of Philadelphia's learned community. In 1787 he was appointed curator of the American Philosophical Society. His museum, located in Philosophical Hall in the heart of downtown Philadelphia, was a fascinating emporium of natural and artificial curiosities from around the world. Peale achieved his initial success painting the portraits of Revolutionary War heroes and early American statesmen. His social circle included Jefferson, George Washington, Alexander Hamilton, James Madison, and Benjamin Franklin, all of whom served on his Board of Visitors and Directors or donated historical relics and specimens to his collection. For a time, Peale's institution assumed the character of a de facto national museum. Unfortunately, despite Peale's persistent urging, the government declined to purchase his collections and create a national institution from them.

Jefferson apparently kept many of the things that Lewis sent to him from Fort Mandan for his personal Indian hall at his plantation, Monticello; most of those objects disappeared after his death in 1826.[1] William Clark retained a number of expedition objects as well, exhibiting them in his Saint Louis council hall and museum. Those, too, vanished after Clark's death in 1838. But Jefferson, Clark, and Lewis also presented Peale with a great variety of expedition materials, most of which had been received as diplomatic gifts. Jefferson sent some to Peale, and the explorers jointly conveyed more upon their return east. Both Clark and Lewis also sent individual shipments to Philadelphia. In November 1809, Peale wrote to his son Rembrandt that he had received a posthumous shipment of unidentified Indian objects from Meriwether Lewis, who had committed suicide en route from St. Louis to Washington (Jackson 1978, 2:469–70).

The following month, Peale entered into his museum ledger book a list of the expedition objects he had received. Although not all of the objects survived the nineteenth century, and the list is incomplete, the Peale memorandum, now in the collections of the Historical Society of Pennsylvania, is the single best record of Lewis and Clark's original collection. It also documents the genesis of the materials now at the Peabody Museum. I reproduce the memorandum here in its entirety—with its original spellings, punctuation, and tribal attributions.

Peale's Memorandum of Specimens and Artifacts

[December 1809]

Articles collected by Merriweather Lewis Esqr. and William Clark Esqr. in their voyage and Journey of Discovery, up the Missouri to its source and to the Pacific Ocean.

A hat manufactured by a Catsop woman near the Pacific Ocian; from whence it was brought by Capts. Clarke & Lewis.

Legings of the Pallatepallers, residing on Lewis'es River, west of the Rocky mountain.[2]

Cap, worn by the women of the Plains of Columbia.

The Tail feathers of the Eagle, much prized by the Indians of North America, who convert them into various ornimental and war-like dresses—these being a pattern for a war-cap would be esteemed by them equal in value to two good horses.

An Indian Pot, found in digging a well at the great Saline near St. Gennevieve, about 17 feet under the surface of the Earth, accompanied by various broken Pots &c. It is supposed that at some former period a walled well had been there.

A Large Mantle, made of the Buffalow skin, worn by the Scioux, or Soue, Darcota Nation.

A small Mantle of very fine wool, worn by the Crow's nation Menetarre.

Legings, ornamented with the hair & scalps taken by the Indian who wore it, and marked with stripes shewing the number he had scalped. Of the Soue Nation.

Two very handsomely ornamented Tobacco pouches, ornamented with Porcupine Quills, and Tin, &c. Of the Saux Tribe.

A Tobacco Pouch, from the Ioway's.

Another, from the Raneird's or Foxes.

A handsomely ornamented belt, from the Winnebagou's or Puount's.

Wampum, of various discriptions, indicating Peace, War, Choice of either, Hostilities commencing, and a disposition for them to ceace &c. From different nations.

Tobacco Pouch not ornimented sent by the Sacks.

Moccosins, worn by the Otoe's.

Do [ditto] from the Soue's.

A Piece of White Buffaloes skin, from the Missouri.

A great number of arrows from different Tribes of Saux. And a Bow.

A handsome Belt worn by the Saux as a garter.

2 ornaments, worn around the neck, by the nations of the Plains of Columbia.

Amulets—taken from the shields of the Blackfoot Indians who attacked Captn. Lewis and were killed by himself and party on the 27th of July 1806. near the Rocky Mountain.

Feathers which were at various times presented to Captn. Lewis and Clarke by the principal Chiefs of the nations inhabiting the Plains of Columbia, whose custom it is to express the sincerity of their friendship by cutting feathers from the crowns of their War Caps and bestowing them on each as they esteem.

Roots, presented to Captn. Lewis on the 24th of June 1806 by Neeshneparkeooh, the great chief of the Pottotepallers as an emblem of the poverty of his nation which he discribed in a very pathetic manner.

Bread, used and formes the principle article of food among the Pattotepallers and other Indian Nations west of the Rocky mountains and is called by them cows [cous]—it is pripared from the bulb of an umbellaferous plant to which they give the same name. these bulbs are pounded between two stones while in a succulent state and then exposed to the sun untill dry when they assume the appearance and consistance of this specimen. This article for many weeks constituted the principal part of the food of Lewis and party, while in that country.

The Roots of Cows. So called by the Pallatepallers with whom it forms a principal article of food.

Bulbus Roots. These bulbs form the food of many Indian Nations residing within and west of the Rocky Mountain. They are called by the Pallotepallers Quaw-mash.

Bread. This is called Passhequo-quaw-mash being only a varied preparation of the quaw-mas bulb.

Bread. This forms a principal article of food among the Enesher, Skillute, Pishquitpahs and others residing near the great falls of the Columbia River, and is called by them Shappellel—it is prepared from the Bulb of an umbellaferous Plant to which they give the same name—these bulbs are pounded between two stones while in a succulent state and then exposed to the sun untill dry when they assume the appearance and consistance of this specimen.

This article for several weeks constituted the principle part of the food of Capt. Lewis and party.

A Bag prepared of grass by the Pishquilpahs on the Columbia River.

A Water-cup of the natives who resides in the Plains of Columbia, made of the same grass.

Spanish Dollar obtained by Captn. M. Lewis, from the Pallotepallers—a nation inhabiting Lewis's River within the Plains of Columbia, who had never previously seen white-men.

A Cap worn by the natives of Columbia River, and the Pacific Ocean.

Stone, Spear points, from the natives inhabiting the Rocky Mountains.

2 Silver Midals, of George the 3d of England, obtained from the Foxes and one from the Socks, or Saukeys.

Four Pipes, or calmets from the nation Saux, of the following tribes, Yankton, on the River La Moine—Teton on the Misuri 1200 miles up—Sharone's 1400 miles up the Misuri—Dacoto's or Sue's.

One of the Puount, called Winebagou. Near Dog Plains, Missippi.

One of the Fox's and one belonging to White Skin, the Chief of the Foxes. Missippi.

3 from the Sauke's on the Missippi.

2 from the Ioway's on the River Lamaine.

One given to the company by White Pigeon.

One made by the Saux, inhabiting the Dog Plains.

A Dress made of Crow or Raven Skins, worn by the Police Officiers of the Saux, nation.

The Dress worn by Captn. Lewis &c.

A number of Minerals &c.

Presented at diferent periods, through the President of the United States, Thomas Jefferson by Govener Meriweather Lewis and General William Clarke, in company, who collected them on their journey &—General Clark. (Jackson 1978, 2:476–78).

Peale was fascinated by the new worlds revealed by European exploration, and he hung portraits of explorers such as Columbus and Vespucci on the walls of his museum. He solicited material from European and American sea captains, explorers, and naturalists through both personal contacts and newspaper advertisements, in the hope of

Captain Meriwether Lewis in Shoshone Costume, by C. B. J. Fevret de Saint-Memin, 1807. This portrait commemorates Lewis's exchange of clothing with the Shoshone leader Cameahwait. When Charles Willson Peale created a wax figure of Lewis wearing Cameahwait's gifts, he replaced the rifle with a calumet, to emphasize Lewis's diplomatic mission.

recreating a "world in miniature."[3] Peale's son Titian served as a naturalist for the government-sponsored Long and Wilkes expeditions, and the museum was a depository for some of the resulting materials. Peale's collection was large—an estimated 100,000 objects by 1820—and broad in scope, including natural history specimens, curiosities from exotic cultures, marvels of science and engineering, and historical relics, particularly those associated with the Revolutionary War and the War of 1812.

In September of 1826, Peale purchased an extensive collection of Native American garments, pipes, musical instruments, and arms from unidentified St. Louis traders (Sellers 1980:250–54). Where the traders acquired this "most complete" collection of Mandan, Cheyenne, Arikara, and Osage materials is a mystery,[4] and it is possible that this collection is a source of some of the Peabody's Peale materials.

Peale commemorated the Lewis and Clark expedition as an event of national significance and as an achievement for Meriwether Lewis, who was honored by Philadelphia's intellectual community by election into the American Philosophical Society. Indeed, Peale seems to have valued the Indian objects that Lewis and Clark conveyed to him primarily as artifacts of the expedition, not as curiosities or specimens. Rather than dispersing the individual objects among his curiosity cases, Peale displayed the group together, emphasizing their interrelationship and forging their identity as a collection. A plan view of his museum from around 1820 reveals a designated "Lewis and Clark Exhibit" situated along one wall in the Mammoth Room, where he showcased his signature attraction, a fossilized mastodon skeleton that he had excavated in New York in 1801 (Sellers 1980:217).

Peale executed individual oil portraits of Lewis and Clark and made both a life mask and a wax figure of Meriwether Lewis. The latter portrayed him as a pioneering cross-cultural diplomat wearing the garments given to him by the Shoshone chief Cameahwait and holding a calumet. This full-sized model was the centerpiece of Peale's Lewis and Clark exhibit. A second wax figure may have represented Cameahwait. Peale kept a gallery of realistic wax models of Asian, Native American, and African people dressed in their customary clothing. The Native Americans were not "types" but named individuals (Blue Jacket, Red Pole) famous for their diplomatic and oratorical skills.

Peale was a pacifist who hoped that the spread of Western principles would unify disparate peoples and end human conflict. In describing his wax model of Lewis to Thomas Jefferson, Peale wrote, "[M]y object in this work is to give a lesson to the Indians who may visit the Museum, and also to shew my sentiments respecting Wars." He recalled the significance of the moment he had memorialized, noting that Lewis had donned Shoshone garments to allay Cameahwait's fear of an ambush by the American party. On an accompanying tablet, Peale displayed the text of a speech that he believed conveyed the spirit of Lewis's reply to Cameahwait. It began:

> Brother, I accept your dress—it is the object of my heart to promote amongst you, our Neighbors, peace and good will—that you may bury the Hatchet deep in the

Showman P. T. Barnum helped drive the staid Peale Museum into bankruptcy and then purchased much of the museum's contents for his own entertainment enterprises. This lively Barnum Museum poster, produced circa 1890, depicts Barnum as the "Sun of the Amusement World."

ground never to be taken up again—and that henceforth you may smoke the *Calmut* of Peace & live in perpetual harmony, not only with each other, but with the white men, your Brothers, who will teach you many useful Arts. (Miller 1988:1055–56)

P. T. Barnum

By the 1830s, the American public was no longer enthralled by ordinary flora and fauna, and the staid Peale Museum found itself competing with a growing number of sensational "dime museums" and specialized natural history enterprises. Peale's sons and heirs struggled to maintain the family enterprise by popularizing the museum's attractions, and they opened new sites in Baltimore and New York. But by 1850, the flamboyant impresario P. T. Barnum had purchased, in whole or in part, the contents of all three establishments. With that purchase, the fragmentation of the Lewis and Clark collection began.

Barnum had launched his spectacular career as a showman in 1841 by purchasing Scudder's American Museum in New York City. The very next year, he enjoyed his first blockbuster exhibit—of an eighteen-inch-long "Feejee Mermaid." The reputed mermaid (a manufactured composite of a monkey and a fish) was loaned to Barnum by Moses Kimball, proprietor of the Boston Museum, who had purchased it from a sailor. Kimball and Barnum became fast friends, and when the contents of Peale's Philadelphia Museum were sold in 1849–50, the two museum proprietors bought most of them in partnership for between five thousand and six thousand dollars (Barnum 1888:82). A catalogue prepared for an 1848 sheriff's sale of the Peale collections lists a case containing a "Wax figure of Captain Lewis, and curiosities," suggesting that many of the Lewis and Clark materials had remained together in a thematic arrangement (Sellers 1980:312–16).

It is unclear how Barnum and Kimball divided the Peale collections of paintings, wax figures, natural and artificial curiosities, mechanical wonders, and historical relics. Parts of the collection went elsewhere (Sellers 1980). Some of Barnum's share may have been transferred to a museum that he operated in the Swaim building in Philadelphia, which he sold during the fall of 1851. If so, those materials must have been lost in a fire that destroyed the museum and most of its contents just a few months later, on December 30, 1851. Barnum may also have sent some of the Peale materials to his principal establishment, the American Museum in New York.

But if Barnum retained any of the Lewis and Clark materials, he did not identify or advertise them as such. During the 1850s and 1860s, he published a series of illustrated catalogues and guidebooks describing "the principal objects of interest in this extensive establishment, and useful to the visitor for purposes of reference, entertainment and instruction."[5] Some of these guidebooks are quite expansive, containing detailed descriptions of thousands of objects found in the various "saloons" or galleries in Barnum's five-story marble building on the corner of Broadway and Ann Street. An example from the 1860s testifies to Barnum's enormous collection of menageries, aquaria, stuffed birds and mammals, reptiles, statuary, paintings, wax figures, cosmoramas, models, antiquities, amusement technologies, and historical relics—reportedly, more than 100,000 curiosities in all. A few of the paintings are attributed to members of the Peale family. There is no mention at all of Lewis and Clark.

Barnum evidently judged that Lewis and Clark's animal specimens and diplomatic gifts from distant tribes held no commercial or popular appeal for museum-goers of the 1850s and 1860s. He seems to have been right, for the Lewis and Clark expedition had been all but forgotten by scholars and the public alike.[6] Barnum apparently sent many of the Peale materials, including some Lewis and Clark objects, to Moses Kimball, his Boston friend. This was a providential decision. In 1865, Barnum's New York Museum burned to the ground, killing hundreds of animals and destroying his entire collection.

The Boston Museum

Like Barnum, Moses Kimball was a self-made man who achieved success in the museum business by emphasizing "rational" entertainment for the public. In 1839, he and his brother, David, purchased the New England Museum, an amalgam of several previous museums dating back to the late eighteenth century. The contents of the New England Museum consisted primarily of curiosities deposited by sea captains and sailors who had ventured to the Northwest Coast, the Pacific, and the Orient; wax figures; portraits of historical figures; and stuffed animals from around the world. The Kimball brothers christened their new enterprise the "Boston Museum and Gallery of Fine Arts" and opened it to the public in downtown Boston in June 1841.

For a twenty-five-cent admission fee, visitors could tour some fifty cases crowded with shells, birds, fossils, stuffed animals, the armaments and clothing of exotic peoples, Chinese curiosities, an extensive portrait gallery, "heathen idols," dioramas, portraits, and historical and moralistic tableaux of wax figures. They could also enjoy, at no extra charge, a range of variety shows, such as comic sketches and scenic spectacles. The museum was so successful that in 1846 the Kimball brothers spent nearly a

BOSTON MUSEUM,

Tremont street...Open Day and Evening.

Admission 25 cts. Reserved seats 50 cts.

This Museum is the largest in America; it comprises no less than SEVEN DIFFERENT MUSEUMS, to which has been added one half of Peale's Philadelphia Museum, the whole collection exceeding HALF A MILLION ARTICLES. Also, a new HALL OF WAX STATUARY, filled with upwards of two hundred Wax Figures, so natural and life-like as to mock reality. Every Evening and Wednesday and Saturday Afternoons, splendid performances given without extra charge.

☞Exhibition Room opens at 6½—performances commence at 7½ o'clock.

The Boston Museum publicized its purchase of "one half of Peale's Philadelphia Museum" in this undated newspaper advertisement touting the breadth of its attractions.

quarter of a million dollars building a four-story, neoclassical building on Tremont Street, in the heart of Boston's commercial district. The centerpiece of the new museum was an enormous interior hall, flanked by Corinthian columns and terminating in a grand staircase, above which hung Thomas Sully's epic painting of George Washington crossing the Delaware.

For the first decade or so, Kimball aspired to create the largest and most complete "Noah's ark" in the country. He offered to the public an expanding array of "splendid, chaste, and amusing entertainments," detailed in "catalogues" of his collection, which were sold to visitors as guidebooks. None of the surviving guidebooks or advertisements mentions the Lewis and Clark objects, which may simply have been integrated into existing cases of Indian curiosities. Unfortunately, no other inventories of the Boston Museum have been located.

The Peabody Museum of Archaeology and Ethnology

Moses Kimball retired from active museum management in 1860. After his retirement the Boston Museum increasingly emphasized theatrical performances over exhibits, and during the early 1890s it began to dispose of its natural history specimens, donating most of them (including a number of Peale's quadrupeds and birds) to the Boston Society of Natural History (Faxon 1915). Kimball died in 1895. When a fire damaged the Boston Museum in April 1899, the Kimball heirs invited the Peabody Museum of Archaeology and Ethnology to select objects from among those remaining in exhibit

Fire-damaged exhibit galleries at the Boston Museum, 1899. After the fire, many of the surviving artifacts were given to Harvard's Peabody Museum of Archaeology and Ethnology.

cases. In May, assistant curator Charles C. Willoughby of the Peabody visited the Boston Museum and chose for the museum more than fourteen hundred "artificial curiosities" from around the world. Among them were Peale Museum objects, some still accompanied by handwritten labels prepared by Peale and his sons.

Probably inspired by the 1904 Lewis and Clark centennial, Willoughby published an article in the journal *American Anthropologist* the following year describing expedition objects at the Peabody Museum. Willoughby's article was the first to publicize their

The Northwest Coast cases at the Peabody Museum, from a glass-plate negative made around 1900. Early anthropological teaching exhibits at the Peabody were arranged to illustrate object types and the features of "traditional" cultures, not their engagement with the Western world.

existence and the first disciplinary gesture claiming them as "ethnological" specimens. Anthropologists of Willoughby's generation viewed culture contact as having despoiled traditional (by implication, "real") indigenous societies and their art. For Willoughby, the Lewis and Clark objects were significant because they were among the few remaining "higher class . . . specimens of the earlier handiwork of the modern tribes" (Willoughby 1905:633). Their collection history was important primarily because it certified their relative antiquity. Many twentieth-century researchers, such as John Ewers, continued to regard the Lewis and Clark objects as "type specimens" that exemplified the formal properties of traditional tribal arts. Their powerful, animate capacity to create and encode the experiences of Indian peoples living in emerging transcultural worlds, their profound meanings as gifts, and their potential to inspire us to reread old histories and to write new ones remained dormant.

Two

THE PEABODY MUSEUM OBJECTS

Chapter 5

From Warriors and Women Traders

Objects Collected by Meriwether Lewis and William Clark

SIX OBJECTS IN THE PEABODY COLLECTION HAVE STRONG ASSOCIATIONS with the Lewis and Clark expedition, matching materials described by Lewis and Clark, objects that museum proprietor Charles Willson Peale received from Jefferson, Lewis, or Clark and catalogued into his collection as expedition artifacts, or both. These six pieces represent the breadth of Lewis and Clark's journey, having been given or traded to the explorers by people living in the midcontinent, on the Plains, and near the Pacific Ocean. The six objects are a quilled otter bag, possibly an ambassadorial gift; two basketry whalers' hats that Lewis and Clark commissioned from women while they wintered at Fort Clatsop; and three raven belt ornaments, almost certainly presented to Clark by a chief or leading warrior.

This raven belt ornament, with its flaring crest, exudes power and presence, attributes appropriate to the soldier society member who probably wore it. (Detail of 99-12-10/53051; see p. 111.)

Otter Bag

Charles C. Willoughby, assistant curator at the Peabody Museum in 1899 when the Boston Museum collection was received, considered this otter bag (99-12-10/53052) to be among the objects having the firmest association with the Corps of Discovery (Willoughby 1905). It arrived at the Peabody accompanied by a paper label reading, "Sioux Tobacco Pouch. Sent to Capts. Lewis and Clarke by the Sock nation. Presented by Capts. Lewis and Clarke." Analysis of the bag and associated documentation seems to substantiate Willoughby's judgment that it was acquired during the expedition.

Lewis and Clark did not write about bags of this type in their journals, but five "tobacco pouches" are listed in the 1809 Peale memorandum, including two described as "very handsomely ornamented . . . with Porcupine Quills, and Tin, &c. Of the Saux Tribe." In addition, an unornamented tobacco bag is listed as having been "sent by the Sacks." Spelling in the Peale documentation is idiosyncratic and inconsistent, and it includes the forms *Saux, Soue, Sock,* and *Sacks* for both Sioux and Sauk. In some cases it seems clear that Peale referred to the Sioux as "Saux." Consequently, the extant paper label may conflate the memoranda indicating quilled Sioux bags with the unornamented example sent by the "Sacks" or "Socks." Willoughby interpreted object 53052 to be a Sauk bag.[1]

How might we determine whether it was the Sioux or the Sauks who gave this bag to Lewis and Clark? The formal properties of the bag are not diagnostic, because in 1800 both the Mississippi Sioux and the Sauks were producing Woodlands-style quillwork that may have included the techniques and patterns evident here. The bag could also have been traded prior to being collected. Three factors might support a Sauk attribution for the bag: the later association of such objects with Woodlands and western Great Lakes peoples, the similarity between the quillwork on this bag and that on Sauk otter bags studied by Alanson Skinner (1923–25:pl. 26), and the relationship of this piece to other items on Peale's list of objects from Lewis and Clark.

A Peale Museum label associated with one of the tobacco bags donated by Lewis and Clark.

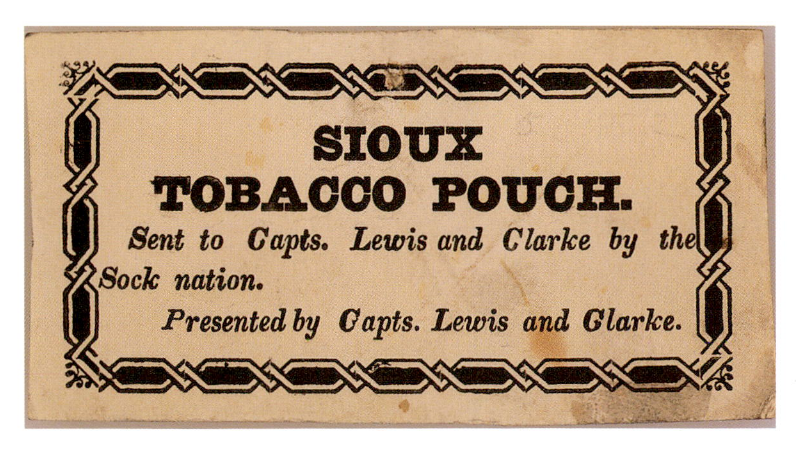

If we presume that this bag is Sauk, then when and where might Lewis and Clark have acquired it? The relationships among the tobacco bags, pipes, and other items acquired from Mississippi River tribes such as the Ioways, Sauks, and Foxes (Mesquakies)

Otter bag

99-12-10/53052

Eastern Plains/western Great Lakes (Sauk?), circa 1800–25

Otter hide, otter skull, deer hide, dyed and undyed porcupine quills, dyed bird quills, sinew, tin-plated sheet-iron tinklers, unidentified plant fiber, dyed horsehair, glass beads, wool fringe

Length 110 cm, width 23 cm

provide one clue. The 1809 Peale memorandum lists the undecorated bag as having been "sent by the Sacks," and the extant label for 53052 describes it as a "Sioux" bag "sent to" the captains by the Socks. Peale also received from Lewis and Clark "2 Silver Midals, of George the 3d of England, obtained from the Foxes and one from the Socks, or Saukeys." The captains' receipt of peace medals originating with a foreign nation that was competing for the allegiance of the Sauk and Fox peoples, along with pipes (one listed as having belonged to the Fox chief White Skin, a treaty signatory) and tobacco bags, suggests that all of these items were presented in an official, diplomatic context. It is less clear how and where Lewis and Clark obtained objects from Mississippi River tribes with whom they had little direct contact. The Corps did not meet formally with a Sauk delegation during the expedition, although it did encounter individuals and small groups: Sauk parties visited Lewis and Clark's winter camp at Fort Wood, Illinois, on March 25 and May 5, 1804 (Moulton 1986, 2:181, 212). Clark noted that the second party consisted of a "Sauckee Chief" with eight or ten companions (Moulton 1986, 2:212).

Lewis and Clark did, however, initiate formal contact with the Sauks and other Mississippi River tribes through intermediaries. On March 15, 1804, Lewis sent a speech to the Sauks and Foxes, apparently via a Sauk chief, which might have been accompanied by medals, a flag, and other customary presentations. Because of translation problems, the message was later reissued by Amos Stoddard, commander of U.S. civil and military affairs in Upper Louisiana Territory, who acted as the expedition's agent at St. Louis (Jackson 1978, 1:197, n. 1). It is possible that the Sauks and Foxes replied to the captains, perhaps including reciprocal gifts. This "reply" might have been delivered during the visit of an intertribal delegation to Washington, D.C., in 1805–6. At the time, White Skin was the leading chief of the Foxes, and the Peale memorandum specifies that Lewis and Clark received a pipe from him.

Once under way, the Corps apparently had no direct contact with the numerous and powerful Mississippi Sauks, but they heard much of their prowess in warfare and observed along the riverbanks the ruins of abandoned Missouri and Kansa towns whose occupants had been driven out by Sauks. The Sauks and Foxes had themselves been displaced from the Great Lakes by postcontact international power struggles stemming from the fur trade. They were generally allied with the British.

As Donald Jackson (1981:203–22) pointed out, the Sauks and Foxes were the first tribes to be directly affected by the Louisiana Purchase and by Jefferson's idea of resettling eastern Indians west of the Mississippi River. Because Jefferson feared the continuing presence of the British and their influence on tribes in Upper Louisiana, he directed Indiana governor William Henry Harrison to acquire Indian land that would provide a buffer zone. As a result, a party of Sauks and Foxes signed a treaty in St. Louis in 1804 relinquishing claims to land in what would become three states. Ensuing conflicts, especially with the so-called British Band, whose members refused to acknowledge the 1804 treaty, made the Sauks and Foxes of considerable diplomatic and military concern to the United States throughout the first quarter of the nineteenth century. Owing to the government's "very delicate standing" with the Sauks and Foxes (Wilkinson in Jackson 1978, 1:266), those tribes made up one-third of a delegation orchestrated by Lewis and Clark that visited Washington and other eastern cities during 1805–6, while the Corps of Discovery was yet in the field.

This delegation, then, is another potential source of the Mississippi River materials that were given to the Peale Museum as expedition artifacts. (A fuller examination of this hypothesis is presented in chapter 8.) Delegation members met with President Jefferson in January 1806. Jefferson addressed them at length about the Lewis and Clark expedition, the new American control over the West, and his desire for peaceful trade with and between the assembled Indian nations. The delegation members presented a formal speech in reply (Jackson 1978, 1:280–89). Diplomatic gift exchanges were almost certainly a part of the proceedings.

Finally, Lewis or Clark might have received some of the Mississippi River materials after the expedition itself. In their post-expedition government positions in St. Louis, both men conducted extensive diplomatic proceedings with Indian peoples, including the Sauks and Foxes, and both sent items to Jefferson and Peale after their return. For example, Lewis, in his capacity as governor of Louisiana Territory, conducted a treaty council with the Sauks and Foxes in August 1808 (Coues 1979 [1893], 1:xxxviii, n. 18). He also forwarded a shipment of ethnographic objects to Peale before embarking on his ill-fated journey toward Washington in the fall of 1809, during which he took his own life.

Otter bags with quilled panels were closely associated with Great Lakes medicine societies, particularly the Midewiwin, and in that context the Ojibwas called them *pinjigosauns* (Harrison 1989:figs. 3, 4). The Midewiwin, or Grand Medicine Society, was a religious institution found among the Ojibwas, Menominees, Winnebagos, Potawatomis, Ottawas, Sauks, Foxes, and other tribes in the Great Lakes and eastern Plains regions (Ritzenthaler 1978). Midewiwin bags often retain the skull of the otter, as does the Peabody bag. Otter bags were also used as tobacco and pipe containers

Still vivid, organic dyes animate the elaborate quillwork on the tail of the Peabody's otter bag, which may have been sent to Lewis and Clark as an ambassadorial gift. (Detail of 99-12-10/53052; see p. 87.)

(Ewers 1979:46–47; Penney 1992:69). Prince Maximilian acquired at least one otter bag during his 1833–34 Missouri expedition, which Karl Bodmer, the artist who accompanied him, illustrated among a group of collected objects (Bodmer 1984:pl. 351). George Catlin painted many otter tobacco bags, including one belonging to the Mandan chief Four Bears, and William Clark listed at least five "otter skin tobacco pouches" in the catalogue of his personal museum.[2]

The ritual role of smoking in diplomatic councils is well known, and such meetings often closed with exchanges of gifts. It is not surprising, then, that tobacco bags made from mammal skins were acquired by many visitors to the Plains. Several Sauk and Ioway otter-skin "medicine bags" bearing similarly quilled panels are presented in Skinner's "Observations on the Ethnology of the Sauk Indians" (1923–25:pl. 25). It seems likely, however, that the five "tobacco" bags acquired by Lewis and Clark were categorized as such by their presenters—not as bags associated with the Midewiwin or other religious societies.

Formally, this otter bag exemplifies the relatively early character of many of the objects in this collection, which display native technologies and aesthetics while creatively incorporating Euro-American materials.[3] The three-part color scheme of the quillwork—whites and yellows, reds and oranges, and browns and blacks—was the classic palette used by Woodlands and Prairie groups at the time of their contact with Euro-Americans. Also typical of early historic Woodlands quillwork are the relatively large, thick quills used here. The subtle yet vibrant hues of the quillwork were achieved by using organic dyes such as bloodroot or buffalo berry for oranges and reds and walnut or wild grape for the darker, wine- or eggplant-colored browns, some of which are so dark that they appear almost black. The white quills are natural, undyed porcupine quills.

This otter bag was created by a woman who had mastered the ritual art of quillworking. She utilized six separate techniques—five for sewing (four of them involving two threads) and one for wrapping: four-quill plait; two-quill triangle; one-quill zigzag band; simple band; simple line; and quill wrapping (see Bebbington 1982; Heinbuch 1990). The complex four-quill diamond pattern down the center of the tail is rarely

JO ESTHER PARSHALL, QUILLWORK ARTIST

Quillwork artist Jo Esther Parshall with her apprentice, Elaine McLaughlin.

Quillwork is a ritual art that was customarily practiced by women and by women's societies in the Woodlands, Great Lakes, and Plains and Prairie regions. After coloring porcupine and bird quills with plant and mineral dyes, women flattened them and then, using sinew thread, wrapped, plaited, and sewed them to garments, pipe stems, and other objects, creating both geometric and curvilinear designs. Senior society members and accomplished artists transmitted both the rights to undertake quillwork and the necessary technical knowledge to younger women. After Native American communities were forcibly settled on reservations and pressured to abandon indigenous cultural practices, few women continued working with quills. In recent decades, Plains Indian artists have spearheaded a revival of this elegant and demanding art form.

Jo Esther Parshall (Cheyenne River Lakota) is one of the most gifted and accomplished of these modern quillworkers. She achieved national recognition in 1993 when her fully quilled horse mask was awarded "Best in Show" at the Northern Plains Tribal Arts Show in Sioux Falls, South Dakota. Many other accolades followed, including best-in-show awards at the United Tribes Art Exposition in Bismarck, North Dakota, and the Denver Cultural Center Art Show. Her major works, which include dresses, vests, and horse regalia, have been acquired by regional and national museums including the Smithsonian Institution and the National Museum of Scotland and by private collectors.

Parshall's quillwork is distinctive for its delicacy, depth of color, and imaginative, often visionary, themes, such as her depiction of the comet Hale-Bopp on a pipe bag in 1997. When she became a quillworker in the late 1980s, she set out to create works that were "visually big and spiritually big, that connected a lot of things and that would draw attention [back] to quillwork." She attributes her success to the way in which she was called to the art:

> I had found some porcupine quills out on the prairie, and I carried them around with me. Finally, a keeper of the quills came to me. That was Carrie Brady (Mandan-Hidatsa), a relative of my then husband. It was nothing that I decided to study, and

Jo Esther Parshall

I never sought it out—it just came about. Carrie Brady wasn't able to show me how to use the quills—her hands were too crippled with arthritis—but she knew that I would honor the quills and pass it along. She's the one that gave me the rules, rights, and medicine to go along with this art form. It was a spiritual blessing for her to see all of this in me. Not until later did I understand what she was passing on to me, and now I see it in others. Since 1997, I have been mentoring students through grants from the North Dakota Council on the Arts. I'm working with an apprentice now who has the same passion that I had in the beginning, which is a good feeling, because sometimes I wonder who will carry it on.

Jo Esther's hands, quilling.

seen; the only other example found during the research for this book was on an Ioway "snakeskin medicine" from a war bundle (Orchard 1982:pl. 8; Skinner 1926:pl. 40). The quillwork on this bag was sewn with sinew onto pieces of tanned deer hide that were attached to the four feet, the anus, and the tail of the otter.

All of the quillwork on the bag is executed in porcupine quills except for an enigmatic section on the underside of the tail. There, a simple band is sewn, composed of bird quill, porcupine quill, and two kinds of brown vegetable material, one of which may be cornhusk. A third thread is visible on the band. The colors are again red, white, yellow, and brown. Gaylord Torrence (personal communication, 2000) observed that this asymmetrical band corresponds to decorations on the pendant tails of Sauk and Mesquakie bear claw necklaces. Ruth Phillips (1987:89–90), relating spatial design to cosmology, identified asymmetry as characteristic of a Woodlands aesthetic.

Handmade sheet-iron tinklers are suspended from the feet and tail of the otter and are filled with both red wool and a material identified as possibly being deer hair. Hide cone attachments are wrapped with red-orange quills, and tassels of dyed red yarn replace cone tinklers on the otter's right front paw (see p. ii). The quilled panels on the hind legs are each edged along the top with four white, tubular "faux wampum" glass beads. White, quill-wrapped loops and red wool tassels are pendant from the otter's ears.

Basketry Whalers' Hats

Anne-Marie Victor-Howe

One or both of these two knob-top conical hats (99-12-10/53079 and 99-12-10/53080; see pp. 94 and 95), commonly referred to as "onion-dome" hats, were very likely among a number of basketry hats acquired by Lewis and Clark in the lower Columbia River region. In his "Memorandum of Specimens and Artifacts" from the expedition, Peale listed three hats. One was "manufactured by a C[l]atsop woman near the Pacific Ocian; from whence it was brought by Capts. Clarke & Lewis." A second was a "cap, worn by the women of the Plains of Columbia," and the third was "a cap worn by the natives of Columbia River, and the Pacific Ocean" (Jackson 1978, 2:476, 478). We know that Clark sent one hat from St. Louis to Louisville in 1806, using Adam Woolford, or Woodford, as a courier. Lewis forwarded another hat to Washington, D.C., via Lieutenant Peters (Moulton 1993, 8:418–19).

In 1805, while on their way down the Columbia River to the Pacific Ocean, the explorers noted Native people wearing basketry hats of many different shapes, ranging from knob-top, rounded-top conical, round, and water-cup-shaped hats to flat-top hats similar to those worn by Europeans. On November 1, 1805, Clark recorded that he had purchased from Native peoples "a hat of their own taste without a brim" (Moulton 1988, 5:371). They were particularly intrigued by several cone-shaped hats with geometric or pictorial figures and some onion-dome knob-top hats on which whaling scenes were depicted. On January 29, 1806, while wintering at Fort Clatsop, Clark wrote a detailed account of onion-dome hats he had seen:

> Maney of the nativs of the Columbia [wear] hats & most commonly of a conic figure without a brim confined on the head by means of a String which passes under the chin and is attached to the two opposit Sides of a Secondary rim within the hat—the hat at top termonates in a pointed knob of a conic form, or in this Shape. [T]hese hats are made of the bark of Cedar and beargrass wrought with the fingers So closely that it Casts the rain most effectually in the Shape which they give them for their own use or that just discribed, on these hats they work various figures of different colours, but most commonly only black and white are employed. [T]hese figures are faint representations of the whales, the Canoes, and the harpooners Strikeing them. Sometimes Square dimonds triangle &c. (Moulton 1990, 6:246)

Basketry whaler's hat
99-12-10/53079
Makah or Nootka (Nuu-chah-nulth),
circa 1790–1830
Spruce root, cedar bark, surf grass,
unidentified mammal hide
Height 27 cm, width 26.5 cm

Basketry whaler's hat
99-12-10/53080
Makah or Nootka (Nuu-chah-nulth), circa 1790–1830
Spruce root, cedar bark, surf grass, unidentified fur, unidentified mammal hide
Height 22.5 cm, width 27 cm

In their journals, Lewis and Clark made several small sketches of the conical hat with an onion dome at its apex (see p. 65; Moulton 1990, 6:141, 143, 247, 250).

Impressed by the rain-repellent qualities of the onion-dome hats, Lewis and Clark acquired a number of them through trade and commission. On February 22, 1806, two women whom the captains identified as Clatsop came to the fort with hats that the men had previously ordered. Lewis wrote that they brought

> a parsel of excellent hats made of Cedar bark and ornamented with beargrass. [T]wo of these hats had been made by measures which Capt Clark and myself had given one of the women some time since with a request to make each of us a hat; they fit us very well, and are in the form we desired them. [W]e purchased all their hats and distributed them among the party. (Moulton 1990, 6:335–36)

Woven hats were such valuable pieces of clothing for the expedition's members that at least one man was still wearing his at the Missouri Breaks in August 1806. By early July of that year, the expedition had split into two groups, one led by Lewis, the other by Clark. The plan was for the two parties to reunite at the mouth of the Yellowstone River. On August 7, while Lewis's party was searching for Clark's camp, the captain sent one of the men, John Ordway, to check the shore for signs of the other explorers. Ordway came back with auspicious news: not only had he discovered "the tracks of two men which appeared so resent that he believed they had been there today," but he also had found a campfire still "blaizing" and, most amazingly, "part of a Chinook hat" that was identified by the Lewis party "as the hat of Gibson" (Moulton 1993, 8:151).

Lewis and Clark probably erred, however, when they attributed the manufacture of onion-dome hats to Clatsop women, if indeed the hats they purchased were of that type. Although the Clatsops were accomplished weavers and the explorers did commission Clatsop women to weave them basketry hats, special skills were required to weave this particular type. In fact, these shapes and designs were made only by whaling peoples on the Northwest Coast, specifically the Nootkas (Nuu-chah-nulth), who lived along the west coast of Vancouver Island from Cape Cook to Port San Juan, and the Makahs, who occupied the mainland of what is now Washington state at Cape Flattery, from the Hoko River to north of Flattery Rocks.

Lewis and Clark's mistaken attribution of the hats to the Clatsops may be explained by the explorers' choice of site for their winter camp in 1805–6 and their relationship with the neighboring Clatsops and Chinooks during those months. The exploring party deliberately chose a site two miles up the coast from the nearest centers of Indian activity, and they maintained a distanced relationship with their neighbors. At Fort Mandan, there had been much peaceful interchange between the expedition and the Mandans, and traders were available to act as interpreters. At Fort Clatsop, relations with the tribes were often tense, and at sunset the Indians were asked to leave the fort. Members of the expedition ventured outside the fort sporadically but generally were not inclined to interact with the villagers. The Indians, for their part, came to the fort strictly for business, not to socialize. As a result, the explorers never closely observed the Natives' daily activities—particularly their indoor winter activities such as weaving—and they lacked the benefit of Native interpreters. Lewis repeatedly complained in his journal that he could not understand the Chinookan language enough even to make simple inquiries (Moulton 1990, 6:97–98).

The knob-top hats Lewis and Clark attributed to the Clatsop women had probably been traded down the coast to the Columbia River area and may have been worn by members of several other coastal tribes along the way. The presence of such "foreign" basketry objects on the lower Columbia in 1805 among people who did not hunt whales can therefore be taken as evidence of regular cultural contact between disparate tribes in the region through trade, exchange, gift networks, raids, enslavement, visits, and perhaps intermarriage. On March 14, 1806, for example, Lewis and Clark were informed by some Clatsops of a recent visit from an Indian of "the *Quin-na-chart* Nation who reside Six days march to the N.W.," no doubt a Makah Indian from Cape Flattery (Moulton 1990, 6:416). The hat might also have traveled from Nootka Sound to the mouth of the Columbia on one of many American trading ships that regularly plied that route (Malloy 2000:133).

Knob-top hats without decorative motifs have been found at three water-saturated archaeological sites on the Northwest Coast: the Ozette Village, Hoko River, and Wapato Creek Fish Weir sites (Croes 1977:44–45; Croes and Blinman 1980:215–17). At Ozette Village, located at Cape Alava near the western tip of the Olympic Peninsula, Washington, where houses were uncovered from beneath a massive mudslide dating between three hundred and five hundred years ago, a hat with a cylindrical knob was found associated with whaling equipment, which indicates that its owner had considerable status (Croes 1977, 3:422, 427). Five examples of knob-top conical-style hats with much smaller knobs than those from Ozette were excavated from the Hoko River site. All five date between 2600 and 3000 B.P. (Croes 1976:6, 1995:134, 136; Croes and Blinman 1980:217). Although these hats were not associated with objects indicating status, it is likely that these wearers, too, enjoyed high social standing. Social stratification certainly existed on the Northwest Coast even in prehistoric times.

It is known that for many years before whales were pushed close to oblivion by commercial whaling, hats such as the two shown here were woven in limited numbers and worn only by high-ranking tribal members, usually harpooners (Cook and King 1784, 2:304, 327; Croes 1977:423–27; Gunther 1972:29–30; Kaeppler 1978:159–60; Mozino 1970:15).

It is remarkable how much the shape of these and other onion-dome hats worn by high-ranking Makah and Nootka chiefs resembles the shape of basketry hats worn by well-educated Manchu and Chinese men during the Qing Dynasty (1644–1911). Such hats had finials at the apex, and the finials of the highest-ranking mandarins had gold-ornamented bases topped by jewels or precious stones and decorated with peacock feathers. Those of lower-ranking officials had only decorated knobs (Garrett 1994:71–73). Winter hats had "a turn-up brim of black satin, mink, sealskin, or velvet, and a padded crown covered in red silk fringing." Summer hats were "made of woven straw for the lower ranks, and split bamboo [covered with silk gauze] for higher officials" (Garrett 1994:71, pl. 29).

Regular cultural interchange had taken place between China and the Northwest Coast for several decades before the Lewis and Clark expedition, because Chinese and other Asians served as crew members on European and American vessels engaged in the fur trade and exploration. Clothing made in China, including hats and other accessories, was traded to Northwest Coast Natives (Quimby 1948:248), and Chinese crewmen manufactured all kinds of trade articles (Meares 1967 [1790]:88; Quimby 1948:248). Some of them even "intended to become settlers on the American coast" (Howay 1940:179; Quimby 1948:249–50). There are reports of at least one Nootka man who spent time in Canton (Meares 1967 [1790]:121). Yet despite the presence of Chinese crew members and Chinese trade articles on the Northwest Coast, the archaeological evidence clearly shows that the resemblance between Chinese and Northwest Coast hats is no more than a coincidence of design. It might have been reinforced, but it was not determined, by cultural interchange.

In a similar vein, it is tempting to suppose that the onion-dome shape of the Makah and Nootka hats was influenced by the ogee, or bulbous, form of dome on the roofs of Russian Orthodox churches. As early as 1799, at least one such church probably existed far to the north of the Nootkas and Makahs in Sitka, the capital of Russian America. Again, the archaeological record proves that supposition wrong.

Onion-dome knob-top hats with decorative motifs were described and drawn by a number of explorers and artist-illustrators who sailed to Nootka Sound, off

Vancouver Island, in the latter half of the eighteenth century. As early as 1774, Juan Josef Pérez Hernández, aboard the *Santiago,* recorded intertribal trade in animal pelts and hats with pear-shaped bulbs at the apex (Gunther 1972:11–12). On his third voyage to the Pacific in 1778, Captain Cook described the inhabitants of Nootka Sound as wearing two different kinds of hats: one in the shape of a sugar loaf, flattened at the top; the other with a knob at the top. "They [the hats] are white and into them are worked in black a description of their Whale Fishery, a man standing up in a Canoe in the action of striking a whale" (Beaglehole 1967:1099, 1325–26, 1411; see also Cook and King 1784, 2:327; Kaeppler 1978: 159–60). John Webber, the official artist on Captain Cook's third voyage, painted portraits of several chiefs and a woman wearing conical, woven, onion-dome knob-top hats with designs depicting whale hunts (these portraits are PM 41-72-10/498 and 41-72-10/496).[4] By the second half of the eighteenth century, although whaling chiefs' hats were no longer worn exclusively by whale hunters, Northwest Coast residents still wore them to display their social status and wealth. By the middle of the nineteenth century, Makah and Nootka weavers had stopped making hats like the ones shown here.

A Nootka chief wearing a whaling hat. Some scholars have suggested that this drawing by John Webber, Captain Cook's expedition artist, depicts Maquinnah, one of the most famous chiefs during the maritime fur trade period.

Hat 53080 (p. 95), woven from cedar bark, split spruce root, and surf grass, has a typical conical base surmounted by an onion-shaped dome. Its geometric, highly representational designs are typical of the Makahs and Nootkas. The weaving is overlay twining, or twining with double strands. The warp is split spruce root; the weft is cedar bark dyed dark brown with an overlay of bone- or ivory-colored surf grass that provides a light background for the dark design (Pat Courtney Gold, personal communication, 2000; Vaughan and Holm 1990:33; Willoughby 1905:65–68; Wright 1991:84).

The hat has two layers, which enhance its rain-repellent qualities. The inner layer, made of coarsely twined red cedar bark, is joined to the outer layer at the brim. The twined cedar bark interior is fitted with a headband made of coarsely twined cedar bark and fur that helped to position the hat on the wearer's head (see p. 12). The hat was secured with two thongs tied under the chin. Its lining extends into the bulb of the hat. Other hats' linings (e.g., 99-12-10/53079; p. 94) start below the knob using various kinds of weaving methods (see also King 1981:82).

Like most other whaling hats, this one shows whales being pursued by hunters in canoes.[5] The zoomorphic motifs are so realistic that one can identify humpback whales (*Megaptera novaeangliae*) with their stout, short bodies; humps; rather long, narrow flippers; deeply notched flukes; dorsal fins bent horizontally at the top; white marks on their sides; and widely spaced ventral grooves from chin to navel.

There are two scenes opposite each other on hat 53080 (facing page), each showing two whales being pursued by three canoes. In two of the canoes in each scene stand harpooners, ready to thrust their harpoons. In the canoe next to the knob—the lead canoe— just behind the harpooner stand seven other crew members: three pairs of paddlers and a steersman in the stern. Each of the seven is represented by a single dark-brown, almost black vertical line. The last line, with a small bar on top projecting outward, represents the stern, rising almost vertically and covered by a broad, flat platform.

Next to the rim of the hat in each of the scenes is another canoe with a harpooner. In one scene, this canoe is somewhat smaller than the lead canoe and has only five crew members; in the other, the leading boat is apparently smaller but has seven crew members. In reality, Makah whaling canoes were all about the same size: thirty to thirty-five feet long. Evidently, in order to fit four canoes around the hat's rim, the weaver had to make two of them shorter. By alternating the sizes of the boats and their crews, the weaver cleverly showed that all the boats were in fact the same size.

In each scene, both whales have lines with inflated sealskin floats running from their sides, as if they have already been harpooned. The whale near the knob has three floats running from its body; the one near the rim is shown with four floats. During an actual hunt, four floats, each with a distinct name, were attached to the line before the whalers set out. More floats were added as needed after a strike (Drucker 1951:29–30). As many as thirteen floats might be attached to one line. The weaver of this hat, however, was limited by space and chose to depict fewer floats. Moreover, during an actual hunt, a smaller last float remained near the surface, marking the whale's position, while all the other floats were pulled underwater when the whale sounded. As the whale swam to the surface, the last float bobbed up long before the

Top view of whaler's hat.
(Detail of 99-12-10/53080; see p. 95.)

whale appeared (Waterman 1920:43). In both the top and bottom images on this hat, the weaver deliberately depicted inflated floats, thus portraying the animals close to the surface. The Makahs and Nootkas believed that visual imagery of whales near the surface made a successful hunt more likely. The whaling scenes represented on this hat are therefore not illusions but real situations; the whale is truly present and "actable upon" (Freedberg 1989:274).

It is also important to note that the stylized scenes on the hat depict several crucial moments during a whale hunt. First, the canoe comes alongside the whale from the rear, outside its field of vision. The whale is just starting to sound, with its flukes still underwater. The harpooner, wearing his chief's hat, is poised to thrust his weapon behind the whale's flipper. The second moment is shown in the upper part of the hat, when the whale is first harpooned and the line is playing out. Finally, the last float on the line is surfacing, signaling the presence of the whale near the surface. The time between the strike and the appearance of the last float near the surface was extremely dangerous for the crew; it was a solemn time when the harpooner needed the help of a powerful guardian spirit and sang his sacred songs, promising the whale great honor to entice it to swim toward shore.

The power of these images resides in the Makah and Nootka belief that the exemplary actions depicted on the hat would help ensure similar actions during a hunt. More importantly, the animal's death must be transformed in order for its spirit to complete its round-trip journey to the whalers' village and back to the ocean. The efficacy of that process is depicted on the hat. Seen from above, it describes several concentric circles: a dark brown one at the rim, a second, broken dark line half an inch above the first one, and a third at the edge of the knob. The alternating black-dyed cedar bark and ivory surf grass form a series of concentric circles converging on a center point at the top. Between the rim and the knob, the whaling scenes progress in circular motion. This circling is emphasized by the symmetry of the scene and the fact that the scenes on both top and bottom are in reality broken circles. The design gives the impression of a spiral pattern converging on a center.

For the Makahs and Nootkas, the circle formed by the hat's brim created a boundary between the worlds that protected animals and crew members. Power flows through the spiral in two directions: outward from the center in the technique of the weaver, and inward from the rim in the temporal sequence of the whaling scene. The evolution of the hat as it is woven balances the counterclockwise involution of the whaling scene; it brings the hunt to the center where the whale can be killed and brought back to life in spirit form, thus allowing it to return to its home in the sea.

Like their designs, the shape of these hats is also significant. J. C. H. King (2000:268–69) suggests that "the knobbed form seems closest to the design of the cod lure—a wood bulb with long backswept fluted projections or wings" that was pushed into deep water at the end of a long pole and allowed to float rapidly back to the surface. The movement attracted the curious cod to the fisherman, who waited at the surface with a harpoon or net (Stewart 1982:56). King also says that the knob may relate to "kelp bulbs," the air bladders of various seaweeds found along the Northwest Coast, or could represent "a simplified harpoon—the rounded knob symbolizing a three-dimensional whaling head" (King 2000:269). King goes on to speculate that "the whole hat symbolizes the instrument used for hunting whales." It is certainly conceivable that the shape of a lure or a kelp bladder on top of the hat was a propitiatory symbol intended to help bring the whale to the surface. Both the lure and the bulb are representations that create a symbolic path for the whale toward the hunter.

An even more likely interpretation is that the onion-dome shape represents a specific kelp species known as bladder seaweed (*Halosaccion glandiforme*), commonly found on rocks in the mid-intertidal zone along the Northwest Coast. At a deeper, symbolic level it may then represent a breast or nipple. Nancy J. Turner and coauthors (1982:24) mention that the Hesquiat Indians on Vancouver Island called *Halosaccion glandiforme* by a Nootka name that means "like a nipple." It was used for medicine, and "when the water-filled sacs are squeezed, water squirts out like milk from a nipple, hence the name." As a breast or nipple, the onion-dome shape would have symbolized maternity and a newborn infant's first source of food. Thus it might have been meant to help the whalers reassure the whale that after staying as an honored, welcome guest in their village it could return, newly born, to the ocean. The onion-dome shape thus might be an example of what Wilson Duff referred to when he wrote that "sexual symbolism is so important in the arts of the world that its virtual absence on the surface of Northwest Coast art permits us to suspect that we might find it in metaphorical forms below the surface" (Duff 1981:214).

The stylized representations woven into these hats thus served the hunter on his adventure beyond the limits of society, clearing the whale's and the hunter's path from animal to human space. The successful interaction between animal and man was not only visually depicted but was physically woven into the hat through the weaver's technique and her choice of materials. Slowly circling the center, from top to bottom, the weaver worked her way out to the hat's rim, ending by cutting the fibers like a mother cutting the umbilical cord of her newborn child. Her main material, cedar, held special cultural significance (see sidebar, chapter 9), and the transformation of the living force of the tree into an object such as this hat made the object a spiritual, living substance with supernatural powers that were directly connected to the cycle of birth, death, and rebirth.

Knob-top whaling chiefs' hats thus served four principle functions. They were utilitarian objects, emblems of the wearer's social status, signs of respect for the whale that was asked to sacrifice itself, and media of communication between the human and animal worlds that helped ensure successful whale hunts.

Raven Belt Ornaments

Three raven belt ornaments are the best documented of all the Lewis and Clark ethnographic materials. Their chain of possession, from Clark to Peale to the Peabody, is supported at each juncture by written documentation. While on the expedition, Clark described seeing members of a Lakota soldier society wearing raven bustles around their waists and raven ornaments affixed to their foreheads. In late September 1804, the Corps was camped uneasily among the Brule Lakotas, who threatened to bar their passage up the Missouri (see Ronda 1984:27–41). On September 26, Clark wrote in his journal of a Brule soldier charged with civil control:

> His duty is to keep the peace, and the whole interior police of the village is confided to two or three of these officers, who are named by the chief. . . . They seem to be a sort of constable or sentinel, since they are always on the watch to keep tranquility during the day and guard the camp at night. His power is supreme. . . . Like the other men his body is blackened, but his distinguishing mark is a collection of two or three raven-skins fixed to the girdle behind the back in such a way that the tails stick out horizontally from the body. On his head too is a raven-skin split into two parts, and tied so as to let the beak project from the forehead. (Coues 1979 [1893], 1:141)

More than a year after returning from the West, Clark sent Thomas Jefferson a shipment of fossilized bones from an excavation commissioned by the president in Big Bone Lick, Kentucky. Included in the shipment was "a Sioux dress for the head and hips when on duty, as a soldier or police man."[6] Jefferson forwarded the raven ensemble to Peale, and it is listed in the 1809 Peale memorandum as a "Dress of Crow or Raven Skins, worn by the Police Officiers of the Saux, nation" (Jackson 1978, 2:478). An 1810 advertisement in the Philadelphia newspaper *Poulson's Daily Advertiser* (March 10, p. 2) heralding Lewis and Clark's donation to the Peale Museum described the "crows skin" objects as part of "a complete dress of the Soux Indian," including leggings.

There is no evidence that Clark acquired the raven materials during the expedition's encounter with the Brules, which seems to have been an inauspicious occasion for acquiring such objects. Clark did, however, refer to the Sioux bustle at the Peale Museum in the context of a conversation about the Brule incident with Nicholas Biddle, the first editor of Lewis and Clark's published journals (Jackson 1978, 2:518). If a Sioux presented these raven objects to Clark, then they might have been affiliated with the Yanktons or Yanktonais. Lewis and Clark spent several days in late August 1804 with a Yankton band whose members were generous hosts. They presented the captains with a pipe following two days of formal negotiations about the regional political economy. Clark noted that four soldier society men escorted the Yankton chiefs to the councils (Moulton 1987, 3:21–37).

At least one other documented incident took place in which an Indian "police officer" or soldier gave his crow insignia to U.S. officials. In 1819–20, Major Stephen H. Long led a combined exploratory and military expedition up the Missouri River, with the object of describing the country and tribes between the Mississippi River and the Rocky Mountains and discouraging the British fur trade. Charles Willson Peale's son Titian served as the assistant naturalist for the party, which traveled in a wood-burning steamboat carved and painted to look like a sea serpent with steam escaping from its mouth. Secretary of War John C. Calhoun directed the principals to read Jefferson's instructions to Lewis and carry out a similar mission (Thwaites 1905b:38, 249). Party members observed veteran warriors among both the Otos and the Omahas painted black and wearing crow belts, which they called "the crow." They first described this accoutrement as worn by the venerable Oto warrior Ha-she-a, or Cut Nose:

> Ha-she-a, called Cut Nose, in consequence of having lost the tip of his nose in a quarrel with Ietan, wore a handsome robe of white wolf skin, with an appendage behind him, called a *crow.* This singular decoration is a large cushion, made of the skin of a crow, stuffed with any light material, and variously ornamented; it has two decorated sticks projecting from it upward, and a pendant one beneath; this apparatus is secured upon the buttocks by a girdle passing round the body. (Thwaites 1905c:235)

During its first summer and autumn, the Long expedition was joined for a time by Indian agent Major Benjamin O'Fallon and his assistant, John Dougherty. In October, Major O'Fallon, who was William Clark's nephew and a patron of the artist George Catlin, held a council with the Otos, Missouries, and Ioways. At the conclusion of the council, presents were distributed to the leaders of the Indian delegations, whereupon "Cut Nose now presented to the agent his crow and bison robe ornamented with hieroglyphicks" (Thwaites 1905c:239). This act exemplifies what I have called the chiefly gift, a mode of diplomatic presentation that honors both the giver and the recipient.

Charles Willoughby (1905:635) characterized the three raven objects that came to the Peabody as "badges of office" and more fully quoted the Clark passage describing the Brule soldier. These objects, now believed to be belt ornaments or bustles, may be the oldest surviving examples of their type in a museum collection, and they are certainly the earliest collected. Although Clark specified that he also sent a piece or pieces "for the head," the printed Peale label that accompanied either catalogue number 99-12-10/53050 or 99-12-10/53051 reads, "Ornament. Worn upon the elbow by the Sioux Indians. Presented by Captains Lewis and Clarke." Apparently Willoughby used that label, as well as his interpretation of historic paintings of Indian soldiers wearing raven ensembles, as the basis for identifying the single raven ornaments, 99-12-10/53050 and 99-12-10/53051, as "armbands." Subsequent researchers followed his lead.

ORNAMENT.
Worn upon the elbow by the Sioux Indians.
Presented by Captains Lewis and Clarke.

A Peale Museum label associated with the Peabody raven bustles.

These raven belt ornaments have tremendous significance as artifacts of the Lewis and Clark expedition and as objects in Native American cultural and art history. Although they are ancestral to the feather bustles still worn by Native American powwow dancers, only nine of them are known to exist in collections around the world. In the following essay, art historian Gaylord Torrence explores the origins and meanings of these rare and important bird forms.

The Raven Belt Ornaments of Lewis and Clark

Gaylord Torrence

The profound relationship between Native American peoples and the birds of their world, both physical and mythic, has been embodied in countless aesthetic forms and visualized through the creation of objects that symbolize, evoke, and manifest this powerful bond. When William Clark described the raven skin belt ornaments he saw among the Brule Lakotas on September 26, 1804, he provided the first written account of objects that have an ancient history in the Woodlands of North America. Clark revealed in his simple statement that these archaic bird forms, which possess such an arcane and powerful beauty, were associated with war, valor, leadership, and sacred power in the lives of the Native American peoples who created them.

In the years that followed, others, too, described such objects. Evidence suggests that raven belt ornaments were never numerous, and surviving examples are exceedingly rare, yet their widespread presence and cultural importance is clearly established. Native American narratives, archaeological evidence, ethnological studies, descriptions by travelers and explorers, images created by artists of the time, and examination of the surviving ornaments themselves all reference these objects in significant ways. It is well understood that the early raven ornaments were the archetypes for later versions of the crow belts and feathered dance bustles with trailers associated with the Omaha Hethu'shka (Heluska) Society and the related Grass Dance and Omaha Dance, as well as for those that became a standard part of the later pan-Indian War Dance costume. They also formed the basis for the feathered bustles with rawhide bird cutouts that were worn during the Ghost Dance, and for the various bustle forms that were adopted by the warrior societies of Plains tribes as part of their officers' regalia. Finally, they are recognized as the original sources for the feathered rosettes worn by male dancers in modern powwows.

Our knowledge of raven ornaments, however, is fragmentary. It does not, and probably never will, completely encompass their origin, distribution, meanings, and evolution. In this essay I want to explore these questions by examining the class of objects as a whole, in order to reveal in some measure the significance of the three raven belt ornaments acquired by Lewis and Clark. As the earliest documented examples, the Lewis and Clark ornaments provide access to all these remarkable works.

The three raven belt ornaments in the Lewis and Clark collection, the earliest known postcontact examples, are descended from a long tradition. The peoples of the Woodlands first created objects of this type, yet none seems to have been recorded or collected in the historic Northeast or Southeast or in the boreal forests of the Subarctic. Instead, all have emerged in historic times from the Midwest—the western Great Lakes, Prairie, and eastern Plains regions—which suggests an origin in the ancient cultures of the Mississippi River valley. This supposition is supported to some extent by archaeological objects from the area.

Perhaps the earliest explicit representation of a feather bustle occurs on the great masterpiece of Adena sculpture, a human effigy pipe found in Ohio that dates sometime between 500 B.C.E. and 100 C.E. It depicts a standing man in flexed position, possibly a dwarf, ceremonially dressed with headdress and ear spools. A bird-skin bustle with fanning tail feathers extends from a wide belt at the back of his waist (Douglas and D'Harnoncourt 1941:72). Other images of elaborate feather regalia or bird-man transformations appear on objects from the Mississippian culture dating

Opposite
Raven belt ornament
99-12-10/53049
Plains/Prairie, circa 1780–1806
Dried bodies of ravens (*Corvus corax*), raven feathers, dyed short-eared owl (*Asio flammeus*) feathers, rawhide, dyed and undyed horsehair, sinew, wood, winter weasel fur, dyed and undyed porcupine quills, dyed bird quills, unidentified plant fiber
Length 70 cm, width 31 cm.

Due to its fragility, this ornament was photographed from the underside only. When worn, the birds' heads would have been visible, as illustrated on page 112.

from 1000 to 1600 C.E. Finely engraved shell gorgets and cups, for example, depict dancing warriors as falcon-men with beaks, feather capes, and hanging feather bustles made from the tails of birds (Penney 1985:115, pls. 133, 134, fig. 20). These objects are generally regarded as the most complete expressions of the Mississippian warrior theme and, like the pipe, must be considered within the larger context of celestial bird imagery that has permeated Woodlands art for more than three thousand years. Clearly, the connection of raptors and other birds with warrior cults or societies is powerfully represented in these ancient works, and it was this association that Clark witnessed in the Brule village.

No generally accepted name for these ornaments exists. They have been variously described as either belts or bustles, with reference to both ravens and crows. The most common name for the feathered assemblages has evolved as "the crow," but all of the terms are somewhat interchangeable. The distinction between raven and crow is often imprecise. The skins and feathers of both birds were used and are similar in appearance, and their symbolic associations were apparently the same. The term "belt," also used at times to describe the entire assemblage, grows out of the tradition held by some tribes in which an elaborately decorated belt, and the designs upon it, functioned as a symbolic element as well as the means of suspension for the attached feathered ornament. This name was further reinforced by the fact that the men who possessed such objects were distinctively known as belt wearers (Conn 1960:12). "Bustle," of course, refers to the position of these ornaments as they were worn, regardless of the presence or absence of a decorated belt. Like "the crow," this designation was commonly used to describe later feather assemblages, particularly those that evolved during the last half of the nineteenth century and the early decades of the twentieth.

Each of the three raven belt ornaments acquired by Lewis and Clark is constructed in essentially the same manner, and all clearly derive from the same tribal group and time period. The most elaborate, catalogue number 99-12-10/53049 (facing page), consists of four distinct raven skins, whereas the other two, 99-12-10/53050 (p. 110) and 99-12-10/53051 (p. 111), like most other examples, are each made from the skin of a single bird. Although the ornaments initially appear to be made from the complete stuffed bodies of the birds, they were actually created by reconfiguring various parts—the head, neck, breast, wings, and tail—into a stylized, abstract representation of the natural form. This primary sculptural image was then elaborated with narrow, quill-wrapped wooden slats attached to the wings and tail, quilled rawhide loops, and dyed horsehair.

Each of the feathered objects is attached to a rectangular base of heavy bison rawhide that forms a support against the body of the wearer, with thongs extending from the back of this base as a means of attachment to the belt that held it in place. The ornaments were worn so that the birds were positioned at the center of the wearer's back, projecting outward from the upper hips. This configuration of projecting wings and tails undoubtedly formed the prototype for later types of bustles in which pairs of slender, decorated feathers, or "spikes," were attached in a near vertical position. These distinctive elements are seen in Catlin's drawings and paintings (see p. 112) and were standard features on later Omaha and Grass Dance bustles (see p. 113).

The Peabody raven skins are attached to their rawhide bases at the back of the bird's head and neck in such a manner that the head and beak rest upon the body, facing backward over the outward-projecting wings and tails. A flaring ruff of feathers surrounds the head (which retains the skull), and a loose stuffing of plant fiber, perhaps sweetgrass, filled the body cavity formed by the breast skin and wings. The wings and tail are held in place with sinew and reinforced with the addition of quill-wrapped slats. A single slat was attached to the inner side of the outer quills of each wing, and a third slat, slightly wider than the others, was joined at the base of the tail and fastened to the shafts of the upper feathers so as to extend the full length.

Each of the slats was wrapped and plaited with porcupine and bird quills or, more rarely, small amounts of plant fibers, forming a progression of intricately banded and checkered patterns running the complete length of the ornament. All were origi-

Raven belt ornament
99-12-10/53051
Plains/Prairie, circa 1780–1806
Dried raven body (*Corvus corax*), raven feathers, rawhide, dyed and undyed horsehair, sinew, wood, winter weasel fur, dyed and undyed porcupine quills
Length 62 cm, width 16 cm

Opposite
Raven belt ornament
99-12-10/53050
Plains/Prairie, circa 1780–1806
Dried raven body (*Corvus corax*), raven feathers, rawhide, dyed and undyed horsehair, sinew, wood, winter weasel fur, dyed and undyed porcupine quills, dyed bird quills, unidentified plant fiber
Length 78 cm, width 15 cm

nally tipped at each end with white weasel skin and both straight-cut and long flowing strands of red-dyed horsehair, some of which was resist-dyed so that a subtle shift of color occurred as the band of natural light hair appeared through the center of the strand. Decorated slats of this type were a common device and often appear attached to the shafts of single eagle feathers worn in the hair or used with Great Lakes head-dresses, as well as those that formed the widespread feather pennants of calumets. Variations of this method of decorating feathers extended throughout the Prairie region and the central and northern regions of the Plains, particularly along the upper Missouri River.

One of the Peabody's single ravens, 99-12-10/53050 (facing page), and each of the birds composing the four-raven ornament are further embellished with single, three-inch loops of quill- or plant-wrapped rawhide affixed to each side of the head, with long strands of dyed horsehair attached at the base. This particular type of decorative loop also originated in the Woodlands and then became a common feature on a variety of Plains objects.

Pash-ee-pa-ho (Little Stabbing Chief), by George Catlin, 1835. Catlin drew this important Sauk chief wearing a raven bustle on the occasion of the artist's visit to Keokuk's village on the Des Moines River. Catlin depicted a number of other prominent warriors wearing similar ornaments during the same visit.

The belt ornament composed of four raven skins, 99-12-10/53049 (see p. 109), is a magnificent object, both dramatic and complex as a sculptural form and wonderfully rich and vibrant in its coloration and the intricacy of its detail and construction. In its composition of four birds, it is also unique among existing examples. The specific significance of the multiple skins is unknown, but the number four carries a multitude of symbolic associations in Sioux religious belief and, by extension, in ritual practice and the creation of sacred objects.

Although it possesses a strange and even startling beauty, nothing about this raven ornament suggests a purely decorative contrivance. Each decision, each element in its creation was informed by the concepts that underlay its meaning and purpose. The object somehow communicates this integrity of belief through its form and material substance alone, and it is this that makes it such a remarkable work of art—it is at once beautiful and emotionally moving, primeval and explicit. The sight of this bustle, with its mass of shiny, black projecting tails and wings and outward-turned beaks, its flashes of colored quillwork and bits of snow-white ermine, and its multiple strands of flowing red-dyed horsehair—all animated by the slightest movement of the black-painted wearer—must have evoked a powerful response in its viewers.

This composite ornament features structural and decorative elements not found on the two single ravens. Most notably, the tails of the four birds were partially stripped of their feather "webbing" (barbs), creating an open, cutwork pattern within each tail feather cluster, which is composed of eleven to twelve feathers, typical for the adult raven. The cutwork is asymmetrical: one side of the tail feather cluster, consisting of five to six feathers, is trimmed, and the other side is not. These cut feathers are partly stripped for a span of about three inches toward the upper middle of the shaft, leaving the lower half of the feather and a three-inch tip at the end intact. Two of the tails have the partly stripped vanes removed from the left side, and the other two, from the right side. This asymmetrical trimming forms a laterally oriented symmetry of design across what were originally the four projecting tails (one has been lost over time). The overall effect is delicate and beautiful and clearly reminiscent of the later cut-feather work seen on headdresses, dance bustles, and peyote feather fans.

In addition, clusters of split, red-dyed owl feathers are tied to the beaks of the two outer birds, echoing the symmetrical structure created by the cutwork of the tails. It is worth noting that the positioning of split owl feathers at the sides of the ornaments and sometimes at the bases of projecting feather "spikes" remained a common feature as bustles evolved in form. Also, the loops at the sides of each of the four heads are wrapped predominantly with cornhusk or similar plant material rather than with quills (a feature sometimes seen on later Cheyenne and Arapaho ceremonial objects), and split bird quills are included in the feather slat decorations. Finally, a small cylindrical roll of hide, invisible when the object was worn, was attached to each of the four birds where they join the rawhide base in order to create additional support.

Mato Wamniomni (Whirlwind Bear), Lakota, wearing a classic Grass Dance bustle. Photographed in 1900 by E. A. Rinehardt, Omaha, Nebraska.

The birds of all three Lewis and Clark belt ornaments were reconstructed so that the skin of either the neck or breast is positioned behind the head and reversed so as to fall forward, forming a flaring crest, or ruff, of short feathers surrounding and partially covering the head and beak. This feature creates an effect that is somewhat different on each of the three but is particularly pronounced on the two single ravens, especially the smaller of the two (53051; p. 111). A few other notable details differentiate these ornaments from one another. The more elaborate of the two singles (53050; p. 110), which was formed from quite a large bird, has finely plaited quill-wrapped rawhide loops on either side of the head that match the designs on the central feather slat attached to the tail. The quillwork on this bird, as on the composite ornament, includes both porcupine and split bird quills and small amounts of plant fiber. In contrast, the smaller of the two singles is the least decorated of the three objects; there is no sign that there were ever quilled loops at the sides of the head, and no bird quills or plant fibers are included in the quillwork. The form of this ornament is also different in that it has no tail or central extension with a quilled slat. This is not a determining feature for comparison, however, because the piece undoubtedly once possessed this component, which has been lost.

Finally, as a group—and in comparison with others of the type—the raven belt ornaments of Lewis and Clark are somewhat austere and archaic in appearance. This may be due to their early date, but it is just as likely the reflection of a distinct tribal or regional style.

Some question has emerged regarding the manner in which the Lewis and Clark ornaments were actually worn. The Peabody's initial identification of the two single raven ornaments as armbands was apparently based on a label recorded by Peale that accompanied the objects when they left his museum. The label states: "Ornament. Worn upon the elbow by the Sioux Indians." This interpretation was affirmed by Willoughby (1905), who described the three objects as an ensemble, "three badges of office such as were worn by certain . . . trustworthy men appointed by the chief." Since Willoughby's time, researchers have classified them as armbands.

There is little to support this interpretation. Neither Clark's first description of raven ornaments in his September 1804 journal entry nor his later citation accompanying the shipment of these objects to Jefferson indicates that the single ravens were armbands. Indeed, he was quite specific in describing ornaments that were worn on the hips and head. Moreover, nothing in the construction of the single ravens allows for a functional means of attaching these large, heavy pieces to the arm. The sturdy rawhide bases to which the skins are affixed are identical to those on the four-raven bustle and an Oto raven bustle collected nineteen years later by Friedrich Paul Wilhelm, duke of Württemberg, which are clearly intended to support the ornament against the wearer's back. Also, the three ornaments do not form an "ensemble" in the details of their construction. Subtle differences in fabrication, in the materials of which they were made, and in the designs on the quilled elements clearly indicate that the three were created by different makers. On the basis of Clark's statements and these other factors, there is every reason to believe that all three of the Lewis and Clark ornaments were made and used as bustles.

On what grounds Peale identified the single ravens as armbands is unknown. It is doubtful that he knew that raven belt ornaments were often composed of single birds.

If he assumed that bustles were constructed of multiple skins, as noted in Clark's journal entry, then he might well have thought that the single ravens he received were intended for attachment to the arms. Without guidance from Clark or Lewis he would have had no way of knowing the truth. And even though Willoughby cited the presence of similar ornaments in paintings by Charles Bird King and by Catlin, I have found no images that explicitly depict armbands of this type. It seems more likely that Willoughby interpreted imprecise imagery in the paintings as the armbands in question. I have discovered no other early descriptions that mention raven- or crow-skin armbands in association with bustles, and none is known to have been collected.

In considering tribal origin, we know that Lewis and Clark acquired these belt ornaments from the Sioux, based on Clark's inventory for the shipment of items to Jefferson more than a year after his return. However, as Castle McLaughlin noted earlier in this chapter, there is no direct evidence that they were obtained from Brules, and they cannot be so precisely identified on the basis of style alone. There exists the strong possibility that the bustles were acquired from friendlier Yanktons or Yanktonais on an earlier occasion, and even the remote chance that they were obtained from an altogether different group of Sioux, providing they were not collected during the expedition.

Still, it cannot be discounted that the acquisition of these remarkable objects corresponds to Clark's elaborate description in his journal, particularly in light of his later comments to Nicholas Biddle. That he was so impressed by the actions and bearing of these soldiers and so intrigued by their appearance indicates that he would have welcomed the opportunity to acquire the striking regalia they wore. Although tensions ran high during the expedition's encounter with the Brules, these Sioux also made efforts to extend at least some degree of hospitality, and the opportunity was certainly created for the presentation and exchange of objects. The Corps of Discovery attended a grand council in the Brule village and was entertained for two successive nights with scalp dances. Other instances in which raven bustles were presented to visiting dignitaries on similar occasions indicate that the tribes regarded them as valued and appropriate diplomatic gifts, and an acquisition from the Brules would have been in keeping with this practice.

I know of only six complete raven bustles apart from the three acquired by Lewis and Clark, a very small sample for comparative study. Three of these six were acquired from the culturally related and closely allied Sauks and Mesquakies (Foxes). This may simply reflect the collecting patterns of early-twentieth-century ethnologists, or it may be noteworthy in revealing that such ornaments were especially common among these peoples. The latter possibility seems to be supported by Catlin's paintings, in which the raven belts of Sauk and Mesquakie warriors are prominently featured, whereas such ornaments are notably absent in his depictions of most other groups. Still, raven bustles from other tribes have been collected and are referenced in Native oral traditions; they are also described in the literature and depicted in early paintings and drawings in relation to other groups. It is apparent that the use of these ornaments extended throughout the central Plains. It is also clear from the earliest surviving examples that a number of stylistic variations were created, which might reflect individual differences within a tribal group, different tribal styles, or an evolution in style over time. Therefore, the question of who specifically made and used these objects must be considered.

One of the difficulties facing any researcher of historic Native American art is the recognition that while a general material culture and many ceremonial forms were produced by every group in a region, other forms were tribally specific. Raven ornaments must be considered in terms of this dynamic. That the ethnographer Alanson Skinner (1913, 1921, 1924, 1926) was intensely interested in medicine bundles and war customs yet did not mention the earliest forms of raven bustles in this context in his extensive studies of the Menominees, Prairie Potawatomis, and Ioways seems highly significant. It is also noteworthy that none of the existing bustles seems to have been associated with or is known to have been collected from the Ojibwas, neighbors of the Dakotas (eastern Sioux) who surely produced them. At the same time, the Omahas, Poncas, Otos, Osages, Nakotas (middle Sioux), Lakotas (western Sioux), and possibly Pawnees—all Prairie and Plains peoples located west of the Missouri River—possessed the crow in varied forms. The apparent conclusion is that during the late eighteenth and early nineteenth centuries, raven and crow belt ornaments were of great importance for most tribes inhabiting the western Great Lakes, Prairie, and eastern Plains regions as a defining feature of their warrior complex and a significant part of their ceremonial structure. Conversely, not all groups in these regions originally created or possessed such objects.

As strong as this cultural pattern might appear, however, traditions defining the production and use of objects were never static. Any pattern discernible to us today might well have shifted over time as a result of tribal migrations, intertribal exchange, and the impact of Euro-American expansion. The creation of raven ornaments must be viewed in the context of the tremendous cultural transitions that took place during the eighteenth century as Great Lakes tribes migrated west and newly developing

Plains and Prairie cultures emerged. This epic transformation occurred in a very brief time, and mythic traditions, ceremonial forms, and artistic production all reflected the forces of continuity and change during this period.

The first of the six known surviving raven ornaments (in addition to the three from the Lewis and Clark expedition) was found among the contents of a war bundle acquired from Sauks in Oklahoma during the early years of the twentieth century by M. R. Harrington (1914:198–99, pl. 27); it is now in the National Museum of the American Indian–Heye Foundation. Another, also contained in a war bundle obtained from Sauks in Oklahoma—in 1922 by Alanson Skinner—is in the collection of the Milwaukee Public Museum (Skinner 1923–25:101, pl. 3). A third, now in a private collection, was acquired with related war bundle material at the Mesquakie Settlement in Iowa in recent years.

Another strikingly beautiful raven belt was collected by Paul Radin from Winnebago Indians in eastern Nebraska and sold to the Field Museum of Natural History in Chicago in 1908 (cat. no. 15495). This belt was undoubtedly transported to the Nebraska reservation from Wisconsin when bands of Winnebagos were forcibly removed to the west during the mid-nineteenth century. A fifth belt, collected by Friedrich Köhler, was accessioned by the Ethnological Museum Berlin (formerly the Museum für Völkerkunde) in 1846 (Bolz and Sanner 1999:84, fig. 64; Hartmann 1979:323, pl. 49). Its tribal origin is unknown, and current scholars usually attribute it to the Dakotas (eastern Sioux). Although this identification is probably accurate, it is by no means conclusive. These five ornaments are all distinguished by the elaborately quilled belt that forms the support for the raven or crow skin or other complex assemblage of materials that includes raven or crow feathers as principal elements.

The last raven ornament, also in the collection of the Ethnological Museum Berlin, was acquired by Duke Paul of Württemberg during one of his several trips to North America; it is believed that he obtained the object from Otos in 1823 (Hartmann 1979:324, pl. 50; Württemberg 1973:381). It was apparently stored in a parfleche cylinder that he also acquired at the time (Torrence 1994:35, 88, pl. 4). Like the Sioux objects acquired by Lewis and Clark, it does not include a belt.

It is beyond the purpose of this essay to describe each of these six bustles in detail, but a few comments comparing the group with the ornaments from the Lewis and Clark expedition as well as with one another may be revealing. The Lewis and Clark bustles and the Oto example collected by Duke Paul stand apart from the others as multiple or single bird-skin ornaments without integral belts. The three Lewis and

Clark pieces were created to replicate somewhat the appearance of the complete bodies of the birds, and their decorated forms remain the primary element of each ornament. The Duke Paul bustle, however, is an assemblage comprising a raven's head facing outward, clusters of raven feathers, a wolf's tail, and other decorative elements. The presumed Dakota bustle is similar in appearance, but the bird plumage ornament is attached to a finely quilled belt that features a delicate geometric design and quill-wrapped suspensions tipped with metal cones hanging at intervals along its bottom edge. The raven's head and cloth-wrapped body, which is surrounded and overlaid with clusters of its feathers, is affixed at the center. A large suspension of split owl feathers is attached at one side of the head, and strips of white swan's down are attached at the other.

The Winnebago belt is one of two with representational designs. Its striking image consists of ten deer within a geometric border; interestingly, those on the right side are rendered upside down. At the center is a single raven skin configured and decorated in a fashion similar to the single ravens of Lewis and Clark; the designs on the quilled slats, however, are exceptionally bold. The bird's body, wings, and tail project outward from the belt, and at the base, where it is attached, two swan's-down pendants tipped with owl or hawk feathers are suspended.

The Harrington bird-quilled belt obtained from the Sauks is predominately a deep red with symmetrically organized rectangular designs in black and white. At its center is the skin of a single crow without the head, which may be missing, with two bunches of red-dyed eagle down attached at the neck. Only one wing remains, and both it and the tail are ornamented with finely plaited quilled slats. In its present condition the crow hangs downward, but it was quite possibly intended to project outward like the others. Attached at the base are a large brass bell, four tiny bundles of roots or herbs, and quill-wrapped thongs with brass cones; to the side is a single, long, red-dyed hawk feather.

Both the Skinner and the Mesquakie bustles feature beautifully embroidered porcupine-quill belts, the first with a repeated geometric design running its full length and the second with an image composed of ten monumental thunderbirds. Hanging from the center of each is an assemblage combining a large cluster of crow or raven feathers and a wolf's tail as the principal elements. The Skinner belt is further embellished with pony beads, suspensions of red wool cloth, quill-wrapped rawhide loops, and quill-wrapped suspensions tipped with brass cones and horsehair. The featherwork of the Mesquakie assemblage is more complex and includes the additional feathers of eagle, hawk, and woodpecker, along with suspensions of copper cones and red-dyed horsehair. Attached at the base of the wolf's tail is a cylindrical roll of hide wrapped with pony beads and purple and white wampum.

In summary, all of these ornaments feature the raven or crow as the primary element. Three include wolves' tails, nearly all include additional feathers of owls, hawks, swans, or eagles as part of the assemblage, and nearly all are embellished with finely quilled slats tipped with ermine and red-dyed horsehair. A few have medicine packets, quill-wrapped rawhide loops, leather thongs with metal cones, and other singular attachments. But even with these subtle variations, all appear conceptually related. All are splendid and radiant in their physical substance—intricately constructed of diverse materials that possess an opulence of color, pattern, texture, and shape. And all possess and communicate a tangible sense of spiritual intent congruent with the concepts underlying their creation.

The six raven ornaments date from the last half of the eighteenth century to the early years of the nineteenth, as determined by either their collection history, the style of their decoration, materials, and construction, or the cultural context in which each was found. The Lewis and Clark raven bustles, which are the earliest documented objects of the type, were probably made within a twenty-five-year period prior to their acquisition in 1804; they are perhaps the oldest of the group, but this is impossible to determine with certainty. The Harrington belt also appears quite ancient; it almost certainly dates from the last half of the eighteenth century and could possibly predate those collected by Lewis and Clark. The others are roughly contemporary with the Lewis and Clark bustles or were made somewhat later, from the last decade of the 1700s to within the first three or four decades of the 1800s.

It is impossible to date these objects more precisely; too few pieces survive from which to determine firmly the characteristics of tribal styles, much less to reconstruct a pattern of stylistic evolution that corresponds to age. It is unlikely, however, that any of the raven ornaments I have discussed, or the types they represent, was produced after 1840. By the mid-nineteenth century, it is doubtful that the Sauks and Mesquakies or other Great Lakes peoples were still creating war bundles, and the raven ornaments of the Prairie and Plains groups were changing significantly in form, evolving into the familiar dance bustles associated with the Grass and Omaha Dances and with standardized warrior society regalia.

From this small group of nine surviving ornaments, the existence of two distinct regional styles can be suggested. Essentially, all those originating with western Great

Lakes tribes, east of the Missouri River, were constructed with an elaborately quilled belt as a major, integral element, and they originated as parts of the sacred war bundles in which they were kept. In contrast, those from Plains and Prairie groups west of the Missouri appear to have been conceived as singular bird-skin ornaments that were held in place with what is presumed to have been an undecorated hide belt—described as a girdle by Clark—that was not a permanent attachment. In these cases the power of the object resided exclusively in the feathered ornament, whereas for Great Lakes tribes, the belt, too, was charged with sacred meaning. It is possible, of course, that the Lewis and Clark ornaments were originally attached to quilled belts, but this seems unlikely. That the three are consistent in their construction and in the absence of a belt, and that Clark mentioned no decorative belts in his description, strongly suggests that they were collected as complete objects. This inference is further supported by the similarity of the Oto bustle acquired by Duke Paul. The object appears to be in excellent condition, and there is little reason to believe that it is not complete in every way, particularly in view of its having been collected within its parfleche container. Nor were the Oto and Omaha bustles observed by Major Stephen Long's party described as having decorated belts (Thwaites 1905a:235).

Encompassing these stylistic differences, all raven bustles consistently embody a set of interrelated religious and military concepts. Within this complex, three principal aspects of function and meaning are signified in the context of the ornaments' use: they were worn as sacred amulets in battle, they were worn in recognition of military achievement, and they were worn as badges of warrior society membership, civil rank, and appointed tribal responsibility. Some groups might have conceived of these three aspects as inseparably integrated; others apparently focused the symbolism in more singular ways. These different facets of meaning and function may represent different regional traditions that had evolved separately long before. More likely, they reflect more recent shifts in emphasis over time and region, from the earliest, most archaic traditions continuously maintained by the western Great Lakes tribes until the end of their active warfare in the mid-nineteenth century to the later practices of the southern Siouan tribes who resided on the prairies west of the Missouri River following their sixteenth- and seventeenth-century migrations. These Prairie groups, descended from Woodlands traditions, developed patterns of life that were similar in many respects to those evolving among the Plains cultures emerging throughout the eighteenth century.

At the time of the Lewis and Clark expedition, raven ornaments were essentially being created in two different cultural contexts, and they reflect the social, political, and military organizations of each. The warrior society complex with its standardized officer's regalia, which defined the use and meaning of raven bustles for Plains and Prairie tribes, did not exist among the western Great Lakes cultures. Conversely, there is no record of the wearing of bustles in war by the Prairie tribes, who conducted much of their fighting on horseback, whereas the early Great Lakes belts—those collected from the Sauks and Mesquakies and, it may be reasonably assumed, the Winnebagos—were powerful amulets worn into battle. They formed inseparable parts of complex war bundles obtained through individual men's vision quests and the blessings of Manitous. Such belts were worn by the men who possessed the bundles, the powers of which enabled them to achieve success as warriors and as acclaimed leaders of war parties, or by members of their parties who, under the protection and guidance of the bundle, had accomplished certain deeds. M. R. Harrington (1914:198–99), describing the Sauk belt he collected, wrote that "because it was considered a powerful amulet and to confer warlike powers on its wearer, it was donned only on the warpath, after the enemy had been sighted, and in the ceremonies connected with the [war] bundle."

Thus, for the Great Lakes tribes, these belts were physical manifestations of visionary power and the guidance of spirit helpers. They were the means whereby the powers of Manitous were focused and transferred to the individual warrior, and they were therefore central to the endeavors of war. This manifestation of spiritual energy lies at the very heart of the creation of these mystical objects and surely reflects the most ancient tradition. Only by extension did the belts symbolize war honors, leadership, and warrior society rank.

At least one dramatic record exists of a raven belt worn in battle. In an autobiographical drawing made about 1830, Wacochachi (Wa-Ko-Ba-Di-A), one of the principal Mesquakie leaders during the early decades of the nineteenth century and an important war chief of the Fox clan, depicted himself killing enemies of his people in two separate episodes (Torrence and Hobbs 1989:26–27, cat. 188). His extraordinary two-sheet drawing (see p. 123), undoubtedly one of the earliest to be executed with commercial materials, is archaic in style; the black figures are rendered as simple silhouettes, devoid of interior details and descriptive features. Still, the visualization is revealing. In both representations Wacochachi has indicated the presence of the raven belt he was wearing, which clearly defines the importance in which it was held. In the uppermost of the two images, he recorded that he was also wearing garters with long pendants (drawn as lines extending forward from the knees) of the type represented by Catlin in a number of his paintings and drawings of Sauk and Mesquakie warriors. That they appear in one episode and not in the other suggests that Wacochachi depicted himself wearing these ceremonial items in relation to specific events in time, and not as pictorial conventions of rank or identity.

Skinner's Sauk informants, too, connected the raven belt closely to active warfare, but they indicated an expansion of the belt's function. In discussing the belt he collected, Skinner stated, "Native opinions differ as to whether it was worn by the partisan [bundle owner and war party leader] before and during battle, or given to the first warrior to count coup to wear during the ceremonial return of the war party, and the subsequent dancing" (Skinner 1923–25:88). While this observation affirms the information obtained earlier by Harrington from his Sauk informants, who described the belt as a war medicine actually worn during fighting, it also references the tradition of wearing the belt in recognition of specific military achievements, the custom noted by others as the defining function and significance of bustles among the Prairie and eastern Plains tribes.

A variation of this Great Lakes tradition—the practice in which a single, honored warrior wore the ornament in recognition of a specific deed during ceremonies accompanying the return of a war party—is revealed in the description of "the dance of discovering the enemy," a ceremony that Long's party witnessed among the Omahas. According to the traditions of this Prairie tribe, the prerogative of wearing the belt was extended; it was associated with the more general accomplishments of war rather than with personal ownership or a specific honor relating to the powers inherent in a sacred bundle. Still, the concept is fundamentally the same:

> [T]he music strikes up, and a warrior advances, who takes a war-club and *crow,* provided for the purpose; the latter of which he belts around his waist. He then dances . . . exhibiting at the same time a pantomimic representation of his combats with the enemy. . . . The warrior then advances to the post, which he strikes with his club, and proceeds to detail one of his deeds of war. This done, the music recalls him . . . that he may continue his chivalric history. This . . . recitation continues until the tale of the warrior is told; when he resigns his *crow* and war-club to another. (Thwaites 1905b:126–27)

Clearly, in both the Sauk and Omaha cases, the wearing of the raven bustle was reserved for distinguished warriors who had demonstrated their bravery through graded deeds of war. It was this concept that underlay the bustle's use and much of its recorded symbolism among Prairie and Plains tribes, particularly in relation to the bustles associated with warrior societies and their regalia and to the "officers" and

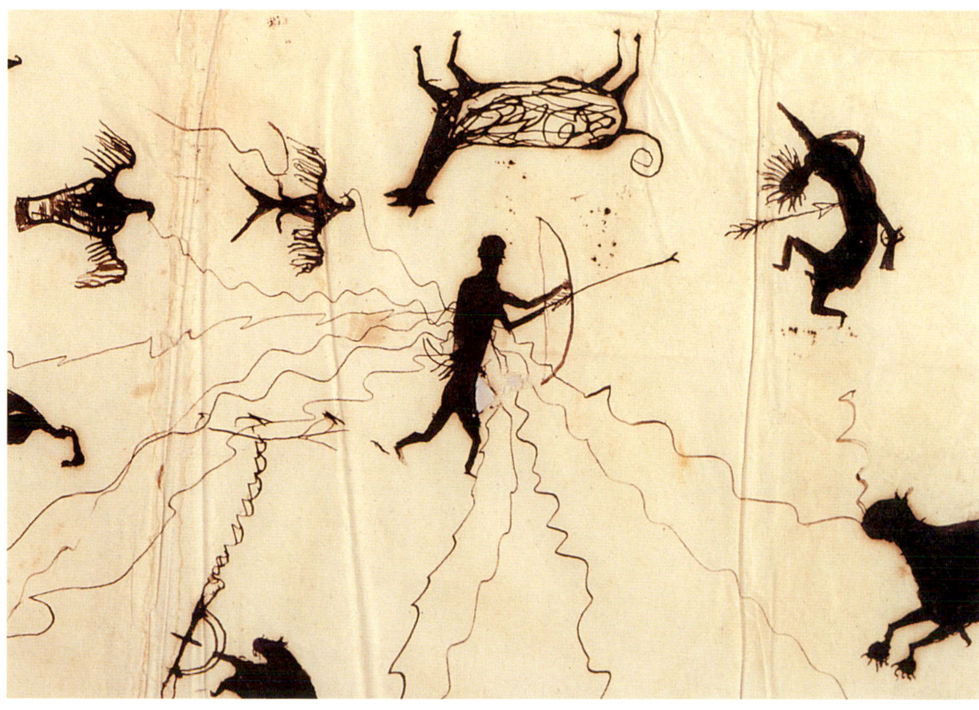

Drawing attributed to Wacochachi, Mesquakie, circa 1830. In this early autobiographical drawing, covering two sheets originally joined by sealing wax, Wacochachi depicted himself wearing a raven belt into battle on two different occasions. Collected by Colonel George L. Davenport.

"soldiers" Clark described. It was from among these distinguished men that such tribal officials were selected. As the most accomplished and respected warriors, they had the right to wear the crow, and this ornament, along with their symbolic black paint and raven headdresses, became synonymous with their rank and duty as tribal police.

It is apparent from all that has been written about these ornaments that they embody a complex of interrelated meanings that have evolved in varied forms over time and in relation to different cultural groups. One of most complete descriptions of the significance of the crow, its meaning, and its mythic origins was provided by Alice Fletcher and her Omaha collaborator, Francis La Flesche, in their monumental study of Omaha culture (Fletcher and La Flesche 1992 [1911]). Although their study was conducted during the last quarter of the nineteenth century, the traditions related to them by Omaha and Ponca religious leaders concerning the bustles derived from an earlier time. The concepts reflected in these traditions surely extended beyond the Omahas and closely related Poncas. They express the most fundamental ideas connected with raven ornaments as they came to be understood and conventionalized over time and within the formalized structure of Prairie and Plains warrior societies.

By the time of Fletcher and La Flesche's inquiry, the crow had changed in form and composition. Now, the entire skin of an eagle, including the head and tail, formed the central element, and two long hide or cloth pendants hung to the ground, covered with eagle feathers suspended in rows. Two spikes, or long, slender, decorated feathers, projected nearly vertically at the top. On the right hip was the tail of a wolf, and on the left, the entire skin of a crow. Thus, the warrior's association with the crow, raven, and wolf embodied in the earliest bustles was symbolized in this later form. Fletcher and La Flesche recorded that the crow and the wolf were believed to be in mythical relation to Wahon'da—the Power That Gives Life—and were pledged to help men in battle and in the hunt; crow ornaments were created to preserve the story of this bond.

The crow and the wolf, together with the eagle, who was connected with Thunder's destructive power, were strongly associated with war. They performed as scouts for war parties and hunters and had the power to reveal future events to the leaders. They were also equated with battle and death, in that they were the first to arrive on the scene following a conflict. Omaha and Ponca bustles were said to symbolize a battlefield after the battle was over, the fluttering feathers on the pendants representing the dropping of feathers from birds fighting over the dead bodies. Finally, it was related that the wolf and crow were not only connected with carnage but also had a mythic relation to the office of "soldiers." The men chosen for this honor were always selected from those entitled to wear the crow, and these regalia were worn as badges of office as well as during certain dances (Fletcher and La Flesche 1992 [1911]:441–46).

One of Fletcher and La Flesche's statements in particular expresses the reason the Omaha crow, the raven belt ornaments of Lewis and Clark, and similar feather bustles have persisted as symbols of mystery, power, and honor to the present day, "namely: That man is in vital connection with all forms of life; that he is always in touch with the supernatural, and that the life and the acts of the warrior are under the supervision of Thunder as the god of war."

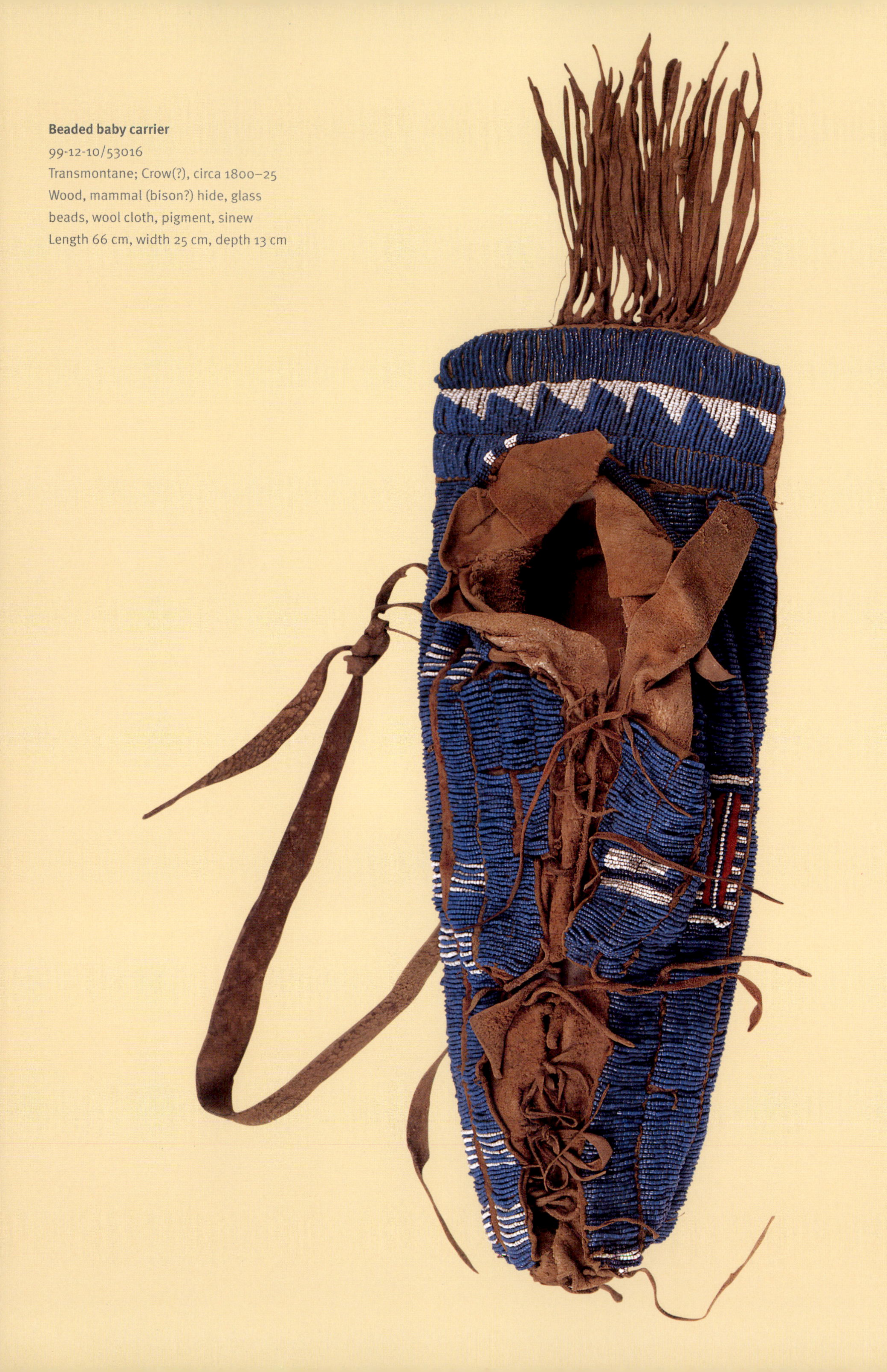

Beaded baby carrier
99-12-10/53016
Transmontane; Crow(?), circa 1800–25
Wood, mammal (bison?) hide, glass beads, wool cloth, pigment, sinew
Length 66 cm, width 25 cm, depth 13 cm

Chapter 6

The Army Moves West

The Curious Collection of Lieutenant George C. Hutter

THE PEABODY'S HUTTER COLLECTION CONSISTS OF SIX OBJECTS that were probably collected along the Missouri River by Lieutenant George Christian Hutter while he served with the Atkinson–O'Fallon expedition in 1825 and 1826. The Hutter family donated a collection of Native American and natural history objects to the Peale Museum in 1828. Because Hutter was married to William Clark's wife's niece and was closely associated with Clark in St. Louis, it is also possible that some of these objects originated in Clark's own collection. In this chapter, I present the extant objects that can be associated with Hutter and discuss their potential relationship to William Clark and the collection he developed in the years following the Corps of Discovery's expedition.

The Hutter-Clark Connection

In June 1828, Colonel Christian Jacob Hutter, a Pennsylvania newspaper publisher and state representative, deposited some thirty American Indian objects and a small assortment of geological specimens with the Peale Museum in Philadelphia. The Peale Museum ledger records that the materials "were collected by Lieut. Hutter US Army and Presented by his Father—C. J. Hutter Esqr."

Portrait of Lieutenant George Christian Hutter. Hutter served on the Atkinson-O'Fallon expedition up the Missouri River in 1825–26 and married William Clark's niece in 1830.

Nonetheless, because of connections between the Hutter family, William Clark, and Thomas Jefferson, scholars have long speculated that the Hutter items originated in either Jefferson's or Clark's collections of expedition materials. Charles Coleman Sellers, who championed this idea, represented a number of Hutter objects as having been collected by Lewis and Clark in his 1980 book *Mr. Peale's Museum.* Because new research confirms that George C. Hutter might well have acquired at least some of these objects himself, I present them here as a separate but related collection.

George Christian Hutter (1793–1879) led an active and privileged life. Born in Pennsylvania, he attended West Point and subsequently spent most of his life in the military. He served on the shifting frontiers of the nation, participating in the War of 1812, the Seminole and Black Hawk Wars, and the Mexican War. He was major paymaster of the U.S. Army at the outbreak of the Civil War and went under a flag of truce to pay off Major Anderson's command at Fort Sumter in Charleston harbor. When Virginia seceded from the union, Hutter resigned his commission and retired to Sandusky, the Lynchburg, Virginia, plantation he had purchased in 1841.

Hutter's life was intertwined with that of the family of William Clark. As a young lieutenant in the Sixth Infantry, he was stationed at Jefferson Barracks outside of St. Louis, where he served with Clark's oldest son, Meriwether Lewis Clark. In 1830, Hutter met and married Harriet Risque of Lynchburg, the niece of Clark's second wife, Harriet Radford. The marriage ceremony was performed at the Jefferson Barracks residence of the bride's uncle, James Kennerly, Harriet Clark's brother and a U.S. Indian agent and sutler, or provisioner to the army post. Both General Clark and his wife were in attendance. Ten years later, Hutter's half-brother, Edward Sixtus Hutter, married Emma Cobbs and moved to her family's Virginia residence, Poplar Forest. The Cobbs family had purchased the estate, which formerly belonged to Thomas Jefferson and his wife, Martha, from Jefferson's grandson Francis Eppes in 1828.[1]

Sellers (1980) speculated that Jefferson had kept ethnographic materials at Poplar Forest and that the Hutter family acquired them after Jefferson's death in 1826, probably through estate auctions. However, no evidence exists for the Poplar Forest collection that Sellers hypothesized, and documentation indicates that the household contents of Poplar Forest were sold *after* June 1828, when C. J. Hutter presented the Indian objects to the Peale Museum (Pond 1984; Stein 1993:457–58, n. 21).[2] Sellers's scenario seems chronologically impossible, considering that Peale received the Hutter donation several years before ties were established between the Hutter, Cobbs, and Clark families. The 1830 marriage date of George C. Hutter and Harriet Risque also undermines a Hutter family story that the painted bison robe given to the Peale Museum in 1828 had been a wedding present from William Clark's brother, George Rogers Clark.

It seems likely that William Clark, then governor of Missouri Territory, inspired George C. Hutter to make his own collection during Hutter's bachelor days in the Sixth Infantry. Following his expedition with Meriwether Lewis, Clark became a towering figure in St. Louis, entrepôt to the American West. Serving as territorial governor and superintendent of Indian affairs, he acted as the chief gatekeeper to the West for the duration of his life. He dispensed advice and granted formal passport to travelers bound for the Indian territories north and west of St. Louis, and he was centrally involved in the cultural politics of the burgeoning U.S. Indian trade. Known among Indians as the "Red-Headed Chief," Clark hosted Native visitors from throughout the West, many of whom traveled to St. Louis to conduct affairs of state and to consult with him about American demands for Indian land. Charged with implementing U.S. Indian policies under several successive presidents, including the taking of tribal land and the removal of Indian peoples farther west, Clark also traveled to Indian communities and to intertribal meeting sites such as Prairie du Chien and Portage des Sioux. His diplomatic roles and wide network of associates enabled him to build a large collection of Indian "curiosities," an activity that he seems to have pursued with growing purpose in the decades of his diplomatic career.

The brick walls of Clark's council chamber, which was attached to his St. Louis residence, were decorated with scores of natural history and ethnographic objects, many presented to him by Indian and Anglo-American dignitaries and friends. Few visitors to his St. Louis home failed to comment on the extent and variety of his collection, which included materials acquired during his trek to the Pacific Ocean with Meriwether Lewis in 1804–6 (Ewers 1967:48–72; Kennerly 1948:41–44; McDermott 1948). Henry Rowe Schoolcraft, who visited in 1818, noted, "We believe that this is the only collection of specimens of art and nature west of Cincinnati, which partakes of the character of a museum, or cabinet of natural history." Schoolcraft described it as being "arranged with great taste and effect" and commented that Clark evinced "a philosophical taste in the preservation of many subjects of natural history, together with specimens of Indian workmanship, and other objects of curiosity, collected upon his expedition" (in McDermott 1948:130). While serving primarily as a diplomatic venue, Clark's council hall seems to have had the eclectic character of a gentleman's private "cabinet of curiosity," mixing mineralogical specimens, portraits of Indian chiefs, animal skins, and mounted horns with calumets and peace medals (McDermott 1948, 1954). Because Clark granted free access to curious "people of respectability," his contemporaries in St. Louis often referred to this hall as a museum (McDermott 1948:129).

Clark's St. Louis "museum" is now considered to have been one of the earliest and finest in the United States and is recognized as having introduced to the West such well-known adventurers and collectors as Prince Maximilian of Wied-Neuwied and George Catlin (Ewers 1967). It is also credited with having served as an example to others, and Clark himself seems to have fanned the "great rage" for collecting Indian curiosities (Feder 1964:10) that developed during the 1830s. George Catlin, Nathan S. Jarvis, and the fur traders James Kipp and Alexander Culbertson were among many individuals whom Clark may have encouraged to assemble Indian collections.

An undated and incomplete catalogue of the contents of Clark's museum remains in the archives of the Missouri Historical Society.[3] According to this catalogue, many of the objects were presented to Clark by Indian leaders as diplomatic and chiefly gifts, given to accompany the consummation of treaties and other political agreements. Others were given to him by traders and U.S. military men. The written accounts of visitors to his council chamber stress the predominance of artifacts of Indian diplomacy, including treaty pipes, peace medals, and portraits of chiefs with whom he had negotiated (Ewers 1967). Clark's display of such objects testified to the national significance of Indian affairs and Indian leaders even as it graphically recorded the process of Anglo-American expansion and the consequent removal of Indian peoples from their ancestral lands.

Indian people were in fact an important audience for Clark's displays, and the portraits and pipes may have provided a familiar and reassuring environment for his Native visitors. Clark himself must have experienced his objects as mementos of real people, lived events, and diplomatic pacts. At the same time, he seems increasingly to have appreciated their value as artifacts, or at least "relics," of the tribes and groups that were being displaced by U.S. policy. He used his collection as a form of "social capital," to be shared and discussed with other members of his class and social circle; indeed, gifting has played an important role in the history of collecting. The sociable and generous Clark gave Indian objects to family members and to interested visitors, such as Duke Paul of Württemberg (Gibbs 1982) and George Catlin (Donaldson 1887; Ewers 1979). He also sent Native American objects to Jefferson in the years after the expedition.

The fate of Clark's collection, which disappeared after his death, remains a mystery (Ewers 1967; McDermott 1948, 1960). A persistent theory is that it was taken to Europe and sold, but portions of the collection may have been dispersed piecemeal. John Ewers believed that a few objects from Clark's collection might be in the Smithsonian, but he was unable to locate them.

The Hutter Collection

William Clark probably became acquainted with Lieutenant Hutter shortly after Hutter's arrival in St. Louis in 1824. The following year, Hutter participated in the Atkinson-O'Fallon expedition, a military venture undertaken by the Sixth Infantry to secure conditions throughout the upper Missouri region for U.S. trade interests. Promoted in part by Indian threats to American river travelers, the

The Missouri River in the 1830s, by George Catlin.

expedition had as its overt diplomatic goal to locate and treat with all principal Indian nations along the upper Missouri, thus reaffirming the allegiances forged by Lewis and Clark.

The Atkinson–O'Fallon expedition proceeded unhurriedly and without incident from St. Louis to the confluence of the Missouri and Yellowstone Rivers and back. Like other early U.S. military expeditions, the party to some extent shared the broad agenda that Jefferson had established for Lewis and Clark. While Indian diplomacy and the protection of U.S. interests were paramount, the mission also included exploration, mapping, and collecting. The anonymous author of the official expedition journal made repeated references to Lewis and Clark's written observations on and maps of the upper Missouri, noting subsequent changes in the riverine landscape (Reid and Gannon 1929). The journalist recorded geographical and natural history information, and while on shore, members of the party collected rocks, minerals, and a human skull (Reid and Gannon 1929:20, 34). Not only did members of the Atkinson–O'Fallon expedition revisit places familiar to Lewis and Clark and conduct similar business, but they were also aided by interpreters who had assisted Lewis and Clark, such as the mixed-blood Garreau family and Toussaint Charbonneau, the husband of Sakakawea.

Diplomatic interactions with tribal groups—which included both formal negotiations and trade—provided the officers and enlisted men with opportunities to acquire Native American objects. The expedition concluded twelve treaties with sixteen tribes; George C. Hutter was a signatory to four of them (Reid and Gannon 1929; Kappler 1975 [1904]:230, 241, 244, 246). Each treaty council concluded with a formal exchange of gifts between Indian leaders and U.S. representatives. All of the Sioux groups that signed treaties (listed as "Sioune" [Blackfoot band, according to Reid and Gannon], "Oglala," "Hunkpapa," "Teton," "Yancton," and "Yanctonies") presented the councilors with pipes, robes, and assorted garments and skins (Reid and Gannon 1929:23–29). The treaty councilors also "made" or recognized chiefs throughout the journey and attended dances and feasts during which gifts were often exchanged.

Enlisted men had opportunities for more informal trade. Writing at the Mandan villages on July 29, the expedition's journalist noted, "The Indians bring in quashes [squashes], corn, robes &c. & expose them for barter before the main ground—our market place at all camps we make before Indians" (Reid and Gannon 1929:35).

Three years later, on June 18, 1828, George C. Hutter's father delivered to Charles Willson Peale a variety of objects that he stated were "collected by Lieut. Hutter U.S. Army." As recorded in the Peale ledger, the items consisted of the following:

A Buffaloe robe on which is painted a battle between the Sioux and Arrickaree Indians, by a Sioux.

A Ladle made of the horn of the mountain sheep.

A Cradle—Crow Indian.

Knife scabbard. Chippaway.

Flute—Winnebago.

Tobacco pouch, made of the skin of a young Beaver, by a Crow Indian.

Mandan Warriors necklace.

Mandan chiefs Buffaloe robe.

Bone whistle—Mandan.

Rattle a musical instrument used in dances—Menomene.

Specimen of Petrified wood from the upper Mississippi.

Do Do [ditto] found on a mountain near the Mandan villages 150 [*sic*] above the water level of the Missouri.

Squaws necklace—Arrickaree.

Head dress worn by the Mandan beaus.

Otter skin tobacca pouch & Indian Tobacca-Omahans.

Mandan warrior's dress.

Sioux squaws dress.

3 Pipes, and 3 Pipe stems used generally by the Missouri Indians.

Specimen of the stone of which the Indians of the Missouri make their pipes.

Six pair of Moccasins from various tribes.

Two pair squaw garters—Mandan.

Three stones from Cannon ball river which empties into the Missouri.

Three Do [ditto, i.e., stones]—from the Big Platt.

Two petrified stones from Do Do [ditto]

Specimen of Lead Ore from Fever River.

A Bundle of various kinds of arrows

Some of the objects donated to Peale by the Hutter family are attributed to tribes that both met with Lewis and Clark and later signed treaties with the principals of the Atkinson-O'Fallon expedition. Others, such as the Crows and Omahas, met only with the Atkinson-O'Fallon party. Neither expedition met formally with Great Lakes and eastern Plains peoples, although Clark later did as superintendent of Indian affairs. The Winnebagos, Potawatomis, and Chippewas were active along the Mississippi River and into the Missouri drainages. Some of their objects might have been acquired by Missouri River tribes through exchange and then traded secondarily to Anglo visitors.

Many of the object types in the Hutter donation correspond to those acquired by Lewis and Clark and donated to the Peale Museum and to objects listed in the Clark museum catalogue. This might be accounted for by their proximity in time and place and by the similar nature of the encounters during which the objects were acquired. Because of this correspondence, a number of the former Peale Museum objects acquired by the Peabody Museum in 1899 *could* have been acquired by either Lewis and Clark or by Hutter. For example, both Lewis and Clark and Hutter apparently collected painted robes depicting battles involving the Sioux and a configuration of Arikara, Mandan, and Hidatsa villagers (see chapter 7). The close association between Hutter and Clark, and the similarity between their collections, makes it difficult to unravel the origins of their objects.

Assuming that George C. Hutter did assemble the objects his father presented to the Peale Museum, as the record states, we still know nothing of his motives for doing so. Hutter's unusual effort to place the objects with Peale's Philadelphia Museum indicates that someone conceived of a higher purpose for them than as simple mementos; it probably again signifies the direct influence of William Clark. Perhaps Hutter collected them as a way of enacting the identity of an enlightened army officer, with Clark as his role model. Or perhaps the objects that Hutter conveyed to Peale had been given by Indian leaders to the commanders of the Atkinson-O'Fallon expedition and were seen as diplomatic artifacts.

But the possibility that some of the Hutter objects originated in Clark's collection cannot be definitively ruled out, given Clark's sense of history, his proclivity for largesse, and the disposition of the objects at the Peale Museum. I have as yet uncovered no evidence that Hutter continued to acquire Indian objects throughout his subsequent military career, which included participation in several Indian wars. Was young Hutter a collector, a courier, or both?

A significant number of objects from Hutter's original donation to the Peale Museum seem to have survived multiple museum transfers, especially if several of the items described in chapter 7 were also acquired by him, and not by Lewis and Clark. Whether this simply represents historical accident or is the effect of past knowledge or intent, I cannot say. Regardless of their origins, Hutter's objects make up one of the earliest documented collections of western Indian materials remaining in North America, one that predates the florescence of collecting in that region by a decade.

Beaded Baby Carrier

This heavily beaded baby carrier (99-12-10/53016; see p. 126) is almost certainly the item Peale received from C. J. Hutter and catalogued as "a Cradle—Crow Indian." It is an early example of the U-shaped baby carriers used throughout the Rocky Mountain and Columbia Plateau regions (see Hail 1993:143), and the simple geometric design elements are typical of early beadwork. It was constructed by encasing a light wooden frame with hide, attaching a hide bag for the baby, and then adding to the surface multiple leather strips covered with beadwork. The beaded strips appear to have been "recycled," perhaps from a woman's dress or other garment. Gaylord Torrence (personal communication, 2000) ventured that the beaded strips might have been removed from a Plateau dress yoke. He pointed out the ease with which these beads might again have been recycled after the child outgrew the carrier. Because of that potential, Torrence feels that the baby carrier might have been collected immediately following its use-life.

Detail of red wool insert on the baby carrier 99-12-10/53016 (see page 126).

To support the baby's head, carrying straps and a collar made of rolled and beaded leather were also added. Bison hide may have been used throughout, and traces of red ochre pigment cling to the leather, which may have been smoked. Laced flaps, rather than the pairs of crossing straps typical of later Crow baby carriers, were designed to secure the infant.

The profuse, simple rows of blue pony beads create an almost tiled effect on the surface of the carrier. This number of blue beads was worth a small fortune in the early 1800s, and their use expresses extravagant regard for the carrier's occupant. At the time of the expedition, the Crow would exchange a horse for one hundred such blue beads (Wildschut and Ewers 1959:45).

Several unusual features make this carrier unique. Because of the reuse of previously beaded strips, the beadwork on each side is different, so that the overall design is asymmetrical. The beaded strips attached to the proper left side of the carrier are dominated by simple rows of large blue pony beads. In the center of the side is an insert of red wool trade cloth (a medium-fine broadcloth), outlined with blue and white bars, a device favored by the Crows, Blackfeet, and Nez Perce. Just above this insert are two rows of transparent navy blue beads bordered by partial and whole rows of white beads. A second, much smaller red cloth insert is obscured by the hide flaps of the head opening (see p. 135). Another area of navy-bordered white beads occurs at the foot of the carrier. On the right side of the piece, many of the beaded strips are dominated by white stripes. Also on the right side, "errant" seed beads—strawberry, white, green, and the same gray-blue beads that appear on the painted side-fold dress illustrated on page 176 (99-12-10/53046)—are lightly sprinkled throughout the blue. A row of white isosceles triangles creates a "sawtooth" design across the top of the frame.

The Atkinson-O'Fallon expedition negotiated a treaty with a Crow delegation at the Mandan villages on August 4, 1825, to which Hutter was a witness and signatory (Kappler 1975:244–46). It is possible that he acquired the baby carrier at that time. The formal features of this carrier, however, are more easily related to a regional than to a particular tribal style. Several contemporary viewers have suggested that it was made by a Shoshone, a Nez Perce, or even a Blackfeet woman.[4] As noted earlier, the wide, crossing flaps that are a hallmark of later Crow baby carriers are absent. The headboard shape, fringe, and tie thongs are more typical of carriers made by Plateau women, and the heavy use of red ochre and the hide color also seem uncharacteristic of Crow work. The sawtooth beadwork design was used by the Nez Perce (e.g., Dubin 1999:358; Peterson and Peers 1993:63). The term "transmontane" describes a shared aesthetic style that linked the Crows to the Nez Perce, Shoshones, and other Plateau and northeastern Great Basin peoples during the nineteenth century.

The apparent recycling of previously beaded strips in this piece conveys an impression of expediency, as though it were made in a hurry. Coupled with the carrier's probable Plateau origin, this impression makes it impossible not to wonder whether the carrier might have been made by or for Sakakawea, who traveled with the Lewis and Clark party from the Mandan villages to the Pacific Ocean carrying her newborn son. That son, named John Baptiste but affectionately called "Pomp" during the expedition, was later raised by William Clark in St. Louis. Clark's museum catalogue lists a "Cradle and belts," which might refer to the beaded strips on this baby carrier, but no such objects are included in the Peale memorandum of expedition materials. Although the baby carrier cannot be linked to the expedition, Sakakawea may have carried the infant Pomp in something similar when she left the Knife River villages and headed west with the Corps of Discovery in the early spring of 1805.

Beaded choker
99-12-10/53017
Upper Missouri River;
Mandan(?), circa 1800–25
Mammal hide, sinew,
glass beads, pigment
Length 17 cm, width 14 cm,
depth 4 cm

Portrait of Mandeh-Pahchu (Beak of the Bird of Prey), by Karl Bodmer, 1834. Bodmer portrayed this young Mandan man carrying a cedar flute and wearing a stuffed and beaded choker.

Beaded Choker

The item listed as a "Mandan warrior's necklace" in the Hutter donation to the Peale Museum is almost certainly this beaded choker (99-12-10/53017). The wide distribution and popularity of stuffed-hide chokers throughout Native North America suggest that they predate the contact era. Lewis and Clark described Shoshone collars as "generally round and about the size of a man's finger; formed of leather or silk-grass twisted or firmly rolled and covered with the quills of the porcupine of different colours" (Moulton 1988, 5:135). Large beaded chokers were very popular along the upper Missouri during the early nineteenth century; John C. Ewers (1968) described them as one of "three ornaments worn by Upper Missouri dandies in the 1830s." Both Karl Bodmer and George Catlin painted men wearing chokers of the same type. Yet despite their popularity, such ornaments are "exceedingly rare in museum collections today" (Ewers 1968:91). Ewers described only two known examples of double chokers, including this one from the Peabody collection.

The Peabody choker is made from two hide tubes that were stuffed with soft material, possibly hair, and then sewn together. A simple thong attached it around the wearer's neck. The bold blue, white, and black design of the attached beadwork was typical of the aesthetic of peoples along the upper Missouri during the early nineteenth century.

Eagle Bone Whistle

The Peale ledger lists a "Bone Whistle—Mandan" among the Hutter donations (99-12-10/53010), and the Peabody inherited a corresponding Peale label for a Mandan "war whistle" (facing page). Bone whistles are part of an elaborate and ancient bone-working tradition throughout the Americas. Prior to the introduction of Euro-American materials, Indian artisans made a great many tools, ornaments, and gaming pieces from animal bone, and this bone industry was highly developed among the horticultural peoples of the middle Missouri River. Bird bone whistles such as this one were used a thousand years before Lewis and Clark were born, and they are still used in ritual contexts today. Later in the nineteenth century, similar whistles were made of metal, and some were even carved from catlinite as trade items.

Single-hole bone whistles were not musical instruments but were sounded during battle and on ceremonial occasions. Prince Maximilian noted, "All the warriors wear small war pipes round their necks, which are often very elegantly ornamented with porcupine quills. . . . As soon as they advance to attack the enemy every one sounds his pipe" (Thwaites 1906a:350). Catlin believed that each end sounded a different note, one used as a signal for battle and the other as a signal for retreat (1973 [1844], 1:242–43). He also recorded that some were made from turkey and deer legs.

This whistle is nearly identical to the one worn by the Hidatsa leader Péhriska-Rúhpa (Two Ravens) in the famous portrait painted by Karl Bodmer less than a decade after Hutter visited the upper Missouri (see p. 258; Bodmer 1984:pl. 330). Mandan and Hidatsa men wore bird bone whistles as society badges, and Bodmer depicted Two Ravens in the regalia of a Dog Society dancer. Both Prince Maximilian and Frances Densmore (1923) described specific men's societies using the wing bones of geese, cranes, and swans for their whistles. Early in the twentieth century, Maxidiwiac (Buffalo Bird Woman), a Hidatsa from Fort Berthold in what is now North Dakota, described bird bone whistles to Gilbert Wilson:

> We also made whistles of wing bones of large birds, heron, geese, pelican, etc. for the Dog Imitators. They were long and sounded louder than others, and low. The Masukawadahi or Reckless Dogs used whistles of eagle's bones, shorter, about five inches long. These whistles hung from a thong from the neck. The thong was wrapped or whipped with bird quills. (Wilson 1911:290–91)

The Peabody whistle was tentatively identified by Carla Dove of the Smithsonian Institution as being made from the ulna of a bald eagle. The neck strap is of the type that Maxidiwiac described as worn by the Reckless Dogs for their eagle bone whistles. It was suspended from the wearer's neck by three doubled lengths of intermittently quill-wrapped leather thongs, ending in six quill-wrapped loops. The leather is colored with red pigment, probably ochre, and only a few remnants of wrapped quill remain. The still-extant quills on the pendant loops are dyed blue-green and are from the split shafts of bird feathers, which are more durable than porcupine quills. Mandan and Hidatsa women often used the quills of gulls (primarily Franklin and California gulls) that inhabited the riverine and wetlands areas of North Dakota (Halvorson 1998:3). As on Two Raven's whistle, blue pony beads are strung around the stem on a length of sinew that binds the wrapped thongs to the bone. A plug of resinous material in the interior shunts the

Eagle bone whistle
99-12-10/53010
Mandan, circa 1800–25
Bald eagle (*Haliaeetus leucocephalus*) ulna, undyed porcupine quills, dyed bird quills, mammal hide thongs, glass beads, pigment, resinous material
Length 27 cm, width 15 cm

Peale Museum label, "Mandan War Whistle."

airflow. According to Maxidiwiac, Hidatsa men used pine or spruce gum for this purpose.

Prince Maximilian and Duke Paul also collected quill-wrapped war whistles from the Mandans and Assiniboins (Schulze-Thulin 1987:92). Warriors may have presented such items as chiefly gifts, underscoring their status, and the military association might have appealed to Lieutenant Hutter.

Wooden Flute

The Peabody flute from the Hutter collection (99-12-10/53006) is one of the oldest surviving Native American flutes in a museum collection. One of several objects included in Hutter's donation to the Peale Museum that are attributed to Great Lakes peoples, this multinote flageolet is unusual and fragile. It was made from two pieces of hollowed wood, probably cedar, that were glued and lashed together with sinew. Seven

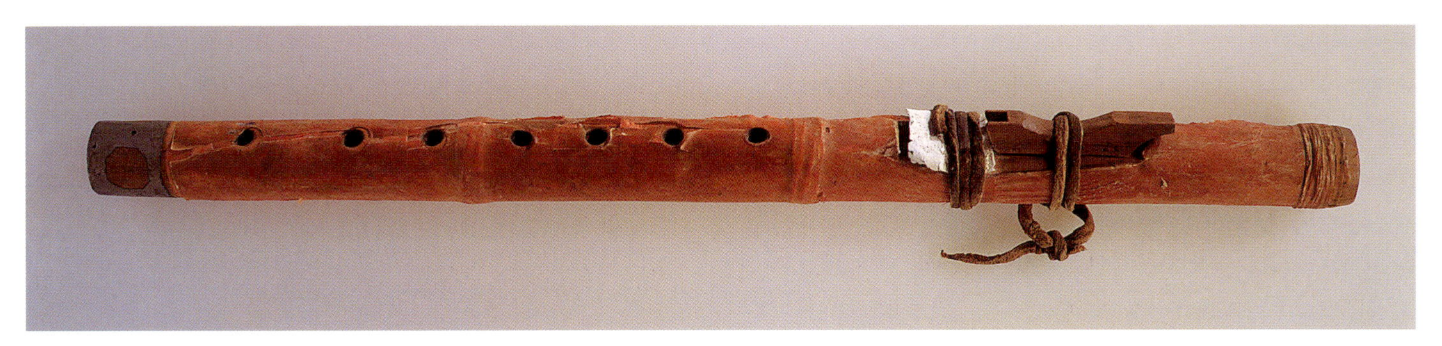

Wooden flute
99-12-10/53006
Western Great Lakes; Winnebago or Winnebago style, circa 1800–25
Wood (cedar?), mammal hide thongs, lead and/or lead alloy, unidentified intestinal(?) membrane, sinew, pigment
Length 41 cm, width 3.5 cm, depth 4.5 cm

holes are bored into the wood at intervals; Christian Feest (personal communication, 2002) notes that most courting flutes have six holes. The flute was painted red, and the entire body of the instrument was then covered with a thin, transparent membrane, probably taken from the intestinal sac of a large mammal. This delicate membrane, which was also pigmented, has now become brittle. Relatively few flutes now in museum collections have such membranes; a Mesquakie example is pictured in Torrence and Hobbs's *Art of the Red Earth People* (1989:no. 151).

Mandan-Hidatsa flute carver Keith Bear (personal communication, 2001) noted that the membranes had both symbolic and functional purposes. "In the old days, they used the heart sac and intestinal tubing from elk and buffalo to cover their flutes. Elk is love medicine, and these flutes are generally used for courting, for touching the heart. Also from the horse—the horse can be strong love medicine. When they wrapped that

MISSOURI MELODIES

Flute Player Keith Bear

Native American flute music has enjoyed a tremendous revival in recent decades and has been widely introduced to non-Indian audiences. Keith Bear, a Mandan-Hidatsa artist from the Fort Berthold Reservation in North Dakota, is one of today's best-known traditional Native American flute players. He also carves and decorates, or "dresses," his own instruments, skills he honed after receiving the right to make flutes from his tribal elders. Since the mid-1980s, Bear has released three compact disks, has appeared in videos, documentaries, and a feature film, and has performed at numerous regional and national cultural events. His CD *Earthlodge* (Makoche Recording Company) was awarded a Native American Music Award in 2001.

More important than these many honors is that young tribal members are beginning to come to him to learn how to make, handle, and play these ancient instruments. Bear says he committed himself to flute playing to keep a musical tradition alive, one he believes has the power to move minds and hearts:

Flute player Keith Bear in a reconstructed earthlodge.

> I was taught to carve these flutes by Carl Whitman. He said, and it was repeated to me even when I was a child, that these flutes are magic, they are powerful, they are medicine. They can take children and turn them into great warriors, and they can take our elder warriors and turn them into the young men and women they used to be. They can take those that are hurt and lost in their hearts and minds and take them to a beautiful place. These are a gift of life. They are the tree of life.
>
> The sound [of the flute] comes from the breath of the Creator, . . . and your breath is the water, the air you breathe is mostly water, so when this song comes out and it washes over you, it washes away pain and it should help people feel stronger, it should help them feel good. . . . I'm not really playing the flute, I'm breathing for the spirit of that flute.
>
> Every one of them has a spirit. I have twenty-two flutes, and each one sounds a little different, even though I try to carve them the same. They can really do some powerful things, so that's how I try to use them. Each one used to be a little branch, growing on a tree somewhere . . . and then they were chosen to do a job—maybe they were a house, or a bed, or a fencepost. I take some from doorjambs, some from windows, some from fenceposts; people have sent me pieces of log from a special tree on their land.
>
> And when I get hold of them, I give them another life. I take a piece of these things and I reshape them, because there's a spirit inside that wood. . . . And all I do is breathe for them, and they play themselves.
>
> Some are courting flutes, some are love flutes, I have one with two barrels, and I have one that was given to me that is a three-barrel flute.
>
> And how all these flutes and all this sound comes together is a medicine.

tubing around the flute, it would 'shrink-wrap' and help hold the pieces of wood together." The rectangular wooden slide or tuner placed above the stop of the instrument is carved to suggest a zoomorphic form.

As Keith Bear remarked, flutes were courting instruments: they were played by men to seduce their sweethearts into a rendezvous. According to Catlin (1973, 1:243), during the 1830s these instruments were called "deer-skin flutes" or "Winnebago courting flutes." The flageolet is one of two forms of love flute used west of the Great Lakes, the other being the more graceful Siouan single-note whistle, often carved in the shape of a bird or animal. Ewers (1986) suggested that both forms might have originated with the Crees or Ojibwas, who were considered masters of "love medicine."

By the early nineteenth century, flutes were popular intertribal trade items, and the "Winnebago" style was much in vogue with suitors along both the Mississippi and Missouri Rivers. George Catlin collected one and wrote, "In the vincinity [*sic*] of the Upper Mississippi, I often and familiarly heard this instrument, called the Winnebago couring [*sic*] flute" (1973, 1:243). Karl Bodmer painted a young Mandan man holding a flute similar to this one (see p. 137). While this flute probably originated in the Great Lakes and may indeed have been made by a Winnebago carver, Hutter could have acquired it in the Mandan villages, where materials from around the world changed hands.

Gourd Rattle

Listed in the Hutter accession at the Peale Museum as "rattle, a musical instrument used in dances, Menomene," this gourd rattle (99-12-10/53018) represents a type with great antiquity in North and South America. Gourds may have been the most widely distributed domesticated plants in human history, and they have been utilized in the Americas since at least 6000 B.C.E. (Heiser 1979:81). People around the world have used gourds to create rattles, drums, and even stringed instruments. Gustavo A. Romero, keeper of the Oakes Ames Orchid Herbarium at Harvard University, believes that the gourd used to create this rattle may be a hybrid of *Cucurbita pepo,* the species commonly known as summer squash (personal communication, 1999).

One of the intriguing aspects of both the Hutter and Lewis and Clark donations to the Peale Museum is their inclusion of objects from peoples living in the region of the western Great Lakes and upper Mississippi River. The Menominees, said to be the source of this rattle, were central Algonquian farmers and hunters who lived in villages

Gourd rattle
99-12-10/53018
Western Great Lakes;
Menominee, circa 1800–25
Gourd (*Cucurbita pepo*), mammal hide thongs, brass bells, dyed horsehair, sinew
Length 32 cm, width 12.5 cm, depth 12 cm

between Lake Superior and Lake Michigan. Like the neighboring Ojibwas, the Menominees became deeply involved in the Great Lakes fur trade conducted by the French and English during the seventeenth and eighteenth centuries. The Atkinson-O'Fallon expedition dealt formally only with Missouri River tribes, and Lewis and Clark interacted primarily with peoples on or west of the Missouri. Great Lakes and Mississippi River objects such as this may have been traded or brought into the upper Missouri region or even St. Louis. Hutter might also have acquired the gourd rattle and the flute while traveling up the Mississippi at some other time. Clark dealt directly with Great Lakes peoples throughout his career as superintendent of Indian affairs and governor of Missouri Territory.

This rattle was created by inserting a carved wooden handle and several small wooden plugs through the body of the gourd. A wrist loop was added to the proximate

end of the handle; on the other end, brass bells and dyed red horsehair were attached with sinew wrappings. Seeds inside the gourd, as well as the attached bells, created sounds when the rattle was shaken. Gourd rattles were used throughout the Great Lakes, the Northeast, and the Prairie region at the time of contact. They remain important elements in the soundscape of Native American culture and in social and religious life. In their study of Northeastern Indian music, Diamond, Cronk, and von Rosen (1994:68) reported that sound symbolism informed the Algonquian word for gourd rattle, *shishikun,* and that the Anishnabeks described that sound as "the sound of Creation."

Knife Sheath

This dramatically designed knife sheath (99-12-10/53027) may be one of the earliest known parfleche (bison rawhide) containers in any museum collection. Other early parfleche objects were collected by Prince Maximilian of Wied-Neuwied during the 1830s and by Duke Paul of Württemberg, who collected from the 1820s through the 1850s. It is also one of few incised parfleche objects believed to have been collected in the eastern Plains (Torrence 1994:85–86). The sheath is thought to be the one listed in the Hutter donation to the Peale Museum as "Knife scabbard. Chippaway," and the Peale Museum label for that object survives at the Peabody.[5]

Knife sheath
99-12-10/53027
Eastern Plains/western Great Lakes; Ojibwa(?), circa 1800–25
Bison rawhide, deer hide, dyed and undyed porcupine quills, dyed bird quills, tin and copper tinklers, sinew
Length 27 cm, width 8 cm

The technique of creating incised patterns in leather may be an ancient one, but it is found on few remaining rawhide objects. A related and more widely continued artistic tradition involves impressing drawn lines into hide and then filling them with hide glue (see "Robes" in chapter 9). Both incising and impressing hide can create alternating negative-positive designs such as the pronged triangular forms on this knife sheath. Both Gaylord Torrence (1994:35) and Ted Brasser (1999:49) have described such elongated triangular forms as characteristic of early-eighteenth-century Great Lakes painting. As Brasser illustrates (1999:figs. 6–8), the same motif was also used in men's body tattoos along the Great Lakes–eastern Plains border and may have been used in the Eastern Woodlands as well.

Norman Feder (1987) included this knife sheath in his discussion of a possible "Upper Mississippi" quillwork style defined by the use of bird quills dyed in a distinctive palette of green-blue, brown, orange, and natural brown or black. The rows of simple quilled bands creating blocks of color on the top of the sheath are composed of yellow-orange and white porcupine quills separated by deep orange, split bird quills. White porcupine quills are embroidered around the edge and are wrapped around the five pairs of hide thongs suspending handmade metal cones. The color scheme exemplifies the red, orange, yellow, white, and brown palette characteristic of early historic Woodlands quillwork. Although many of the objects in the Lewis and Clark and Hutter collections are decorated with metal cones or tinklers, the copper tinkler on this sheath is the only example of the use of that material and may link the sheath to the Great Lakes fur trade network.

The quilled portion of this knife sheath is reminiscent of nineteenth-century western Great Lakes–eastern Plains sheaths (e.g., Feder 1987:fig. 6), but overall it has a slightly more western character. The construction pattern of adding an inner "pocket" for the knife, thus creating a double "edge" from which a keyhole-shaped wedge is removed to accommodate a belt, is seen on later knife sheaths from the northern Plains (e.g., Hanson 1994:pls. 22, 24). A somewhat similar western Great Lakes or Dakota knife sheath collected by Duke Paul was pictured on the cover of *American Indian Art* magazine in the spring of 2000.

Pictographic bison robe
99-12-10/53121
Upper Missouri River; Mandan or Yankton (?), circa 1780–1825
Bison hide, deer hide, dyed and undyed porcupine quills, dyed bird quills, unidentified plant fiber, sinew, pigments
Length 259 cm, width 239 cm

Chapter 7

Enigmatic Icons

Objects Probably Collected by Lewis and Clark or by Lieutenant Hutter

AMONG THE MOST FAMOUS AND SPECTACULAR OBJECTS IN THE Peabody's "Lewis and Clark collection" are numerous items that cannot be firmly attributed to the Corps of Discovery: early-nineteenth-century robes and garments that were probably collected either during the Corps of Discovery's expedition in 1804–6 or by Lieutenant George C. Hutter in 1825–26. Both Lewis and Clark and Hutter acquired bison robes, women's dresses, and men's shirts from the Indian people they met along the upper Missouri River. Both exploring parties spent time in the Mandan villages, and each apparently returned from the upper Missouri with a painted robe depicting a battle between some configuration of Sioux, Arikara, Mandan, and Hidatsa warriors. Lewis, Clark, and Hutter deposited many of these objects in Charles Willson Peale's Philadelphia Museum.

In this chapter, I present evidence suggesting that the two side-fold dresses, the two robes, and the warrior's shirt that follow—all of them among the best and earliest known examples of their kinds—were acquired by either Lewis and Clark or Hutter. I also discuss why they have become icons both of the expedition and of Native American art history, and why they remain enigmatic today.

The battle scene on this robe includes sixty-four pictographs of warriors mounted and on foot. It was probably painted by as many as three different artists. (Detail of 99-12-10/53121; see p. 146)

Pictographic Bison Robe

The Peabody's elaborate pictographic bison robe (99-12-10/53121; see p. 146) has long been thought to have been acquired from Mandan Indians during the Lewis and Clark expedition. In the twentieth century it became one of the most recognizable objects in Native American art history. In his landmark exhibit catalogue *Sacred Circles,* Ralph T. Coe (1976:163) described it as "the earliest datable Plains artifact," one that "marks a beginning of our knowledge of Plains art." It has been represented in countless publications as quintessential of Plains painted robes, the Lewis and Clark expedition, and Plains Indian lifeways at the moment of contact with Euro-American society. Individual pictographs on the robe have been abstracted and reproduced to symbolize traditional Plains Indian culture and art. John Ewers, for example, isolated a pictograph of a mounted warrior for use on the frontispiece of his 1968 book *Indian Life on the Upper Missouri.* Peabody curator Charles Willoughby (1905:638), among others, felt that there could be "no question" about the robe's provenance.

My research, however, suggests that this robe might equally likely have been acquired by George Hutter in 1825. Both Lewis and Clark and Hutter donated robes to the Peale Museum depicting battles between the "Sioux" and the semisedentary Mandans, Hidatsas, and Arikaras. Clark also sent Jefferson a "pied buffalo robe" from Kentucky in the fall of 1807, in the same shipment in which he sent the raven objects described in chapter 5 ("pied" generally refers to a black-and-white spotted hide, like that of a paint or pinto horse). In 1826, Charles Willson Peale purchased a large collection of ethnographic materials from the upper Missouri, including "a buffaloe skin upon which is painted a record of the fight of Colonel Leavenworth with the Aricaree Indians executed by themselves" (Sellers 1980:252)—but this is unlikely to be the Peabody robe, which does not depict Anglo-Americans.

The documentary evidence seems to weigh slightly in favor of Hutter as the collector of this robe, although, as I discuss later, he could have donated a robe acquired by Clark. Unfortunately, neither the formal qualities of the robe nor the available documentation clarifies its collection history or the cultural identity of the artists who painted it.

During their winter with the Mandans, Lewis and Clark obtained a robe that they described on a packing list the following April as "painted by a mandan man representing a battle fought 8 years Since by the Sioux & Ricaras against the mandans, *menitarras* & Ah wah har ways (Mandans &c. on horseback)" (Moulton 1987, 3:331).[1] That robe, along with other ethnographic and natural history objects (including six other bison robes), was crated and shipped to Thomas Jefferson. The shipment arrived in Washington, D.C., in August and was transported to Monticello the following March.

Jefferson may well have retained most of the ethnographic items sent from Fort Mandan. In a letter to C. W. Peale, Jefferson indicated that he was forwarding to Peale the natural history specimens, but "there are some articles which I shall keep for an Indian Hall I am forming at Monticello" (Jackson 1978, 1:260). Correspondence between the two during subsequent months discusses only natural history objects, although Jefferson apparently did send Lewis's Fort Mandan inventory to Peale (Jackson 1978, 1:263, 264, 267). In his first letter to Lewis after learning of his safe return, Jefferson invited him and the Mandan chief Sheheke, then traveling east with Lewis, to visit Monticello so that they could "see in what manner I have arranged the tokens of friendship I have received[,] from his [Sheheke's] country particularly" (Jackson 1978, 1:351). Written accounts by people who visited Monticello after Jefferson's 1809 shipment of material to Peale, such as George Ticknor (1812) and Baron de Montlezun (1816) (both in Peterson 1989), indicate that as much as a full wall was devoted to the display of expedition materials and note the presence of painted battle robes.[2] Unfortunately, Jefferson's personal Indian collection has not survived (Chew 2003).

The Peale memorandum of December 1809, which itemizes the bulk of items acquired from the expedition, specifies only two robes: "A large mantle, made of the Buffalo skin, worn by the Scious, or Soue, Darcota nation," and "A small Mantle of very fine wool, worn by the Crow's nation Menetarre." Neither is described as Mandan or as painted. Jefferson could have turned more expedition robes over to Peale at a later date, but no such transfers were recorded.

In 1828, C. J. Hutter presented to Peale ethnographic materials, including a battle robe, on behalf of his son, George. According to the Peale ledger, the robe donated by the Hutters depicted "a battle between the Sioux and Arrickaree Indians, by a Sioux." As discussed in chapter 6, Hutter may have acquired his materials while serving on the Atkinson–O'Fallon expedition, but we cannot rule out the possibility that some of them originated in William Clark's St. Louis collection.

Recovering the history of this robe is further complicated by the knowledge that Clark personally retained Indian objects from the expedition and often gave objects to others. Recently published correspondence from Clark at Fort Mandan to his brother Jonathan in Kentucky reveals that at the same time Lewis and Clark shipped expedition materials to Jefferson, Clark had five "Buffalo Robes of Different figures" sent home. Two of them went on behalf of York, Clark's slave, to York's family, and another on behalf of John Shields to his wife. Clark himself sent his brother a Cheyenne robe and a Mandan robe, both of which were apparently painted with pictographs. Clark described the Mandan robe as representing "three actions between the Sioux & Mandans near this place, the Mandans on horseback and sucksessfull. the scalps represented [symbols resembling *C* and *O*] on the Sious of the robe" (Holmberg 2002:86).[3] Clark might have decided by 1828 that the battle robe should be in the Peale Museum, and he could have given Hutter items to deposit with Peale. But because Hutter, too, had occasion to collect objects directly from Indians, and his father conveyed them to Peale as such, the evidence suggests the existence of two different robes. If that is the case—and considering the uncertain fate of Clark's Mandan robe once it left Fort Mandan, in contrast to the high rate of survival of the Hutter materials—then this may well be the Hutter robe.

The robe collected by Hutter and the one on Lewis's Fort Mandan packing list chronicle battles between, respectively, the Arikaras and the Sioux and the Arikaras and Sioux versus the Mandans and their Hidatsa allies. These robes graphically depicted the unstable political economy of the upper Missouri region, a circumstance that was of keen interest to Lewis and Clark (Ronda 1984:42–112). Bands of Sioux, particularly Nakotas (Yanktons and Yanktonais) and Lakotas, both raided and traded with the semisedentary Mandan, Hidatsa, and Arikara villagers from the eighteenth century until the reservation era. The villages of these three peoples were centers of intertribal exchange and key nexuses in the English, French, and American fur trade.

Awatixa Hidatsa village on the Knife River, by George Catlin, circa 1832. Toussaint Charbonneau and Sakakawea are thought to have lived here when they joined the Corps of Discovery. Catlin drew the village shortly before it was destroyed in a Sioux raid.

Sioux-Arikara relations were particularly complex, turning on the differently structured positions of these groups in the regional political economy. Sioux groups were well armed with guns and ammunition, which eastern Dakota bands obtained directly from British traders on the Mississippi River in exchange for pelts. The guns were redistributed to Nakota and Lakota warriors at annual "rendezvous" held on the Minnesota and Des Moines Rivers, to which the western bands of Sioux brought produce, horses, and goods obtained from the Arikaras. Western Sioux bands needed Arikara goods with which to buy the guns that were in turn sought by the Arikaras, and they used force to intimidate the Arikaras from trading with others and to prevent Euro-American traders from reaching Arikara villages. The Arikaras supplied the Sioux with garden produce and trade goods and often participated in their military raids against the other village tribes while remaining themselves an object of Sioux aggression (Jablow 1994 [1951]:51–58).

Moving up the Missouri during the autumn of 1804, Lewis and Clark saw the remains of fortified villages abandoned by Mandan and Arikara communities that had been weakened by successive smallpox epidemics and driven upriver by Sioux attacks (see Mike Cross's essay in chapter 2). During the following winter, Mandan leaders such as Sheheke recounted for the captains the story of that tribe's northward retreat and reduction from nine villages to two (Moulton 1987, 3:233). Like their French and Spanish precursors, Lewis and Clark felt that the development of an American presence in the region depended upon neutralizing Sioux threats and undermining their "uneasy symbiosis" (Ronda 1984:48) with the Arikaras.

As volatile and intricate as were relations between the Arikaras and the Sioux, so, too, were those between the Arikaras and their fellow earthlodge villagers, the Mandans and Hidatsas. Lewis and Clark arrived during a period of enmity between them, which the captains tried to arbitrate, much to the displeasure of the Sioux. Lewis and Clark discussed the region's turbulent intertribal relations at length with their Mandan hosts, and it is possible that the pictographic robe they collected was presented (and even created) in response to such deliberations. Indeed, it might have been Black Cat (Posecopsahe), chief of the Nuptadi Mandans and one of Lewis and Clark's strongest supporters, who gave them the battle robe. On October 31, Lewis and Clark were invited to hear Black Cat's response to their general council of October 29, during which they had underscored the importance of peace with the Arikaras. Clark recorded in his journal that upon arriving at Black Cat's lodge to receive his thoughts on the prospect of intertribal peace, the chief "threw a Robe highly decoraterd over my Sholders" (Moulton 1987, 3:217), a gesture often indicating a gift presentation.[4] In another journal passage Clark was more explicit, stating that Black Cat "gave me a roabe" (Moulton 1987, 3:219). Later that evening, Black Cat and his sons paid a return visit to Clark, wearing clothing presented by the captains.

The robe that Lewis and Clark packed the following spring would certainly have depicted a battle in which the Mandans and Hidatsas were victorious. Although warfare was endemic, in 1833 Toussaint Charbonneau recounted for Prince Maximilian the story of a large battle that took place during the same year as the one depicted on the Mandan robe. Charbonneau, who accompanied Lewis and Clark to the Pacific and back with his wife, Sakakawea, had lived in the Hidatsa villages since before 1800. He told Prince Maximilian that in about 1797 an estimated thirteen hundred to fourteen hundred Sioux and seven hundred Arikaras attacked the primary Mandan village but were repulsed by a combined force of Mandans and Minnetarees (Hidatsas), who killed one hundred of their attackers, including the son of an Arikara chief (Thwaites 1906a:230).

After pipes, robes represent the largest category of objects that Lewis and Clark brought back from their encounters with Indian people (not all of which went to Peale). Robes were traditionally appropriate and highly valued gifts, the precursors to the quilts that are so widely exchanged on the Plains today. During the early contact period, Siouan-speaking communities in the Great Lakes and Plains regions often greeted non-Indian dignitaries by carrying them to council on buffalo robes. Nicholas Perrot (Hall 1983:3), Father De Smet (Moulton 1987, 3:n. 23), and Lewis and Clark (Moulton 1987, 3:22, 115–18) were all so honored; the Brule Lakotas carried Clark into camp on an "elegent painted B. robe" (Moulton 1987, 3:117–18). Guests were often given the robes on which they were seated in councils. The leaders of the Atkinson-O'Fallon expedition were honored in this way at the conclusion of their treaty councils along the Missouri River (Reid and Gannon 1929:27). While Lewis and Clark generally wrote sparingly, if at all, about the objects they received from Native people, they did record receiving robes from the chiefs of several tribes.

By the time Lieutenant Hutter visited the upper Missouri region twenty years later, the American fur trade had been established, and fancy robes had become market commodities. In 1833, Prince Maximilian observed that the American Fur Company paid six to ten dollars for painted robes (Thwaites 1906a:102), which were then resold to traveling Euro-Americans or to retail merchants in towns such as St. Louis. Lorenz Alphons Schoch, a Swiss dry goods merchant living in St. Louis during the 1830s, traded for and purchased a number of Plains and Great Lakes objects while visiting the Shawnees, Delawares, Kansas, Kickapoos, and Potawatomis during July 1837. His companions on that trip, which was staged from Westport, Missouri, included E. A. Johnson, Joseph Parks, one of three Findlay brothers, and John A. Sutter, all of whom were involved in the fur trade.[5]

Schoch acquired a painted warrior's shirt bearing pictographs very similar to those on this robe (Thompson 1977:figs. 85, 86). In his notes, Schoch described it as "a shirt of the Sacs, with the history of the campaigns of a chief"; Norman Feder attributed the Schoch shirt to the Santee Sioux (Thompson 1977:158). Schoch annotated a number of the objects he purchased as "de Cere," a possible reference to Pascal Cerre. The Cerre family was active in the fur trade for many years, beginning during the Spanish regime, and intermarried with the powerful Choteau clan. Prince Maximilian noted during his Missouri River journey that a "Sere" was trading with the Sioux (Thwaites

1906a:237). If the Peabody's painted robe and Schoch's painted shirt, now in Berne, Switzerland, were the work of the same artist or artists, presumably someone connected to the trade network, then perhaps this robe was collected by Hutter from the Sioux in 1825.

Exploit robes such as this one recorded feats of masculine valor and are one manifestation of a Plains art genre that has been called "biographical art" (Keyser 1987) and "warrior art" (Maurer 1992:188). Although biographical garments foreground individual men's military accomplishments, they were made and worn in social settings and so also served to record and evoke group history. The event-oriented, narrative style of historical Plains biographical art is believed to have developed from a prehistoric tradition of ceremonial rock art. The captains encountered such artwork on at least one memorable occasion. On July 25, 1806, returning east from the Pacific Ocean, William Clark scaled a "remarkable rock" along the Yellowstone River near present-day Billings, Montana, and came face-to-face with petroglyphs of animals incised into the wall (Moulton 1993, 8:225). Clark added his name and the date to the sandstone face and named the location "Pompy's Tower" in honor of Sakakawea's son. (Evidently Clark's editor, Nicholas Biddle, changed the name to "Pompey's Pillar" to give it a more classical cachet.)

William Clark viewed the painting on the pictographic robe that he acquired as testimony to the importance of warfare in Plains Indian culture, which he juxtaposed with "civilized" society. Nicholas Biddle, who published the first account of the expedition, included commentary on the robe, presumably given by Clark, in his 1814 *History of the Expedition under the Command of Captains Lewis and Clark.* Biddle incorporated this statement into his account of the expedition's activities on April 4, 1805, the day on which Lewis and Clark packed ethnographic and natural history specimens to send to Jefferson before they departed from Fort Mandan. Because the exegesis that Clark provided to Biddle represents the only known evidence that Clark speculated about the significance of Indian art and material culture in what might be called "proto-ethnographic" terms, I quote it in full:

> . . . also, a number of articles of Indian dress, among which was a buffalo robe, representing a battle fought about eight years since between the Sioux and Ricaras against the Mandans and Minnetarees, in which the combatants are represented on horseback. It has of late years excited much discussion to ascertain the period when the art of painting was first discovered; how hopeless all researches of this kind are, is evident from the foregoing fact. It is indebted for its origin to one of the strongest

Two mounted warriors, one riding a paint or "calico" horse, depicted on the Peabody's bison robe. (Detail of 99-12-10/53121; see p. 146.)

> passions of the human heart; a wish to preserve the features of a departed friend, or the memory of some glorious exploit: this inherits equally the bosoms of all men either civilized or savage. Such sketches, rude and imperfect as they are, delineate the predominant character of the savage nations. If they are peaceable and inoffensive, the drawings usually consist of local scenery, and their favorite diversions. If the band are rude and ferocious, we observe tomahawks, scalping-knives, bows, arrows, and all the engines of destruction. (Biddle 1904 [1814], 1:240)

John C. Ewers, a pioneer in the study of Plains Indian art styles, considered the Peabody's painted robe a "type specimen," a single object that is believed to exemplify the key features defining a class or type of object. Ewers felt that this robe typified "traditional Plains Indian painting, an aboriginal art untouched by stylistic influences from the white man's world" (Ewers 1968:8). Like Clark's understanding, Ewers's interpretation of the robe tends to essentialize Plains culture, emphasizing a distinction between precontact, or "traditional," cultures and those shaped by subsequent contact with non-Indians. Since Ewers's work, anthropologists and art historians have recognized that the term "traditional" cannot be limited to precontact culture and have refined their understanding of the evolution of Plains painting styles. The term "early biographical art" (Keyser 1987, 1996) is a better register for the genre typified by this robe.

Early biographical or narrative Plains art is typically "relatively flat, and semi-abstract in style" (Berlo and Phillips 1998:121), employing few of the techniques associated with Western realism, such as perspective and dimensionality. Artists depicted human and animal figures as simple geometric forms with few details and little variation between individuals, but they highlighted specific features such as hairstyles, shield symbols, and tattoos that denoted tribal and personal identities. As Ewers pointed out, many of the artistic conventions evident on the Peabody's painted robe are typical of protohistoric and early historic Plains Indian painting. These include the representation of the human upper torso and arms as a W or as linked V shapes, the depiction of horses' hooves as hooked (probably denoting tracks), and the use of dashed lines symbolizing bullet and arrow trajectories to link actors and convey the progression of action. Humans are presented frontally, and most figures are composed of triangle and line arrangements. These conventions also appeared in prehistoric rock art in the Great Lakes region and on the northern Plains. They gave way to more naturalistic painting styles as the nineteenth century progressed.

Among the Mandans, whom Prince Maximilian (Thwaites 1906a:263–64) considered to be exceptionally fine artists, specialists were sometimes commissioned to paint such robes (Bowers 1991 [1950]:92). The varying representational painting styles on this robe suggest that as many as three (male) artists may have participated in its creation. A woman or women would have executed the quillwork.

Evan Maurer (1992:188), Joseph D. Horse Capture (1998), Colin F. Taylor (1990:65–73), and others have ventured interpretations of this robe, all on the assumption that it was painted by Mandan artists. Careful consideration of the sixty-four pictographic figures reveals features such as the denotation of leadership roles and society membership, the pattern of action, the presence of ancient weapons such as the bow lance as well as trade muskets and powder horns, and other aspects that aid in understanding the story the robe recounts. If the cultural identity of the artist were known, these features could be more closely interpreted through correlation with ethnographic information.

If we could determine the cultural identity of the artists who painted this robe—Sioux? Mandan?—we would also know whether Lewis and Clark or Hutter brought it back from the Missouri. Unfortunately, determining the cultural identity of its painters on the basis of its formal elements is hampered by a number of factors. First, a review of early Plains narrative painting in diverse media such as garments and "winter counts" (Native calendars) reveals that many Plains Indian peoples used overlapping repertoires of visual symbols (see Howard 1976; Keyser 2000; Mallery 1972 [1893], 1987 [1886]). Tribal distinctions are particularly hard to discriminate for peoples who interacted as closely as did the Yanktonai Sioux and the Mandans, who shared many aspects of material culture. Consequently, while artists from different tribes employed common symbols and forms, the meanings of those symbols and forms may have varied. For example, the conventions that an artist used to denote "others," such as enemy groups, may represent either categories (e.g., all earthlodge dwellers) or specific peoples (e.g., the Arikaras). Even seemingly specific denotational devices such as hairstyles may function in this way. Indeed, one might argue that because Plains narrative art had a communicative function, the use of tribally specific formal vocabularies might have been counterproductive. We simply have too few well-documented examples of early-nineteenth-century Plains art to permit us to specify the relationships between regional, tribal, and individual components of the visual vocabulary. As artist Butch Thunder Hawk notes, "The clue to whether this robe was painted by a Lakota or a Mandan is in there somewhere. It's just a matter of finding it."

Plains Indian artists often depicted shamans and powerful warriors wearing their hair gathered into a knot above the forehead. (Detail of 99-12-10/53121.)

Objects that might instruct us through comparison are few, and those that exist only highlight such difficulties. Two other autobiographical garments from the Plains, the Berne shirt collected by Schoch near St. Louis in 1837 (Maurer 1992:186; Thompson 1977:figs. 85, 86) and another early robe in the Musée de l'Homme, Paris, for which there is no collection information

(Horse Capture 1993:102–3), seem to share the pictographic vocabulary used on this robe. Commonalities include representational styles such as conventions for depicting human forms, equestrians, and relationships; content elements—hairstyles, shield decorations, and armor; and color palettes. The design field of the Paris robe is also quite similar to that of the Peabody robe, and both robes situate the war leader or partisan in the same position. Unfortunately, the tribal origins of both the Berne shirt and the Paris robe are unknown, and in fact the shirt has been attributed to both the Sioux (Thompson 1977:158) and the Mandans (Maurer 1992:186). Prince Maximilian collected a "Dakota" (Sioux) coat with similar anthropomorphic figures (Hartmann 1979:figs. 11, 12), and a robe given to Maximilian by the Hidatsa leader Péhriska-Rúhpa (Two Ravens; see p. 258) bears similar "calico" (pinto) horses (Bolz and Sanner 1999:fig. 60).

Both the skinning techniques used to shape this robe and the execution of the pictographs indicate that it was created in the upper Missouri region during the first quarter of the nineteenth century. Ken Woody, of the National Park Service, studied and replicated the Peabody robe for the Lewis and Clark National Historic Trail Interpretive Center in Great Falls, Montana. He noted the presence of the following skinning and finishing techniques: double crescent head seams, sewn and quilled shoulder seams, the retention of dew claws, inverted ears, stake holes, and sewn nostrils (Woody 1998). The quill-wrapped deer hide fringe added to the neck is reminiscent of the fringing on early Mandan robes but, given our limited understanding of tribal styles, cannot be considered diagnostic. Woody also detected fringe remnants behind the left front leg. The rear edges of this robe are punctured with small, square holes, an embellishment also often in evidence on early Mandan robes but which could be an artifact of display. The head of the robe is split into two triangular portions. The robe is made from a single piece of hide—it was not taken from the animal in two parts and then resewn, as many were.

While visiting the upper Missouri tribes in 1832–33, Prince Maximilian remarked that the use of quilled center strips on bison robes (which hid the seam on two-piece robes) was "now old-fashioned, and was worn before the coloured glass beads were obtained in such numbers from the Whites" (Thwaites 1906a:263). The quillwork on this robe incorporates orange and white porcupine quills, blue-green and black bird quills, and an unidentified brown plant fiber. Although Mandan and Hidatsa women became noted for using bird (often gull) quills and plant fibers (notably hairy grama grass) in their quillwork, the Yanktonais, Suhtaios, Cheyenne, Assiniboins, and others

BUTCH THUNDER HAWK, PAINTER

Hunkpapa Lakota artist Butch Thunder Hawk grew up on the Standing Rock Sioux Reservation. A teacher of tribal arts at United Tribes Technical College in Bismarck, North Dakota, he has long used traditional Native American pigments in his work—including some of the same materials found in early-nineteenth-century hide paintings.

According to Thunder Hawk, "traditional art forms have been kind of forgotten. There are a lot of Native artists doing contemporary work, but not many understand traditional art forms and techniques. It's important for Native artists to bring them back to life. It's not just about making objects, it's about the values and spirituality behind the objects, and how to respect nature and use materials from Mother Earth.

"I teach to honor those older people that taught me. They shared with me, and I have to share with others. When I was young, one old man and one older lady showed me some painting techniques, taught me about the symbolism, and told me about the pigments: where to find them and how to use them. Then I went out on my own and tried to learn more. Some of the pigments that we used to use, such as a blue mud, are harder to find since the Missouri River was dammed in the 1950s. I found these yellow ochre pigments in the Black Hills and in Montana. A friend of mine in Arizona gave me the cobalt blue. One of the red earth paints is from northern California, and I got the other one in Hawaii—the Native peoples there use it for face painting and for painting tribal designs."

Thunder Hawk examined the Peabody's painted bison robe during a visit to the museum in 2001. "What caught my eye first about this robe were the shields—the green and red circle design is repeated several times. The things that are painted or colored are the important warrior society markings: the horses, the shields, the feathers coming off a lance, the sash. The rest of the figures are hardly colored at all. The main character is probably the man on the black horse wearing the feather bonnet. This might have been his story."

Butch Thunder Hawk with a shield that he created for Monticello, based on his personal vision.

Butch Thunder Hawk's earth pigments.

also used these materials. The quilled strips on the spine and shoulders of this robe are plaited in two- to four-quill diamond bands and simple bands. Fringe attached to the neck is wrapped with orange porcupine quills. The shoulder seams terminate in quilled loops, from each of which two thongs wrapped with orange and white porcupine quills are suspended.

Prince Maximilian observed Mandan robes painted with "representations of their heroic deeds . . . in black, red, green, or yellow figures" (Thwaites 1906a:263). According to analysts at the Canadian Conservation Institute, the Peabody robe was painted with a combination of indigenous earth paints and trade colors (Miller, Moffatt, and Sirois 1990). The brown and red-orange pigments are hematites, the yellow is iron oxide, the green is a copper salt compound, and the red is vermilion. The green pigment was probably prepared locally, perhaps from corrosion on a copper object. Vermilion was a trade item of long standing by the time of the Lewis and Clark expedition and one the captains themselves distributed (as well as red lead and verdigris). A protein solution, probably hide glue fixative, is also present on the surface of this robe.

Whether collected by Lewis and Clark or by Hutter, this robe remains one of the earliest extant examples of Plains Indian biographical art, and certainly one of the most elaborate. It powerfully evokes the worlds that Lewis and Clark entered—and in which they lingered—midway through their journey to the sea.

Biographical War Shirt

In 1980, Charles Coleman Sellers published a photograph of this early Plains Indian shirt (99-12-10/53041; see pp. 167, 169) accompanied by a caption suggesting that it might have been worn by William Clark (Sellers 1980:180–81). Since then, that association has become conventional wisdom. Images of the shirt, often described as Clark's, have been widely circulated in the burgeoning secondary literature on the Lewis and Clark expedition; Stephen Ambrose (1998:51) pictured it as Clark's "uniform" during the expedition. It has been replicated for several Lewis and Clark exhibits—by Larry Belitz for Monticello and by Ken Woody for the Lewis and Clark Interpretive Center in Washburn, North Dakota—and it was one of six objects illustrated in a Lewis and Clark slide set sold in the Peabody Museum gift shop.

Sellers's claim was based on a label at the Peabody Museum that may have originated at the Peale Museum, but the connection that he drew between the label and the shirt was an act of interpretation informed by Sellers's conviction that the objects donated to the Peale Museum by George C. Hutter had actually originated in Jefferson's private collection. Hutter's 1828 donation included a piece described as "Dress, Mandan warrior's." This undoubtedly referred to an *item* of dress rather than a woman's garment, and it might well have described a long hide shirt. Sellers made the assumption that this shirt was donated by Hutter but collected by Lewis and Clark.

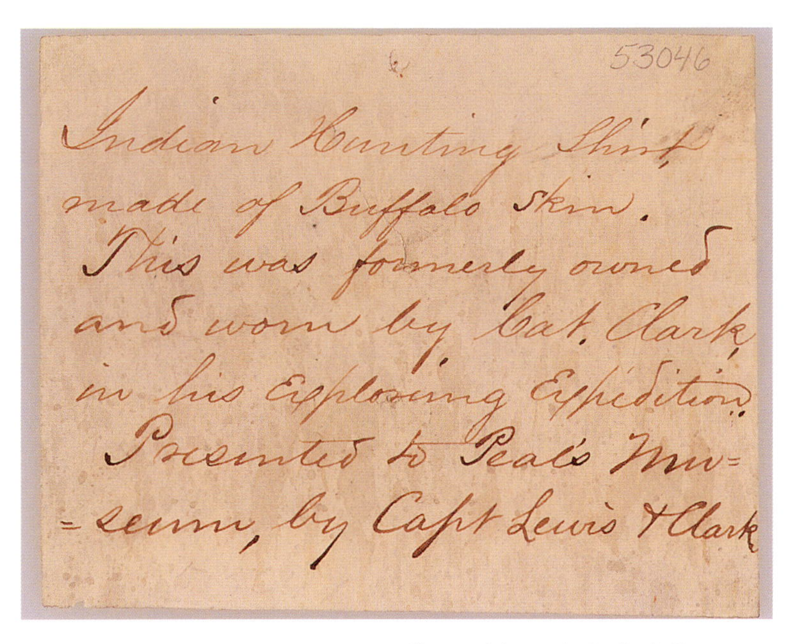

53046
Indian Hunting Shirt
made of Buffalo skin.
This was formerly owned
and worn by Cat. Clark
in his Exploring Expedition.
Presented to Peals Mu=
=seum, by Capt Lewis + Clark

Since arriving at the Peabody Museum in 1899, this Peale Museum label has been associated with several different objects, the biographical war shirt 99-12-10/53041 among them.

The handwritten label at the Peabody that caught Sellers's attention reads, "Indian Hunting Shirt made of Buffalo skin. This was formerly owned and worn by Cat. Clark in his Exploring Expedition. Presented to Peal's Museum, by Capt Lewis + Clark." The spidery handwriting has been tentatively identified as that of Franklin Peale, one of Charles Willson Peale's sons (though if it was his, then the misspelling of the family name seems curious). A second label in the same handwriting (see p. 172) reads: "Indian Hunting shirt formerly owned by Capt. Lewis.—Presented to Peal's Museum by Capt. Lewis + Clark." These two labels have also been associated with a pair of side-fold dresses at the Peabody. (For more on the labels, see the discussion of dresses 99-12-10/53046 and 99-12-10/53047 later in this chapter.) It may be that whoever catalogued the dresses mistook them for shirts—or perhaps, because only one shirt remained in 1899 when the Peabody received the Boston Museum materials, the cataloguer assumed that the labels belonged with the two similar garments. Understandably, many researchers have resisted associating the "hunting shirt" labels with the dresses, especially because the labels describe garments worn by Lewis and Clark. It is now impossible to say to which objects the hunting shirt labels were originally attached. In the absence of documentation, we do not know whether this shirt was collected by Lewis and Clark, by Hutter, or by another party who donated it to the Peale Museum.

The term "hunting shirt" is a Euro-American category, not a Native American one. Apparently "hunting shirts," inspired by Native American patterns (themselves influenced by European models) and generally made of linen, were designed for American Revolutionary War soldiers (Baumgarten 2002:69). On his "List of Requirements" for the expedition, drawn up in the spring of 1803, Lewis included "20 Fatigue Frocks

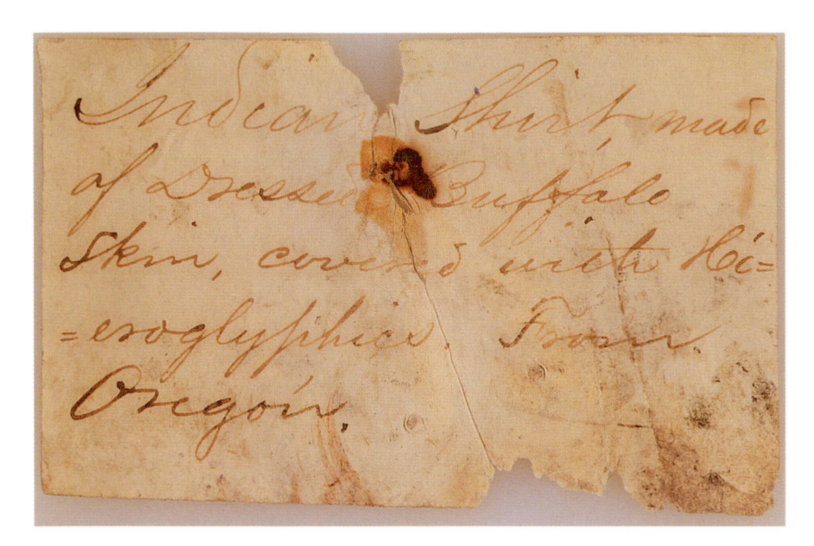
Indian Shirt, made
of Dressed Buffalo
Skin, covered with Hei=
=eroglyphics. From
Oregon.

Another Peale label that may originally have been linked to the Peabody's biographical war shirt.

or hunting shirts" (Jackson 1978, 1:70). Lewis and Clark, like other contemporary explorers, also used the term in their travel journals. Early curators at the Peabody Museum seem to have taken their cue from the old museum labels, for they also catalogued other Indian-made hide garments as hunting shirts.

Although Sellers inspected all of the Peabody's Peale Museum labels in order to match them with objects (and even commissioned personnel at Harvard's Fogg Museum to remove labels still affixed to objects in order to inspect their reverse sides for writing), he apparently disregarded a third Peale label that might refer to this pictographic shirt. That label, which is also handwritten, probably by a member of the Peale family, reads: "Indian Shirt, made of Dressed Buffalo Skin, covered with Heieroglyphics. From Oregon." The place name "Oregon" was already in use by the time of the Lewis and Clark expedition, and although the label does not so specify, it is possible that the "hieroglyphic" shirt, too, was collected by the exploring party.

In their journals, Lewis and Clark described themselves wearing Indian and frontier-style hide shirts during their rigorous journey. Such garments had obvious practical value. Members of the party both made and traded for durable clothing as they traveled. Dressing in Indian garments also expressed an identification with the people the Corps encountered. This was demonstrated in dramatic fashion during an incident in August 1805. After several failed attempts, Lewis and a small detail of men had finally made contact with the Lemhi Shoshones, who they hoped would help them cross the Continental Divide. Lacking an interpreter, Lewis struggled to win the confidence of Cameahwait's warriors, who suspected the Americans might be luring them into an ambush. When Cameahwait draped Shoshone tippets over the shoulders of the Americans to mask their identities in case of attack, Lewis understood the gesture and immediately reciprocated. In recounting the incident in his journal, he remarked, "I put my cocked hat with feather on the chief and my over shirt being of the Indian form my hair dishivled and skin well browned with the sun I wanted no further addition to make me a complete Indian in appearance" (Moulton 1988, 5:104).

On another occasion, Clark and Tunnachemootoolt, or Broken Arm, a Nez Perce headman, exchanged shirts while the Nez Perce were hosting the expedition members during the spring of their return journey. The Nez Perce cared for the expedition's horses while the Corps journeyed to the Pacific Ocean and back. When two young Nez Perce men returned some of the animals, Lewis wrote that he and Clark gave them ribbons, wampum, and vermilion, whereupon "one of them gave me a hansome pare of legings and the Broken Arm gave Capt. C. his shirt, in return for which we gave him a linin shirt" (Moulton 1991, 7:249). In presenting their own shirts to Lewis and Clark, Broken Arm and Cameahwait were enacting the diplomatic protocol expected of great men and leaders; these were "chiefly gifts."

Lewis and Clark might well have kept such shirts as mementos of their experiences and the relationships they forged with individuals like Broken Arm. But they might not have considered worn garments appropriate museum donations. It is also unlikely that they would have customarily worn a remarkable shirt such as this one. Presumably, Lewis did bring a hide shirt back from his journey. Upon his return to the East, he both had his portrait painted and was replicated in wax wearing the Shoshone tippet given to him by Cameahwait, over what appears to be a hide shirt. Charles Willson Peale executed the wax figure of Lewis and displayed it in his museum. An entry in the 1809 Peale memorandum listing "The Dress worn by Captn. Lewis &c." may refer to that Shoshone ensemble (see chapter 4). No other entries in Peale's museum ledger describe shirts, and the list of objects sent to Jefferson from Fort Mandan in the spring of 1805 notes only "some articles of Indian dress," without specifying their nature. This shirt might have been among those items or among the "Indian dresses" that Lewis sent to Peale before his death. Clark might also have sent it to Peale after the expedition, possibly through Lieutenant Hutter.

The construction technique, painting style, and quillwork on this shirt testify that it was created early in the nineteenth century. We can also be certain that it was the garment of a powerful and eminent man who had led several successful war parties. Determining the precise cultural origins of early men's shirts is more difficult, because shirts were standard trade items in intertribal exchange networks, and the shirt-making histories of particular groups are poorly understood. Norman Feder identified a similar biographical shirt collected by Lorenz Schoch as "Santee Sioux," and Barbara Hail (1993:74) discussed a regional "Teton/Assiniboine or Hidatsa/ Mandan type." The Peabody shirt has previously been identified as Mandan, as Sioux, and as Kiowa. It shares a number of features with other shirts that have been identified as Mandan or as "upper Missouri," such as fringed and crenulated edges, the use of the "quirt/man" symbol, the style of the quillwork, and the profusion of colored hairs attached to the sleeves (e.g., Horse Capture and Horse Capture 2001:55–59). Although few of the "Mandan" shirts available for comparison are well documented, the character of this shirt suggests that it belonged to someone who lived in the upper Missouri region, perhaps a "Mandan warrior" who conveyed it either to Lewis and Clark or to George Hutter in 1825.

The Peabody shirt exemplifies one of the earliest recognized patterns in Plains Indian men's shirts, the simple "binary" poncho style made by piecing two hides together, front to back (Hail 1993:68; Wissler 1915). Ted Brasser (1976:31) and others have written that the poncho style probably originated south of the central Plains, although southern Plains men seem to have preferred capes to shirts until well into the nineteenth century. The elaborate fringing on this shirt has also been suggested to have been a "southern" trait, which might explain an old Peabody identification of the shirt as "Kiowa." Early Plains shirts made of untrimmed hides such as this one are often called "deer leg" shirts. The unmodified hides of mammals of the order Artiodactyla, such as deer, antelope, elk, and bighorn sheep, were preferred for such garments and probably had symbolic significance. William Clark recorded a detailed description of the long, quill-decorated hide shirts worn by Shoshone men, noting that men's shirts and leggings often retained not only the legs but also the tail, neck, and hooves of the animal (Moulton 1988, 5:126–27). Such shirts were worn from the Columbia Plateau to the eastern Plains during the nineteenth century.

The extent and distribution of shirt wearing on the Plains at the time of Lewis and Clark's travels are unclear. Shirt wearing is believed to have come into vogue only in the late eighteenth century. When it did, shirts became vehicles for social symbolism formerly invested in body tattoos and painting. They also became favored exchange items and circulated widely through intertribal trade and gift giving. Initially they were worn by men who had achieved recognition in war. Among some central Plains groups such as the Lakotas, shirt wearing was a prerogative reserved for recognized leaders throughout much of the nineteenth century. When shirt wearing became more widespread, shirts remained markers of male social identity as defined by military prowess and membership in ceremonial associations. Because they heralded men's martial achievements, garments like this Peabody shirt are often called "war shirts."

This shirt was made from the hides of two animals, almost certainly deer (for the pattern, see Wissler 1915:51–52). The faint smell of the hides indicates that they were possibly smoke tanned. Holes remain where the hides were staked down for the removal of hair and flesh, and the necks and tails were not removed. A woman joined the two hides together by tying and sewing them across the shoulders, leaving the hindquarters of the animals pendant along the lower hemline. The two sides of the shirt are not sewn but rather joined by knotted thongs reinforced by light sinew stitching; such lacing was an early construction technique. A tail remnant dangles in the front and back. The hair, or grain, side of each hide forms the exterior surface of the shirt, so that the flesh sides of the animals' necks form flaps on both the front and back of the shirt's neckline. The trader François Larocque saw similar necks on Crow shirts in 1805 (Wood and Thiesson 1985:215); on later shirts, separate collars were added. Two epaulet-like strips of a much thicker hide embroidered with quillwork were then sewn with sinew to the arm sockets. A single blue bead was added with each stitch, creating a beaded border.

Strips of delicate quillwork, blue beads, and strands of colored horsehair on the shoulder of the war shirt. (Detail of 99-12-10/53041.)

Using thin sinew and fine stitches, the maker next attached the front quarters of the deer to the shoulders of the shirt, forming sleeves. With an awl, she punched a double row of holes through the hide along the outer edge of the sleeves, extending from the shoulder to the wrist. Two rows of long fringe were attached through the holes, one made from bundles of what are believed to be both human and horse hair and one of long, thin hide strips. These fringes were wrapped with sinew and quills, and the hair bundles were tied to one another inside the sleeves. The sleeves were partially formed into a tube shape by joining them with a threaded hide thong. The woman who made the shirt fringed all of the exterior edges, either by cutting them or, as at the tops of the shoulders, tying on short strips of hide. Clark Wissler (1915:53) considered such edge fringing an "early" treatment. Some pre-1850 Mandan garments, including robes, have crenulated edges. Later in the nineteenth century, northern Plains peoples such as the Blackfeet continued to fringe the edges of shirts.

A man or several men would have painted the dramatic war exploits depicted on both sides of the shirt, working in the semiabstract "early biographical" style characteristic of the period 1750 to 1850 (Keyser 1987). James Keyser and Timothy Brady (1993) hypothesized that different artists painted the front and the back of this shirt.

The human figure on this shirt is probably the shirt's owner, and the red sash signals his membership in a warrior society. (Detail of 99-12-10/53041.)

Opposite
Biographical war shirt
99-12-10/53041
Upper Missouri River; Mandan(?), circa 1780–1825
Deer hide, dyed and undyed horsehair, unidentified (human?) hair, dyed and undyed porcupine quills, dyed bird quills, unidentified plant material, glass beads, sinew, pigments
Length 119 cm, width 77 cm

A brown earth or ochre pigment was used to apply all of the primary images, which represent the shirt owner's honors in war. Stylistically, the painting is consistent with other works attributed to the Mandans.

The front of the shirt is dominated by a circular central image filled with dots and surrounded by "ball-headed" radiating lines moving both toward and away from the interior circle (facing page). (Structurally there is no difference between the front and the back of this shirt, but Peabody conservators T. Rose Holdcraft and Scott Fulton believe that the two sides can be distinguished by wear patterns.) Four concentric rings of short dashed lines encircle these radiating spokes, creating the appearance of a field of action. In other Plains pictographic paintings, such circles have been glossed as rifle pits, as war lodges, and as besieged villages (Keyser 2000:38–39). Beneath the central circle stands the solitary figure of a frontally facing warrior, perhaps the owner of this shirt. (Interpretations of anthropomorphic figures in Plains warrior art sometimes cast them as "ego," or the artist/wearer, and sometimes as vanquished foes. It is likely that artists used both strategies.) The figure is connected to the action above him by a loop of dashed lines. He has "spiked" hair—a style sometimes used to denote Caddoan peoples or earthlodge dwellers—and is armed with a quiver, bow and arrows, and a lance. A trailing red and white sash identifies him as an officer in a military society, and the horizontal bars drawn across one shoulder indicate his coup strikes or other war honors. Excepting the red pigment on the sash, all of the figures were painted with a thick, dark brown earth paint. On both sides, the upper third of the shirt was darkened with blue-black paint, a color widely associated in Plains societies with the powers of warfare and death.

On the reverse side, two rows of contracting vertical bars dominate the lower half of the shirt. The fifteen figures in the upper row, with attached heads, torsos tapering below the shoulder, and single feet, appear to be anthropomorphic and may represent

Opposite
Some of the tally marks on the reverse side of the shirt resemble abstract human figures, whereas others may be horse quirts (whips). Quirt symbols were used to indicate either horses captured or horses given away. (Back of 99-12-10/53041.)

vanquished foes. The similar figures beneath them, half of which are oriented upside down, may denote either human forms or quirts (horse whips). Upside-down anthropomorphic figures in Plains pictographic art are often interpreted as tallies of dead enemies, and quirts were sometimes used to represent horses stolen or given. In contrast to the footed anthropomorphic forms above, each of these figures flares out into a triangle at the proximal end, terminating in loosely drawn single strands (lashes) rather than feet. Variations on these quirt/men figures appear often on garments and robes identified as Mandan (e.g., Feder 1965:no. 16) or Hidatsa, including items collected by Prince Maximilian, and they also are painted on the side of a Hidatsa horse in a work by Karl Bodmer (Thomas and Ronnefeldt 1976:frontispiece). Above these figures, on the shoulders of the shirt, are semicircles, which might represent human heads or horse tracks, and another set of horizontal coup lines.

Unlike later shirts, which generally have quilled or beaded strips across both the shoulders and sleeves, this shirt has quilled strips sewn only over the shoulders. These strips are strikingly Woodlands in character, underscoring the relatively early date of manufacture. Thin, delicate porcupine quills were intricately worked into five flat rows of solid orange that change into solid yellow two-quill diamond patterns. In the second and fourth rows, the quillworker left fingers of yellow extending into the orange field and punctuated both color blocks with intervals of black and white diamonds. The technique she used to create the diamonds, now known as the "double-quill zigzag," involves superimposing one quill over the other to form opposing rows of triangles. The double-quill zigzag is a relatively uncommon technique (Orchard 1916:24). Peabody researchers have found it on a few early historic Woodlands and Great Lakes moccasins, pouches, and knife sheaths utilizing the same red or orange, black, and white color palette.

There are three possible explanations for the quilled strips on this shirt, and they are not mutually exclusive. The strips might have been acquired through intertribal trade, they might reflect the migration of Woodlands peoples onto the Plains, or they might indicate that the shirt itself is eastern Plains (Dakota?) in origin. The quilled strips are each bordered with a single row of blue glass beads; each bead is one to two millimeters long and two to three millimeters in diameter.

Porcupine quills, bird quills, and an ochre-colored plant material were used to wrap the bundles of hair cascading from the quilled strips and sleeves of the shirt (many of these quilled wrappings have been lost). Prince Maximilian observed of the Mandans: "A man may have performed many exploits, and yet not be allowed to wear tufts of hair

on his clothes, unless he carries a medicine pipe, and has been the leader of a war party" (Thwaites 1906a:349).

An intricate color symbolism may have informed the choices of hair and quilled hair wrappings on this shirt. There are three colors of what appears to be human hair—black-brown, dark brown, and light auburn—and at least two colors of horse-hair, gray and red. (There may also be small areas of black horsehair, but the hair on this shirt was not sampled for identification.) None of the hair appears to have been dyed. The hair bundles are wrapped primarily with white quill, blue-green bird quill, ochre quill, and a yellow ochre plant material. On the right sleeve, one bundle of (presumed) human hair is wrapped with quills dyed a deep red, while on the left sleeve, one bundle is accompanied by a split thong wrapped with orange quills.

Wolf Chief, a Hidatsa man interviewed at Fort Berthold in 1916, noted: "If the [war] party killed an enemy or enemies and captured horses also, the leader was let wear a man's hair or in place thereof, black horse's hair on his coat [shirt]; and white or colored red or yellow horse's hair for capturing of horses" (Wilson 1916b:13–14). Wolf Chief elaborated: "If a horse was taken, a bit of white or red or yellow horse tail hair was fastened to the shoulder—never black however for taking a horse; black meant that an enemy had been killed" (Wilson 1916b:30). Filling both sides of a shirt with hair required leading multiple war parties. The human hair used on men's garments was in some cases taken from the heads of slain enemies; in other cases it was donated by a warrior's compatriots or relatives.

Unfortunately, we do not know who the valiant man was who owned this shirt or in what community his achievements were heralded. But the garment seems to have been well worn, and red pigment remaining on the interior of the shirt recalls the wearer's presence.

Side-Fold Dresses

Like the pictographic robe and the warrior's shirt, two rare side-fold dresses are among the most famous but least understood items in the Peabody's Lewis and Clark collection. Charles Willoughby's 1905 article profiling the collection presented the dresses as expedition artifacts that Lewis and Clark had obtained from Cree Indians. Like the robe, they have subsequently been treated by many authors as type specimens and as chronological markers. Early in the twentieth century, they were marshaled as evidence in support of arguments on topics ranging from the pre-Columbian exploration of North America (Jackson 1917) to the chronology of trade goods (e.g.,

Side-fold dresses became "old-fashioned" a few decades into the nineteenth century. Lewis and Clark described seeing hide dresses that had been impressed with heated sticks. The impressions on the yoke of this dress were filled with "sizing" such as boiled hoof or hide glue to keep them defined. (Detail of 99-12-10/53047; see p. 181.)

Hanson 1975; Olsen 1963) and the evolution of Plains Indian garment design (Wissler 1915). They have influenced the identification of like garments in other institutions as well as generations of scholarship on Plains women's clothing.

But Willoughby's temporal and cultural classifications of these dresses are open to question. Neither the 1805 packing list itemizing cultural materials sent by Lewis and Clark from Fort Mandan nor the 1809 Peale memorandum listing objects received from the expedition includes explicit references to women's dresses. The Fort Mandan packing list indicates that "some articles of Indian dress" were folded into bison robes and shipped, but it provides no further details. In November 1809, C. W. Peale wrote to one of his sons that he had received a shipment of tribally unidentified materials following the unexpected death of Meriwether Lewis, including "Indian dresses" (Jackson 1978, 2:469–70). But the term "dress" was often used as a synonym for "garment." In his 1809 memorandum, written the following month, Peale did not specify the receipt of women's dresses from Lewis and Clark, and all of the items are accounted for by tribe. In that fairly detailed list, Peale seems to have used the term "dress" only in the general sense of "garment," in the entry reading "The Dress worn by Cap't Lewis &c." That entry may refer to the hybrid Shoshone and frontier ensemble in which Lewis was painted by C. B. J. Fevret de Saint-Memin (see p. 74) and modeled in wax by Peale.

We know, however, that the 1809 Peale memorandum was an incomplete inventory of the expedition objects held at that institution. We also know that William Clark did, in fact, acquire at least one, and probably two, Plains woman's dresses during the expedition. A card recently discovered in an 1809 scrapbook kept by the family of President James Madison reads, "Mr. Clark's compliments to Mrs. Madison, and request her to accept [one] of the greatest curiositys he met with in Louisiana—part of an Indian dress, such as the Soues woman ware in their dances."[6] Clark also sent a dress to his brother Jonathan from Fort Mandan in April 1805. In his letter to Jonathan, Clark described it as a "*Shirt* worn by the mandan & [word crossed out] Indian women of this countrey made of the Skins of the Antilope or goat" (Holmberg 2002:84). Nothing is known of the fate of the Dolley Madison dress, but it is tantalizing to speculate that she might have presented it to Peale, a family friend.

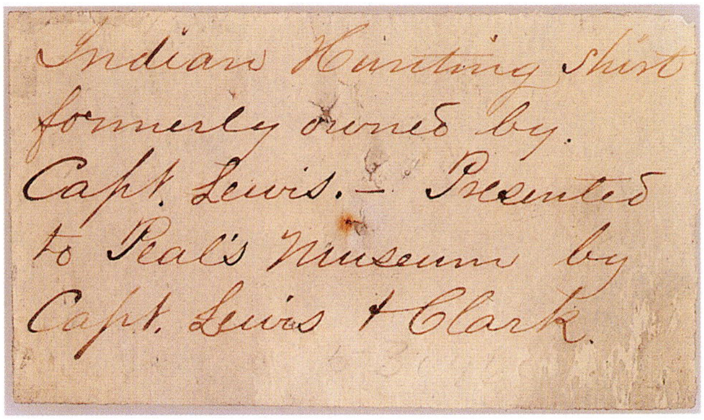

Indian Hunting shirt formerly owned by Capt. Lewis.— Presented to Peal's Museum by Capt. Lewis + Clark.

Peale Museum label for an unknown garment donated by Meriwether Lewis.

Peabody Museum documentation associated with these two dresses further problematizes their origins. In the accession files are two handwritten paper labels believed to be from the Peale Museum that have long been associated with the garments. One label reads, "Indian Hunting shirt formerly owned by Capt. Lewis.—Presented to Peal's Museum by Capt. Lewis + Clark"; it bears the later notation "53046." The second label (see p. 161) reads, "Indian Hunting Shirt made of Buffalo skin. This was formerly owned and worn by Cat. Clark in his Exploring Expedition. Presented to Peal's Museum, by Capt Lewis + Clark." It has been annotated "53047." Based on comparison with other labels at the Peabody identified by Franklin Peale's daughter Anna as having been penned by him,[7] it has been hypothesized that these were written by the same hand. Again, though, it seems curious that a member of the Peale family would have misspelled the family name; perhaps these labels were generated while the objects were in transit between institutions.

There is also evidence of past confusion at the Peabody about which dress each label refers to and whether one of the labels might be associated instead with the man's biographical war shirt described earlier in this chapter. Though it is apparent today that these are not shirts but women's dresses, it is impossible to reconstruct the origin and nature of this error. Were these dresses indeed once thought to be male shirts? Or did someone associate them with the two old labels because they were the only remaining pair of like garments? If the former were the case, it would not have been a singular mistake. Norman Feder (1984:51) examined a side-fold dress from a European private collection that had been inscribed "Male Upper Dress." We could be looking at a problem of terminology, because early European men seem to have often referred to Plains women's leather garments as "shirts." The explorer David Thompson, who visited the Mandans in 1797–98, described "the women's dress" as "a shirt of antelope or Deer leather, which ties over each shoulder, and comes down to the feet . . . the sleeves separate" (Tyrrell 1916:233). Lewis and Clark themselves described Arikara women's garments as "shirts" (Moulton 3:164), and, as noted earlier, Clark used this term to describe a Mandan woman's garment. If Lewis and Clark presented dresses to Peale but described them as "shirts," this language use could well be responsible for the subsequent confusion with regard to their documentation.

In 1828, the Hutter family deposited at the Peale Museum a "Sioux squaw's dress." Because many of the objects that Hutter donated to Peale seem to have been transferred to the Boston Museum and then to the Peabody, it is possible that one of these two Peabody dresses was collected by Hutter. If so, it is likely to have been 99-12-10/53047 (p. 181), which is formally related to other side-fold dresses reliably attributed as "Sioux" that were collected during the early 1830s. Because Clark is also known to have collected a Sioux dress, and given that a dress very similar to 99-12-10/53046 (p. 176) was collected before 1800, it is also possible that Lewis and Clark acquired both of the Peabody dresses.

Side-fold dresses are a rare and early style about which little is known, but they are thought to reflect the westward movement of Indian peoples from the Woodlands to the Plains. Clark Wissler (1915), John Ewers (in Feder 1964:22), and Norman Feder (1984) all postulated a historical relationship between the side-fold design and the early, perhaps Algonquian style known as the (shoulder) "strap dress," which often had detachable sleeves (see Gelb 1993:139; Feder 1964:fig. 22; Taylor 1997:figs. 10–16; Thompson 1977). The trader Alexander Henry saw Cree women wearing strap dresses in 1775 that he thought were made of bison or elk skin. Henry observed: "The . . . garment covers the shoulders and the bosom: and is fastened by a strap which passes over the shoulders. . . . The arms, to the shoulders, are left naked, or are provided with sleeves, which are sometimes put on" (in Bain 1901:247). Wissler (1915)

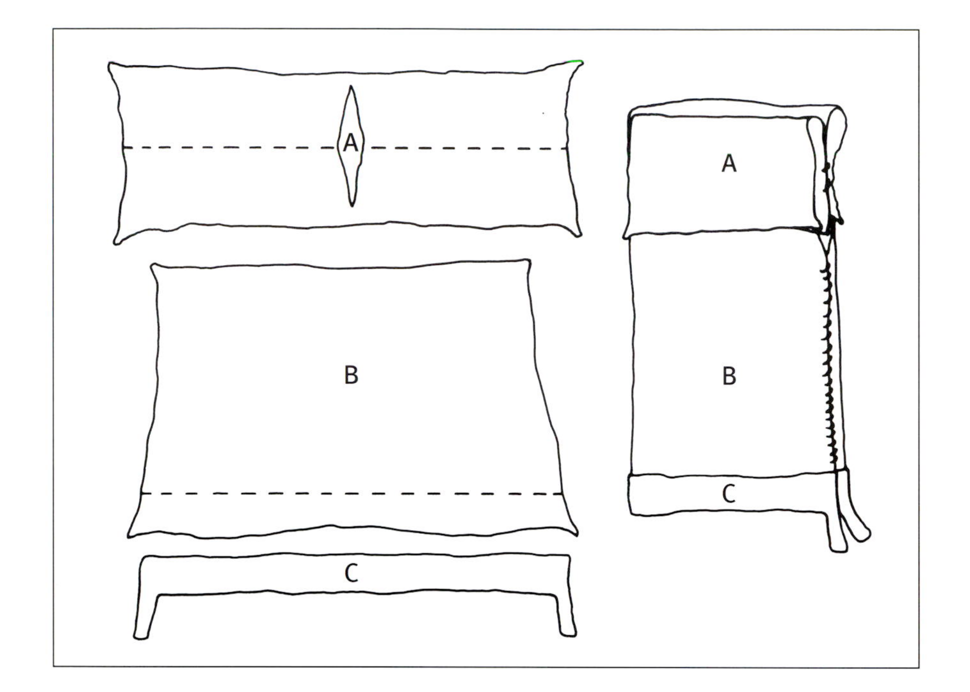

Pattern for a side-fold dress made from three pieces of hide: *a*, yoke; *b*, body or skirt; and *c*, hem piece.

and Rosemary Lessard (1980) suggested that the side-fold pattern, defined by a folded breast flap or yoke and a single side seam, was inspired by the shape of a wrapped body robe. Barbara Hail (1993:89) noted that the side-fold style might also have developed from the eastern wraparound skirt, the earliest form of Native women's garment observed by European colonists. It remains unclear whether this dress type represents regional or tribal styles, a stage in the evolution of dress design, membership in a particular society, or some combination of these factors.

Side-fold dresses are so little understood that it is not entirely clear how they were worn. In seeming contradiction to the account of side-fold dresses given to George Bird Grinnell (1923, 1:57) by Cheyenne women, Feder (1984:49–51) concluded that the vestigial shoulder strap on several examples (including 99-12-10/53047) was worn on the right shoulder, with the open seam of the yoke to the left of the wearer. The two Peabody dresses seem to support this interpretation.

Feder (1984) summarized the existing information on the ten examples of side-fold dresses then known in museums worldwide, the majority of which are in European collections. Provenance is sketchy for most examples. Two may have been collected in the 1780s, and Prince Maximilian (Feder 1984:fig. 4; Hartmann 1979:fig. 29) and Nathan S. Jarvis (Feder 1964:fig. 11, 1984:fig. 8) each collected one during the early 1830s. Tribal affiliation is less well understood. Three of Feder's examples seem to be Cree, and a doll in a London museum dated about 1780 is thought to be wearing an Ojibwa version of a side-fold dress (Feder 1984: fig. 7). On the basis of the Prince

Maximilian and Jarvis dresses, which Feder judged to be "well documented as Sioux," he considered several other examples to be Lakota or Nakota as well (Feder 1984:55). A graphic record also documents side-fold dresses among the Lakotas. Karl Bodmer, Prince Maximilian's artist, painted a Lakota woman at Fort Pierre in what is now South Dakota on June 1, 1833, wearing a tabbed side-fold dress decorated with beads, brass buttons, and iron tinklers (Bodmer 1984:pl. 192). That dress, although partially obscured by a wearing robe, looks very similar to 99-12-10/53047.

Plains women's dress patterns were undergoing transition at the time of Lewis and Clark's travels. Although Prince Maximilian observed at least one woman still wearing a side-fold dress in the early 1830s, other documentation from that decade makes it clear that most Plains women already preferred the later two-skin dress pattern, made from matching hides of mule deer, elk, or bighorn sheep (see Hail 1993:88–89). Side-fold dresses span the late eighteenth and early nineteenth centuries, and so the two Peabody examples might have been acquired by either Lewis and Clark or, several decades later, by Hutter.[8]

Because personal clothing is a point of mediation between the self and society, worn garments are among the most elegiac of museum objects. Women's dresses are integral aspects of their social identity and lived experience. We cannot fully reconstruct the contexts in which side-fold dresses were worn, know what meanings they held for the women who made and wore them, or understand all they may have signaled in the social lives of which they were a part. But clearly, both of the Peabody dresses were made with a great deal of care and attention and were meant to be worn on special occasions. They evoke collective historical and cultural experiences and bear the imprints of individual lives. Traces of red ochre, perhaps residual body paint, linger on the collar area of the quilled dress (53047) in the Peabody collection, a poignant reminder that this beautiful garment was once animated by a woman's body and spirit.

Painted Side-Fold Dress

This unique and enigmatic dress (99-12-10/53046; p. 176) is one of only two known side-fold dresses that are painted. Of those two, it is the more elaborate. Willoughby may have considered it to be Cree because of its affinity with Naskapi Cree hide-painting traditions and because the related strap-style dress persisted among the Crees and Assiniboins, some of whom visited the Mandan villages while Lewis and Clark were wintering there. But as Feder (1964) points out, this dress seems quite unlike others of Plains Cree origin (e.g., Brasser 1976:pl. 96). It is most similar to a side-fold dress probably accessioned into a European collection in the 1780s and now

Painted side-fold dress
99-12-10/53046
Central or northern Plains,
circa 1790–1825
Large mammal hide (elk?), pigment,
glass beads, dyed porcupine quills,
unidentified plant fiber, sinew
Length 121 cm, width 66 cm

in the Musée de l'Homme (Feder 1984:fig. 9). Unfortunately, the Musée de l'Homme dress is undocumented. Both it and 99-12-10/53046 are geometrically painted in red and brown linear bands accented by forked curvilinear designs, and each has a quill-wrapped fringe hem.[9] Although he doubted Willoughby's Cree attribution for this dress, Feder (1984:55) could venture only that the two dresses might represent a common but as yet unidentified tribal style.

The cultural origins of both Peabody side-fold dresses have generated a good deal of speculation. In a letter to the Peabody, the late John Ewers also expressed skepticism regarding their Cree attribution. Ewers suggested that they might have been obtained from the Mandans or Hidatsas, noting that women in those cultures had worn strap dresses.[10] When queried during the research for this book, both Ted Brasser, former curator at the Canadian Museum of Civilization, and art historian Gaylord Torrence suggested that this dress might be of Siouan origin. Both Brasser (1999) and Torrence (1994) have called attention to the widespread traditions of abstract painting (and tattooing) on bodies, hides, and garments that were being practiced by Native peoples in the Northeast, western Great Lakes, and Midwest when Europeans arrived in North America. Few three-dimensional examples of this aesthetic have survived in museum collections, and geometric hide painting persisted as an important art form primarily on the Plains and in the boreal forest.

On the basis of their analyses of early painted objects in the collections of European institutions such as the Musée de l'Homme, Brasser and Torrence emphasized that many of the design principles and elements now associated with later geometric painting traditions in the American North and Northeast were also present in the repertoires of earlier, more southern painters such as the Illinois peoples. Implicit in their argument is the notion that existing museum collections, most of them assembled after the mid-nineteenth century, as well as continuous traditions of hide painting (such as among the Naskapi), have obscured our awareness of earlier styles and their distribution. Of this particular dress, Brasser (personal communication, 2000) stated, "The painting of the Peabody dress and the Paris example both represent the same style. The closest similarity is with the painting on some skin coats, collected in the Minnesota-Manitoba region, 1780–1800. The style is intermediate between the northeastern boreal forest ('Naskapi') and the central Plains."

Gaylord Torrence emphasized that this dress had been painted as elaborately and carefully as a robe, utilizing a similar visual syntax of organization and vocabulary of elements. He was struck by how well thought-out the design seemed to be, suggesting well-established conventions of spatial organization. Like Brasser, Torrence (personal communication, 2000) felt that the linear bands of color forming the major designs on the dress were an early and widespread pictorial element that might reflect diverse influences on historic Plains Indian painting:

> Because the Missouri River villages were a major trade center and, as a result, a confluence of various artistic styles deriving from so many different contacts, I am inclined to believe this dress may have come from one of these groups—Mandan, Hidatsa, or Arikara. I also consider the strong possibility that it may be from the Plains Cree or Assiniboine, or the lesser possibility that it might come from the Arapaho or Cheyenne—but if so, I am perplexed by the beadwork.

Torrence pointed to the unusual delicacy of design and subtlety of color characterizing this dress as atypical of central Plains tribes such as the Sioux, Cheyenne, and Arapahos. In particular, central Plains groups generally produced beadwork in a more contrastive color palette. Regarding the sea-green and blue-gray beadwork on this dress, Torrence (personal communication, 2000) noted: "This coloration of closely related middle tones is related more to the work of the Crow and Missouri River tribes, rather than to that of the Sioux, Cheyenne, or early Arapaho. In this respect it is particularly consistent with later pieces of beadwork and quillwork collected during the last half of the nineteenth century at Fort Berthold from the Mandan, Hidatsa, and Arikara (see, for example, Penney 1992:pls. 91, 131, 132)."

This dress is made from three primary hide pieces, possibly elk, pieced together with the hair side out: a skirt, a yoke or breast flap, and a fringed hem. In addition, a narrow beaded appliqué strip is attached to the lower skirt. As on many side-fold dresses, pendant tabs hang from the seamed corner of the hem. Sinew thread was used throughout. Two pigments were used to paint the dress: a red vermilion and a brown earth or ochre paint. The painting, which has been impressed into the hide, does not seem to have been executed with a stylus, as the lines vary in width and course. The hem fringes were originally wrapped at the top with dark brown plant fiber and (now) orange porcupine quills; only fragments of these materials remain.

The beadwork on this dress is unexpected in several respects. "Lane stitch" beadwork, or short rows of beads, appears along both sides of the collar, on the breast, down one sleeve, and across the lower skirt. The rows of bar and band designs, derived from earlier quillwork motifs, are characteristic of the early "pony bead period" (Hail

This unusual painted dress may once have had a shoulder strap. (Detail of yoke, 99-12-10/53046.)

1993 [1980]:54; Lyford 1940:66). However, most of the beads used (which range from 1.5 to 3 mm in length) would be classified as the smaller "seed" beads. Conventional wisdom has long held that seed beads arrived on the Plains later than pony beads, but this chronology is now recognized as oversimplistic. Lewis and Clark themselves distributed a variety of bead types and sizes, including yellow and white seed beads (Moulton 1987, 3:495–98).

Torrence points out that the analogical color scheme of the beadwork on this dress, which relates beautifully to the muted hues of the painting, is unusual on a garment of this vintage. A single row of lane-stitched seed beads stretches across the shoulders below the collar. Against a field of blue-gray beads are alternating designs of green horizontal and vertical bands organized into groups of three. Two small, blue and green concentric circles are beaded on the yoke, which is also edged with beads; an "overlay" or spot-stitch was used to anchor the concentric circle motifs on the yoke.

Near the hem is a double row of lane-stitched beads that form green hourglass shapes separated by vertical bars, a design element widely used in early Plains beadwork. According to Carrie Lyford (1940:66), the hourglass was a "much used motif" in early Sioux beadwork; among the Cheyenne, the hourglass was a women's design (Coleman 1980:56). The beadwork on the yoke was applied directly to the brain-tanned dress leather, whereas that on the hem was first sewn to an additional, attached strip. In places, the beadwork seems to have been applied before the dress was painted, because some of the beads are smudged with red paint.

Portrait of Chan-Chä-Uiá-Teüin (Woman of the Crow Nation) by Karl Bodmer, 1833. Bodmer painted this Lakota woman, wearing a side-fold dress and a "box-and-border" robe, at Fort Pierre.

"Sioux-Type" Side-Fold Dress

As discussed earlier, the existing documentation on the Peabody's side-fold dresses is inadequate for determining with certainty either the culture of origin or the collector. In terms of formal characteristics, however, this dress (99-12-10/53047; facing page) is more familiar than the painted dress, with clear affinities to examples in other museum collections. It is most similar to a side-fold dress collected by Prince Maximilian, identified as "Dakota" and now at the Ethnological Museum Berlin (Bolz and Sanner 1999:fig. 54; Feder 1984:fig. 4). Both the Berlin dress and the Peabody dress are included in a group of side-fold dresses that Feder (1984) tentatively termed "Sioux-type." The best documented exemplars of this style are the Berlin dress and another acquired by Nathan S. Jarvis, both collected during the 1830s (Feder 1984:figs. 4, 8). Another is the dress worn by Chan-Chä-Uiá-Teüin, the Lakota woman painted by Bodmer at Fort Pierre in June 1833 (Bodmer 1984:pl. 192). Prince Maximilian described the woman as the principal wife of a man he visited. His written description of her dress, which he tried unsuccessfully to purchase, could easily apply to 99-12-10/53047:

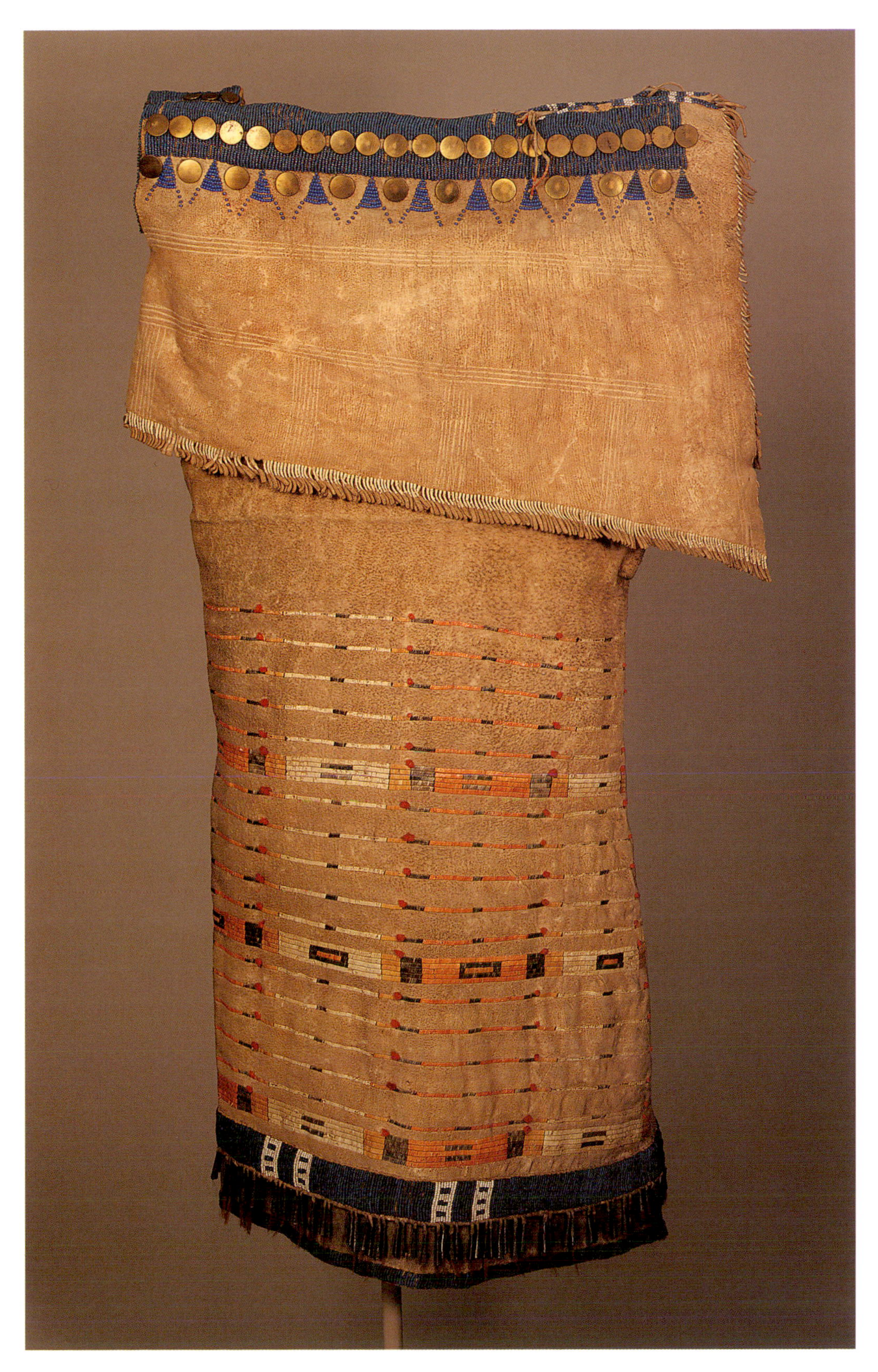

"Sioux-type" side-fold dress
99-12-10/53047
Central Plains; Lakota or Cheyenne (?), circa 1800–25
Large mammal hide (elk?), dyed and undyed porcupine quills, dyed bird quills, gilt brass buttons, cowrie shells (*Cypraea moneta*), glass beads, tin-plated sheet-iron tinklers, horsehair, unidentified plant fiber, woven cotton tape, wool cloth, sinew, glue sizing, pigments
Length 125 cm, width 75 cm

> . . . she wore a very elegant leather dress, with stripes and borders of azure and white beads, and polished metal buttons, and trimmed as usual at the bottom with fringes, round the ends of which lead is twisted, so that they tinkle at every motion. Her summer robe, which was dressed smooth on both sides, was painted red and black, on a yellowish white background. She estimated all these articles very highly. (Thwaites 1905d:321)

As a group, Feder's Sioux-type dresses are united by the presence of a shoulder strap, decoration of the skirt with horizontal rows of dyed porcupine and bird quills accented by red wool tufts, the use of metal tinkler cones and pony beads, decoratively impressed yokes, an added "flounce" on the skirt, and, generally, pendant tabs on the hem. An additional example of this style, collected by the artist George Catlin during the 1830s and now at the Smithsonian, was unknown to Feder when he published his survey (Taylor 1997:fig. 28).

Feder (1984:55) acknowledged the inadequacy of the small and poorly documented universe of known side-fold dresses for defining tribal styles. Both Colin Taylor (1997:34) and Wingfield Coleman (personal communication, 2000) believe that this dress, 99-12-10/53047, may be Cheyenne or Suhtai. But since several similar dresses were acquired from Sioux women, it could be the "Sioux squaw's dress" that George Hutter deposited at the Peale Museum in 1828.

Irrespective of tribal origins, this dress can be read as a site of worldwide intercultural contact and the creative reconfiguration of culture. In pattern, sensibility, and design the dress is indigenous—even "traditional" or conservative relative to the "deer tail" style that came into vogue on the Plains after the first quarter of the nineteenth century. But into this traditional pattern have been incorporated new materials from distant parts of the world. These places were not, as yet, in direct contact; the maker of the dress obtained its exotic elements from traders, and the woman who wore it was doubtless unaware of their ultimate origins. Likewise, the global reach and long-term consequences of the fur-trade economy remained obscured to local actors early in the nineteenth century. This dress suggests the relative balance and latitude that prevailed in early contact relations on the Plains, when Indian women could to some extent rework that engagement on their own terms.

The design of this dress affirms that the wearer was both socially prominent in her native community and had access to commercial exchange networks; she might even have been the wife or relative of a trader. Novel materials such as glass beads and metal buttons must have been interpreted initially in terms of existing frameworks of meaning, and they were probably selected and arranged on the basis of qualities that were culturally salient, such as color and reflectivity. Cowrie shells—which may have been equivalent to the elk teeth used on Plains dress yokes—beads, and metal buttons and trade tokens were also key elements in what Hail (1993:92) has termed women's "prestige" ornamentation on the Plains. The woman who wore this dress was able to marshal and display an exceptional quantity of valuable materials. During the 1830s, George Catlin painted several Indian women wearing similarly ornate dresses, including Tchon-su-mons-ka, or Sand Bar, the Lakota wife of the trader Francis A. Chardon. The yoke of her dress, Catlin wrote, was "almost literally covered with brass buttons" (Catlin 1973, 1:224).

Portrait of Tchon-su-mons-ka, or Sand Bar, by George Catlin, 1832. Sand Bar, the beautiful wife of the Fort Clark trader Francis Chardon, wore a dress emblazoned with brass buttons when she sat for this portrait.

The Peabody's "Sioux-type" dress is made from four pieces of brain-tanned mammal hide, possibly elk: an upper yoke, a skirt, and two additional strips that lengthen the hem (a strip of beadwork is also sewn to the skirt). In contrast to the painted side-fold dress, this dress appears to have been made with the tanned flesh side out. The yoke or cape fold is fringed, and the lower two-thirds of the yoke is decorated with perpendicular rows of linear impressions that were filled with translucent white hoof glue, or "sizing." On their return voyage down the Missouri, Lewis and Clark encountered a party of Cheyenne visiting the Arikara villages. Clark noted that the women's (possibly side-fold) dresses "are als[o] frequently Printed in various regular figures with hot sticks which are rubbed on the leather with Such velosity as to nearly burn it[.] this is very handsom" (Moulton 1993, 8:319).

Horizontal lanes of orange, white, blue, brown, and yellow porcupine and bird quill embroidery, applied in simple bands, fill the skirt of the dress. As on the quilled bison robe described in the next section, each lane of quillwork is punctuated by red wool tufts wrapped with sinew. Every sixth lane of quillwork is a compound of five rows that form multicolored rectangular boxes filled with bars or smaller boxes; these alternate with blue squares. Dark brown "guide lines" drawn on the hide are visible in several places. The quillwork was sewn with three sinew threads and seems to have been worked onto the dress from both directions. The blue and brown-black quills are from the split feathers of birds, and some of the brown areas in the quillwork appear to be plant fiber. A thin hide thong attached to the neck is wrapped with brown and orange porcupine quills.

The top third of the yoke is covered with a profusion of trade goods: blue and white glass beads, English gilt brass buttons, and cowrie shells from the Indian or Pacific Ocean, all of which are arranged horizontally. Most of the buttons are attached to the dress with leather thongs; those on the shoulder piece are held by a warp-faced, plain-weave cotton tape, of the type that English and colonial American women wove on wooden hand looms for household clothing production. Such garment tapes ("gartering") were a staple in the fur trade, and in preparation for the expedition Lewis had purchased several varieties (e.g., "binding" "Dutch tape," and "gartering"), as well as thread and ribbon, to be used as gifts for Indian women (Moulton 1987, 3:492–505). Early cotton garment tapes appear on a variety of Indian-made objects in this collection, including pipe stems.

This is a special-occasion dress, worn for dancing. A row of handmade, conical metal tinklers, each encasing a fringe of horse hair, is suspended above the hemline. Indian people made brass, copper, tin, and iron tinklers from sheet and scrap metal obtained from Euro-Americans and from trade objects such as kettles and cans. Often said to have replaced the carved hooves used as rattles prior to contact, metal tinklers have been found in archaeological sites in northeastern North America dated before 1650 (Quimby 1966:111–12). Snuff cans became a source of material for metal tinklers during the early twentieth century. Metal tinklers are still widely used, particularly on women's "jingle" dance dresses, where they unify sound and motion.

Lewis and Clark reported to Jefferson that throughout their travels among Indian nations, blue beads (which, using a Chinook phrase, they called "chief beads") were desired above all other trade items. Both blue beads and brass buttons appear, with remarks about their value, on the "List of Requirements" Lewis prepared for the expedition before he left Philadelphia (Jackson 1978, 1:69–74). Advising Secretary of State Henry Dearborn about which items Lewis recommended be carried by future explorers, Jefferson wrote, "He says that were his journey to be performed again, one half or 2/3 of his stores *in value,* should be of these [blue beads]" (Jackson 1978,

Detail of tinklers at the hemline of a side-fold dress. Alexander Henry, observing an Assiniboin dance in 1776, wrote, "The sound of bells and other jingling materials, attached to the women's dresses, enabled them to keep time." (Detail of 99-12-10/53047.)

1:375). Indeed, European colonial expansion and resulting trade so fueled the demand for glass beads that companies sprang up in the Netherlands, France, and elsewhere to challenge Italy's domination of their manufacture. The blue and white drawn-glass, blue pound or pony beads sewn to the breast and hem of this dress are of three or four sizes, varying between two and four millimeters in diameter. Such irregularity is common in beads of this vintage, which were cut from hand-blown glass canes. Most of the blue beads are variations of the semitranslucent, blue-green color found on many of the objects in this collection.

The beads on this dress have primarily been sewn in rows of lane stitching to create large blocks of color. The lowest row of beadwork on the yoke consists of blue triangular shapes with extended sides, resembling upside-down bison heads; these alternate with brass buttons.[11] On the front of the dress, these triangular design elements are executed in beads of a distinctive, opaque, "true blue" color. White beads two millimeters in diameter create vertically oriented box designs on the hem and shoulder that repeat the horizontal motifs in the quilled bands wrapping around the skirt of the dress. A single lavender seed bead appears on the shoulder strap.

Two sizes of disk-shaped, one-piece English gilt brass buttons are sewn to the yoke on the front of the dress. English buttons were originally produced for men's garments in Europe but found such favor in the Indian trade that the Hudson's Bay Company was buying them in bulk by 1748 (Woodward 1948:5). According to Meriwether Lewis, brass buttons were "more valued than any thing except beads" (Jackson 1978, 1:375). When Lewis left Philadelphia in the spring of 1803, he carried with him at least thirty-two dozen buttons in two sizes, as well as twelve pounds of brass strips and a wide variety of other brass objects (Jackson 1978, 1:92, 94). Traveling across the continent, the Corps of Discovery observed brass buttons and recycled brass fragments already in use as garment and hair ornaments by tribes throughout the Plains and Rocky Mountains. Lewis and Clark's supply of buttons proved insufficient, and the captains resorted to cutting off their own coat buttons in exchange for Nez Perce roots and bread as the party prepared to recross the Rockies on their return from the Pacific Ocean (Moulton 1991, 7:328).

Gilt brass buttons such as the ones on this dress were produced in Birmingham, England, after 1768 (Woodward 1965:25). Beginning about 1800 (Luscomb 1967:17–18; Olsen 1963:31), they were generally stamped on the reverse with "quality marks" touting their thin mercury and gold-amalgam coating. American companies produced similar buttons after about 1830, and they are often found in contact-era archaeological sites along the middle and lower Missouri River (e.g., Bass, Evans, and Jantz 1971:121–22; Lehmer, Wood, and Dill 1978:326–30; O'Shea and Ludwickson 1992).

Unfortunately, the chronology of brass button forms is so poorly understood that they can provide us only with relative, not absolute, dates of manufacture.[12] The brass buttons of both sizes on this dress were made in England between about 1795 and 1830, probably during the earliest days of machine stamping (Luscomb 1967; Joyce Saler, personal communication, 1999, 2000). The smaller buttons, which measure twenty millimeters in diameter, are of the type made for men's waistcoats (vests) and sleeves and are stamped "Superfine Standard" on the reverse. "Orange Colour" is stamped on the larger (25 mm in diameter) and perhaps slightly later frock coat buttons.

Most remarkable on this dress are the forty-eight cowrie shells sewn in a line across the top of the reverse yoke, framed by rows of blue beads. These are *Cypraea moneta,* generally called "money cowries" because of their ancient and ubiquitous use as a standard of value and a medium of exchange throughout Asia, Europe, India, and Africa. Cowries' widespread use and exchange belie the fact that although they occur

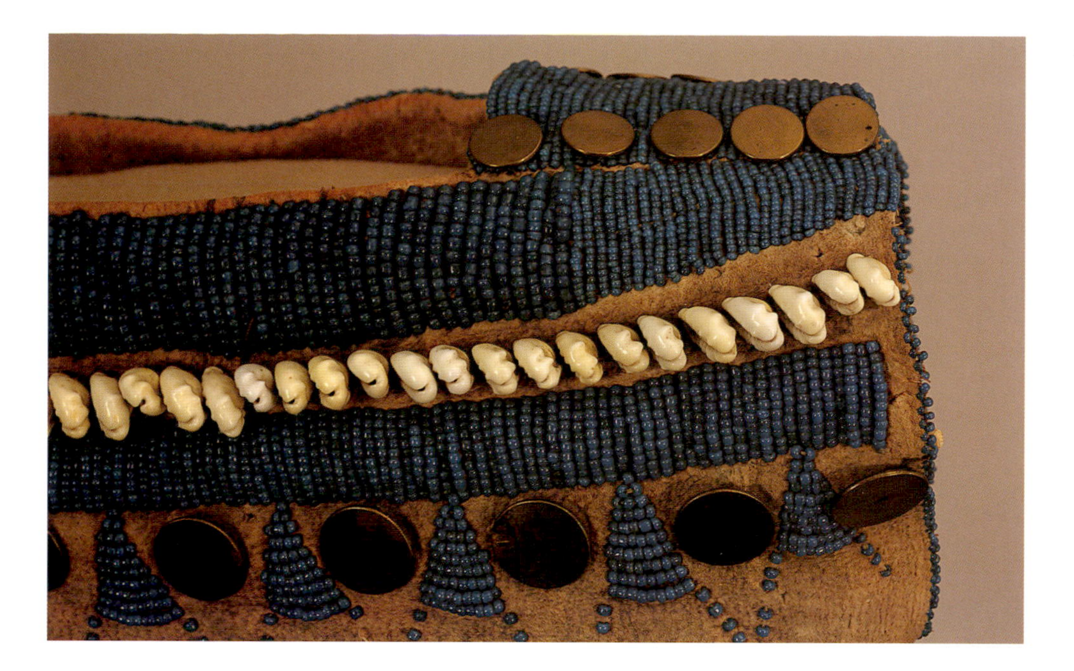

A line of forty-eight cowrie shells on a side-fold dress. (Detail of 99-12-10/53047.)

throughout the Indian and Pacific Oceans, the isolated Maldives Islands have been the primary source of money cowries since the pre-Christian era (Hogendorn and Johnson 1986). Money cowries were apparently acquired by some Native Americans prior to direct contact with Europeans, and their reputed presence in prehistoric archaeological sites in North America has excited a host of speculations. During the early twentieth century, the presence of money cowries on this dress was cited by the conchologist Wilfrid Jackson (1917:188) as evidence for diffusionist theories of pre-Columbian voyages to the New World.

The profusion of cowries on this dress is indeed unusual for a garment made before 1850, although the shells were often used this way in the late nineteenth century (e.g., Taylor 1997:fig. 33). However, both the African slave trade and the fur trade are now recognized as having been vectors for the importation of *Cypraea moneta* into North America, and the Hudson's Bay Company probably imported cowries for the American Indian trade. Unfortunately, most fur trade records apparently include cowries only under broad collective categories such as "seashells" or "porcelain" beads, rather than listing them in their own right.

An old Peabody Museum photograph of this dress alerted my colleagues and me to the fact that at one point, a long decorative dangle was suspended from a brass button on the left shoulder. Peabody records describe the dangle as a metal ring to which a dyed yellow cotton cord was attached, which was strung with animal teeth, glass beads, and a wooden bead. A length of dark brown hide with hair was also attached to the ring. Unfortunately, the dangle is now missing, and we do not know whether it was original to the garment or not. However, Feder's photograph (1984:fig. 4) of the similar Berlin dress reveals a plain leather dangle that falls to the skirt of the dress, attached near the collar of the garment.

Quilled Bison Robe

No evidence directly links this spectacular bison robe (99-12-53120; facing page), decorated almost entirely with bird quills, to either Lieutenant Hutter or to Lewis and Clark. In contrast with the other objects presented in this chapter, it has not been widely associated with the expedition, nor is it well known. Among those who have written about the expedition, only Paul Russell Cutright (1969:454) mentioned a "buffalo robe decorated with quillwork" as being among the Peabody Museum objects thought to have been collected by Lewis and Clark—a possible reference to this robe. Internal documentation at the Peabody, however, indicates that staff members have long speculated about the robe's possible association with either Lewis and Clark or Hutter. Either of these sources is possible, although it might also have been given to the Boston Museum or to the Peale Museum by another donor. In addition to its possible history, this robe is significant by virtue of its early vintage and unusual type. It evokes a primeval North America, a time when Plains Indian peoples were intertwined with bison in a relationship sustained by ritual communication.

Lewis and Clark sent seven robes to Jefferson from Fort Mandan (Jackson 1978, 1:234–36). Beyond designating one as "Minetarre" (Hidatsa) and two as Mandan, including one depicting a battle, the captains provided no written description of the robes. The only glimpse we have of their formal qualities is found in correspondence relating to the Fort Mandan shipment, which spent five months in transit to the East Coast. Upon their arrival in Washington, Henry Dearborn, Jefferson's secretary of war, had the containers opened and their contents checked and dried. He wrote to Jefferson, then at Monticello: "The Buffalo robes are good skins, well dressed, and highly embellished with Indian finery" (Jackson 1978, 1:254–55). In 1806, Clark shipped an additional four Mandan robes home to Louisville, as well as a "Vulter's quill with a buffalow coat" (Moulton 1993, 8:418–19). Because the bird-quilled robe shown here was designed to be worn vertically, like a hooded cloak, rather than wrapped horizontally like most other bison robes, Clark's description of a "buffalo coat" is particularly intriguing.

Quilled bison robe
99-12-10/53120
Plains, circa 1800–25
Bison (bull?) hide, dyed bird quills, undyed porcupine quills, wool cloth, carved ungulate (juvenile bison?) hooves, possible bast fiber thread, sinew, pigment
Length 239 cm, width 168 cm

Opposite
The hood and straps on this robe suggest that it was made to be worn vertically, with the hood covering the wearer's head. (Detail of 99-12-10/53120.)

The 1809 Peale Museum memorandum lists only two robes from Lewis and Clark, one described as "A Large Mantle, made of the Buffalow skin, worn by the Scioux, or Soue, Darcota Nation," and the other described as "A small Mantle of very fine wool, worn by the Crow's nation Menetarre." After the Peabody received the Boston Museum materials, Peabody curator Charles Willoughby annotated his copy of the Peale memorandum. He underlined the entry itemizing the "Large Mantle," indicating that he thought he had a match, and in the Peabody ledger this robe was identified as "Sioux."

The 1828 Hutter donation to the Peale Museum included both a pictographic battle robe and a "Mandan chief's Buffaloe robe."[13] A photograph of an early-twentieth-century exhibit at the Peabody reveals that this quilled robe was displayed as Mandan in a case devoted to objects from that tribe. Following that exhibit, the robe seems to have attracted little scholarly attention until, much to our astonishment, my colleagues and I encountered it while working through the Boston Museum accession.

This robe is strikingly unusual and beautiful, the dark, still-luxuriant winter bison hair presenting a rich contrast to the vivid natural hues of the quillwork. Several features distinguish it from the more typical "striped" robes seen throughout the Plains, on which rows of dyed porcupine quills are laid out in horizontal lanes, often punctuated by wool tufts or downy feathers. Such robes, which came to be associated with women, were widely used throughout the nineteenth century and into the reservation period. In contrast, this large, heavy winter robe may have been too massive for use by women, and it is quilled almost entirely with split bird quills (probably gull and raven), materials generally found on objects made before 1850. On this robe, lanes of quillwork run in two directions, perpendicular to each other. In addition, the robe is elaborated with a "hood" of quill-wrapped rawhide fringe surrounding the head of the bison, a quilled "brain disk," quilled ear panels, quilled, slatted shoulder straps, and a quilled, slatted tail. These "ornaments," as Alfred Kroeber (1983 [1902]) referred to such pieces, are all sewn onto the robe with sinew. We solicited information about this robe from a number of Indian and non-Indian people knowledgeable about Plains materials, but none was certain of its meaning or origin.

Winfield Coleman, however, a student of Cheyenne culture, identified it as a "buffalo-calling robe" (personal communication, 2000). Such robes were probably first used by hunters on foot who "impounded" bison—drove them into concealed traps or corrals—long before the advent of the horse and the development of equestrian hunting techniques. The ancient technique of bison impounding was associated during the historic era primarily with Algonquian-speaking peoples of the northern and central Plains, though it was also practiced by Siouan speakers along the Missouri River. Some Plains groups who impounded bison (and other animals) did so by ritually "calling" them into the trap, a shamanic practice. Other bison-calling rituals were performed to lure animals closer to the village during times of hunger. Still other variants of buffalo ceremonialism sought to attract bison more generally. The Plains Crees, Plains Ojibwas, Blackfeet, Assiniboins, Atsinas, Cheyenne, Suhtais, Arikaras, Mandans, Awatixas (Hidatsas), and Dhegiha Siouans are among the groups believed to have practiced some form of bison calling (Coleman 2000; Harrod 2000; Howard 1974). The documentation of bison calling by other groups, such as the Sioux, may simply be missing.

While the Lewis and Clark party was encamped at Fort Mandan in January 1805, the Mandans staged winter bison-calling ceremonies in order to entice the animals closer to the villages (Moulton 1983, 3:268). Members of the Corps attended the four-night Red Stick ceremony at Mitutanka, which the explorers referred to as the "first village" (the Nuptadi Mandans had a similar rite, known as the Snowy Owl ceremony; see Bowers 1991 [1950]). As part of these ceremonies, the wives of young men often had intercourse with the elders of their husband's father's clan, who assumed the personae of bison bulls. In this way, the essence of the bulls was transferred to the young men, who could then propitiate the herds. Among the Mandans, the Red Stick ceremony was hosted by the owner of a medicine bundle that contained bison-calling power. Bundles, special robes, songs, myths, pipes, and food were integral to the ritual process in both the Red Stick and Snowy Owl ceremonies. The brief account of the Mandan ritual witnessed by members of the Lewis and Clark party notes its function of luring bison to the village but does not mention impoundment.

Little is known about animal-calling practices on the Plains, but the fragmentary information that exists suggests that they were ancient and widespread. Recently, Howard L. Harrod (2000:75–104) synthesized data on Plains hunting beliefs and rituals, and Winfield Coleman (2000) addressed associated robes and other ritual objects. The core concept of animal calling, that of evoking primordial relationships between humans and other animals through ritual transformation and then invoking animal sacrifice, was widely shared. Specific calling practices, however, were grounded in the religious thought and origin myths of individual communities (e.g., Brasser 1984; Coleman 2000; Harrod 2000; Webber 1983). Alika P. Webber (1983), who collected information about magical robes used by Montagnais-Naskapi caribou hunters, noted that until the 1960s, in order to preserve the robes' efficacy, their use and meaning had been hidden from outsiders.

Coleman's identification of 99-12-10/ 53120 as a buffalo-calling robe is based on the robe's formal properties, including its size, its quilled ornaments, its hood, and the design of its quillwork. Like the painted caribou robes worn by Montagnais-Naskapi shamans and hunters to become and attract caribou (Webber 1983:60–69, 75–77) and the magical hunting hoods worn by the James Bay Cree (Brasser 1984:57), buffalo-calling robes were apparently worn with the head of the bison over that of the wearer. Coleman believes that this is a male shaman's robe, worn to lure bison into a pound. As such, it would have been worn vertically, possibly with the hair side out (Coleman 2000:11). During rituals to impound or call bison to their villages, members of the Arikara Buffalo Society wore their robes fur-side out (Howard 1974).

Bison robes with quilled ornaments (often including a brain disk, eyes, and ears) are clearly related to the "banners" that were attached to Plains Cree and Plains Ojibwa backrests—frames people leaned against as they sat on the ground—as described by Brasser (1984). Such banners, or robe fragments, are made of hide or cloth and are decorated with bison symbolism, including quilled or beaded brain disks, eyes, and ears. Many carry symbolic designs representing bison pounds and drive lanes. Brasser learned that the backrest banners (and probably horned and ornamented bison fur caps) were used by Cree and Ojibwa shamans to bring in bison, and he speculated (1987:54) that the banner form evolved from robes or hides. Although he stated that "only the Arapaho and possibly the Cheyenne appear to have used something [formally] similar" (Brasser 1984:61), he noted a corresponding association between bison hunting and backrest covers among the Crows and Hidatsas.

If Lewis and Clark or George Hutter collected this robe, then existing documentation suggests that it was probably received from either Sioux or Mandan people. Given patterns of trade and intermarriage on the Plains, however, we cannot assume that a member of either of those two groups made and used this robe. Robes were key commodities in the intertribal trade network. The Mandan villages were an important nexus in the robe trade, and a Mandan chief might have had access to robes from many sources. The formal properties of this robe suggest at least two possible life histories.

Early-nineteenth-century Plains robes with quilled ornaments are uncommon. Other known examples have been identified as Mandan, Hidatsa, Atsina (Gros Ventres), and Sioux. Like 99-12-10/53120, most of these robes are simply quilled in multiple lanes or stripes, but at least one, a robe collected by Prince Maximilian

Pale blue-green, orange, and brown bird-quill ornaments highlight the ears and tail of the robe and the area where the brain once was. (Detail of 99-12-10/ 53120.)

(Schulze-Thulin 1987:fig. 13), is also painted with battle scenes. The production of quilled ornaments, particularly rosettes, is strongly associated with a group of affiliated Plains Algonquian peoples: the Arapahos, Atsinas, and Cheyenne. These groups, particularly the Arapahos, used quilled rosettes on robes, tipis, backrest hangings, and baby carriers (Feder 1980; Kroeber 1983 [1902]). Among the Cheyenne, Sutaios, Arapahos, and Atsinas, quilled ornaments were considered sacred and were made only by members of the women's ritual sewing societies (see Grinnell 1923; Kroeber 1983 [1902]; Marriott 1956). Historically, quillworkers in those communities also used both bird quills and plant materials in their embroidery.

Consonant with the explicit bison hunting symbolism seen on Cree and Ojibwa backrest banners, some Cheyenne and Arapahos considered the quilled stripes on robes to be bison paths (drive lanes?), and the Cheyenne regarded the periodic wool "tufts" as the bison (Coleman 2000:13; Kroeber 1983 [1902]:34). According to Coleman (2000:13), the Cheyenne referred to the "hoods" of buffalo-calling robes, to which were attached quill-wrapped slats, as "rainbows"—the same word used to designate a trap. When Cheyenne people evoked the power of their sacred arrow bundle to call buffalo, they butchered the animals so as to leave the head, backbone, and tail attached (Harrod 2000:86). The design of ornamented robes highlights the same anatomical articulation, creating a physical manifestation of the ritual relationship between people and bison.

Winfield Coleman (2000; personal communication, 2000) suggests that the production of quill-embroidered, striped, and anatomically "ornamented" buffalo-calling robes was initially restricted to the Cheyenne, Suhtaios, Arapahos, and Atsinas, all of whom were known to have calling rituals. He points to the design of the quilled rosettes on the head of the Peabody robe, which begin in a "radial burst," as indicative of Atsina or Arapaho work (personal correspondence, 2000). Prince Maximilian saw striped robes among the Atsinas and wrote, "They ornament their large buffalo robes in a peculiar manner, with narrow parallel transverse stripes of porcupine quills, and many little pieces of scarlet cloth fastened to them in rows. This way of adorning their robes is said to be likewise usual among the Arapahos" (Thwaites 1906a:75). If the Peabody robe is Atsina or Arapaho, and if it was acquired by Lewis and Clark, then perhaps it is the robe presented to Clark by a visiting Cheyenne chief when the Corps of Discovery stopped in the Arikara villages on their return voyage (Moulton 1993, 8:314, 1996, 10:270).

A second hypothesis, suggested in the literature, is that this robe was made by the Yankton (Nakota) or Santee (Dakota) Sioux. Coleman (2000:20) argued that the Lakotas, Nakotas, and Dakotas did not call bison, but James H. Howard (1976) reported that a number of Sioux "winter counts," or pictographic calendars, record bison-calling rituals, at least one of which was performed by a Sans Arc Lakota shaman. A robe very similar to 99-12-10/53120, probably collected before 1815 and recorded for posterity in a painting by the British military officer C. H. Smith, was identified as "worn by the Tetons [Lakotas] and worked by women of the Sioux tribe" (King 1994:62). According to Smith's depiction, the robe he saw, like the Peabody robe, had a "hood" of quill-wrapped fringe, eye and ear ornaments, shoulder straps, and perpendicular rows of quillwork. Interestingly, that robe may have belonged to the physician Samuel Latham Mitchill (1764–1831), an avocational naturalist and a member of the American Philosophical Society who served in both the U.S. House of Representatives and the Senate and was a close acquaintance of Jefferson, Madison, and Lewis and Clark. In 1805, Mitchill entertained an Indian delegation visiting Washington, D.C., at the invitation of Lewis and Clark, and he was said to have acquired natural history specimens from the expedition for his own collection (Janson 1935 [1807]:232–34; King 1994:62).[14]

In his 1987 survey of bird quillwork, Norman Feder cited this robe as an example of a quillwork style that he identified as characteristic of the upper Mississippi region, particularly the Fort Snelling, Minnesota, area, before 1850. Hallmarks of this early bird quill style include the use of a "unique blue green color that is more green than blue, or a dull blue" (Feder 1987:52), as the primary or background color, often in combination with brown, orange, and black. The earliest documented examples of this style discussed by Feder were collected by Duke Paul in the 1820s and were later attributed to either the Yanktons or the Santees (e.g., Feder 1987:fig. 8; Schulze-Thulin 1987:pl. 3). Two later collections made in the Fort Snelling area, one by Nathan S. Jarvis (Feder 1964) and the other by Bishop Henry Whipple (Casagrande and Ringheim 1980), also contain examples documented as Yankton, Santee, or possibly Ojibwa. Bill Plitt, a private conservator in Santa Fe, New Mexico, reports having repaired an early tobacco bag identified as Yankton that was executed in this distinctive palette (personal communication, 1999). The quilled strip on a pair of Dakota leggings collected in 1820 (Grimes, Feest, and Curran 2002:cat. no. 76) also illustrates this style.

The movement of goods between groups, the complexities of stylistic "signaling" in social relations, and the poor documentation of most early museum collections complicate efforts to associate styles of material culture with specific Plains Indian peoples. Affiliated groups, such as the Sioux and Cheyenne, often exchanged not only objects but associated concepts, stylistic inspiration, and design elements. Whether or not the three divisions of the "Sioux proper" (Lakota, Nakota, and Dakota) impounded or called bison, they did use the "striped" quillwork pattern, which appears, often in red lanes, on robes, baby carriers, dresses, and storage bags. They also utilized bird quills—notably the Yanktons. Although there seem to be no early ornamented robes firmly documented as Lakota, Nakota, or Dakota in museum collections (but see Hanson 1996:fig. 49), Richard Conn (1986:93) illustrated an ornamented Lakota robe made during the 1930s from Hereford hide.

I might also note robe 53120's relationship to two other items in the Peabody collections that are potentially associated with Lewis and Clark. First, Feder (1987:53) included a knife sheath (99-12-10/53027), probably acquired by Hutter, as illustrative of the "Upper Mississippi" bird quill style discussed in his survey. Second, the quilled robe is stylistically linked to a quilled side-fold dress (99-12-10/53047) that may also be either Algonquian or Siouan in origin. In turn, that dress has affinities with a side-fold dress in the Jarvis collection, identified as Yankton Sioux and embellished with green, brown, and black bird quills—the "Upper Mississippi" palette (Feder 1964:24, 42).

The Peabody's quilled robe was made from a high-quality winter bison hide, probably that of a bull, and remains in remarkably good condition. Unlike most other historic bison robes, which are pieced together from two hides, this robe, like 53121, was made from a single hide, a method that requires much more careful skinning and processing. On the body of the tanned, flesh side of the robe are twenty-seven lanes of one-quill simple bands, two rows of which (the first and the sixteenth) are compound rows of three lanes. The lanes, predominantly executed in bird quills, are alternatively pale blue-green and orange-brown, banded by shorter lengths of white porcupine quills, with blocks of color divided by dark brown-black bird quills. Occasionally, darker blue bird quills appear in the lanes, and in some locations, blue-green quills alternate with the white pith of other quills in a "candy-stripe" pattern. Eight perpendicular rows of red wool tufts bisect the quilled lanes. These tufts, probably made from English stroud cloth, have the character of miniature tobacco bundles. The neck of the hide is quilled with five rows of simple bands that are oriented perpendicular to the field of twenty-seven; these rows, too, are accented by red wool tufts.

Three lengths of quill-wrapped rawhide fringes are sewn to the sides and front of the head or "hood" of the robe. On the fringes, geometric shapes, including the widespread "E" design and linked, open squares, appear in orange-brown against the field of blue-green. Three hide ornaments—a round brain disk and two ear coverings—quilled in rows of blue and orange simple bands are attached to the head. An abstract thunderbird dominates the center of the brain disk, while each of the ear attachments bears an E design element outlined in dark brown. Two shoulder straps and a pendant tail are wrapped with long blue, orange, and black bird quills, seamed by multiple rows of sinew thread and divided into slats in the center. The shoulder straps presumably served to secure the robe to the wearer during use. Similar straps are attached to the robe painted by C. H. Smith (King 1994), a robe at the Smithsonian, and a Mandan or Hidatsa robe in a private collection (Coleman 2000:figs. 5, 21).

The pendant hooves that edge the hood of this robe are an ancient device in North America, predating the later use of metal tinklers. (Detail of 99-12-10/53120.)

White, porcupine-quill-wrapped loops pendant with carved ungulate hooves (from juvenile bison?) are attached to the hood fringes, the ear ornaments, and the shoulder straps. Sinew threads (used as needles) protrude on the hair side of the robe, and a plant fiber thread may have been used to secure some of the quillwork. The quilled side of the robe and the hood fringes are stained with red pigment and may also retain remnants of a blue-black pigment. Small holes around the perimeter of the robe indicate that it was once tacked with square nails to a wall.

Chapter 8

The Language of Pipes

Pipe ceremonialism, particularly the use and exchange of pipes in diplomatic rituals, shaped Lewis and Clark's relationships with the Indian peoples they met. Because pipes were such important diplomatic gifts, Lewis and Clark acquired or kept more of them—and donated more of them to the Peale Museum—than any other category of object they received from Native people. The 1809 Peale memorandum listed fourteen pipes, all of which were attributed to tribes or individuals. But because the Corps of Discovery did not meet directly with all of the tribes Peale listed, the origins of some of those pipes are unclear. In this chapter I suggest a number of possible sources for the pipes that Lewis and Clark gave to Peale and present the pipes and calumets now at the Peabody that may match those listed in the Peale memorandum.

Eagle-feather tips on a fan calumet. (Detail of 99-12-10/53099.2; see p. 222.)

Indian peoples across much of the North American interior used pipes both to communicate with the supernatural and to mediate relationships among people. Pipe ceremonies were key mechanisms for alliance building within and between Native groups, and they became equally essential for Indian-white diplomacy after the arrival of Europeans. "The pipe," wrote William Clark, "is the Semblem of peace with all."

Pipes were used, for example, to conduct adoption ceremonies, creating the fictive kinship ties that facilitated intertribal and then intercultural trade on the Plains. Both Lewis and Clark (Moulton 1987, 3:235–36) and Prince Maximilian (Thwaites 1906a:319–21) chronicled adoption ceremonies among the Mandans, which involved the ritual exchange of two calumets adorned with eagle feather fans (Bowers 1991 [1950]:329–31). A related practice called "trading on the pipe" enforced balanced reciprocity in the value of exchanged goods, obviating negotiations (Jablow 1994 [1951]:46–47). The act of pipe smoking also solemnized diplomatic proceedings and, in the context of intercultural agreements such as treaties, enacted what Dorothy Jones (1988:186) termed "sacred collective obligations." Lewis noted the solemnity of pipe-smoking rituals in his description of the Shoshone custom of removing moccasins before smoking: "[T]his is a custom among them as I afterwards learned indicative of a sacred obligation of sincerity in their profession of friendship given by the act of receiving and smoking the pipe of a stranger" (Moulton 1988, 5:79).

The social dimensions of pipe ceremonialism became, in a sense, like a pidgin language between cultures on the early North American frontier. French, English, and U.S. officials found that learning the ritual language of pipes was as indispensable for conducting business in Indian country as it was for intertribal trade, diplomatic relations, and adoption. Early French explorers and missionaries, who called pipes *calumets* ("reeds"), found that pipes were such respected symbols of peace that they could serve as "passports" through enemy land. In a well-known example, the Illinois people gave Father Jacques Marquette a calumet to ensure his safety while he descended the Mississippi River in 1763 (Hail 1993:234). Ritual pipe smoking accompanied political negotiations between Indian peoples and foreign nations from the earliest European trade and treaty agreements to the end of the American Indian Movement's occupation of Wounded Knee, South Dakota, in 1973.

Gift exchange was another essential component of social relations in Native North America, and during the colonial era pipes were the diplomatic gift par excellence. Generations of Indian leaders presented pipes to European and American traders, diplomats, military leaders, Indian agents, and heads of state (e.g., Ewers 1981; King

1977; Viola 1981). For example, when a group of Kaskaskia Indians called on Virginia governor Thomas Jefferson in 1781, Jefferson smoked the "pipe of peace" with chief Jean Baptiste du Coigne, after which the two men exchanged diplomatic gifts. Jefferson pronounced the pipe ritual "a good old custon handed down by your ancestors, and as such I respect and join in it with reverence" (Wallace 1999:73). In 1793, when members of a second delegation died during a visit to President George Washington, du Coigne gave the president a "black pipe" in their memory, after which a second pipe was smoked by the dignitaries, including Jefferson and Secretary of War Henry Knox (Jackson 1981:71–73).

By presenting the pipes used to create and cement relationships, Indian leaders made manifest not only their own pledges but also the reciprocal responsibilities of the recipients. Pipe symbolism was so well understood that often pipes were simply sent, rather than personally presented, for the purpose of forming alliances, declaring war, or initiating peace. Dissatisfaction could be signaled by returning a pipe to its sender (e.g., Ewers 1981).

The pipes given to Lewis and Clark were the most symbolically potent and meaningful objects they received. Lewis and Clark and the Indian leaders with whom they met in council were initiating the first formal relations between those Indian nations and the new federal government. The pipes they shared joined the futures of their descendants in ways that they could not have imagined and that continue to reverberate today. As embodiments of the founding pacts made between Jefferson's ambassadors and tribal leaders, the pipes given to Lewis and Clark might today be considered powerful national symbols of the history of Indian-white relations.

Where and how did Lewis and Clark acquire the pipes they presented to Peale? No pipes were listed on the invoice that accompanied crates of objects shipped to Jefferson from Fort Mandan in April 1805 (Moulton 1987, 3:329–31), but two pipes and some red pipestone were among articles Jefferson shipped from Washington to Monticello on March 10, 1806, along with other objects Lewis and Clark had sent to him from Fort Mandan the previous spring.[1] Some of the pipes might have been acquired during the expedition's return trip in 1806 and sent to Peale by Jefferson. Both Lewis and Clark also sent pipes to Jefferson and Peale in the years immediately following the expedition. In the fall of 1807, Clark included "4 stone pipes with stems" and a warrior's raven regalia in a shipment of fossilized bone that he sent to the president, and an unspecified number of pipes were among the items Peale received following Lewis's death in 1809 (Jackson 1978, 2:469–70).

Intriguingly, many of the pipes listed in the Peale memorandum are recorded as having been given to Lewis and Clark by representatives of tribes from the Mississippi River valley, the western Great Lakes, and the eastern Plains-Prairie edge—areas the expedition did not visit. As I discuss later, one possible explanation is that some of the pipes originated with members of an intertribal delegation that visited Washington in 1805–6, while the Corps of Discovery was wintering near the Pacific Coast. Others may have been acquired by the captains after the expedition or sent to them as ambassadorial gifts.

The Peabody Museum received a number of pipes from the Boston Museum in 1899, some of which must have come from the Peale Museum. Unfortunately, we cannot identify which of these magnificent pipes might have been presented to Peale by Lewis and Clark. In addition to the fourteen pipes that the Peale Museum attributed to Lewis and Clark, Lieutenant George Hutter donated "3 Pipes, and 3 Pipe stems used generally by the Missouri Indians." Hutter's pipes may have been diplomatic gifts received from Sioux and Cheyenne people by the leaders of the Atkinson-O'Fallon expedition. Other fur traders, explorers, military men, collectors, and Indians also donated pipes to the Peale Museum throughout the early nineteenth century.[2] Peale's 1826 purchase of a large collection of Missouri River materials also included pipes (Sellers 1980:250–54).

Because it cannot be known for sure which of the Peabody pipes might have been given to Peale by Lewis and Clark, I present in this chapter the fifteen candidates that are of the right age and type to have been associated with the expedition. Many of these pipe stems seem to have come from the same tribes and regions as those described in the Peale memorandum.

Pipe Diplomacy during the Expedition

In his expedition journals, William Clark recorded receiving at least three pipes, one from the Yankton or Yanktonai Sioux, one from the Shoshone leader Cameahwait, and one from the Nez Perce leader Broken Arm (Tunnachemootoolt). In late August 1804, the Corps of Discovery entered Sioux territory with some trepidation. At the mouth of the James River, near present-day Yankton, South Dakota, the party encountered a camp of about seventy Yankton Sioux. The Yanktons received them hospitably, and the Americans spent nearly four days in their company, conducting formal diplomatic business and socializing in each other's camps. Lewis and Clark were treated to feasts

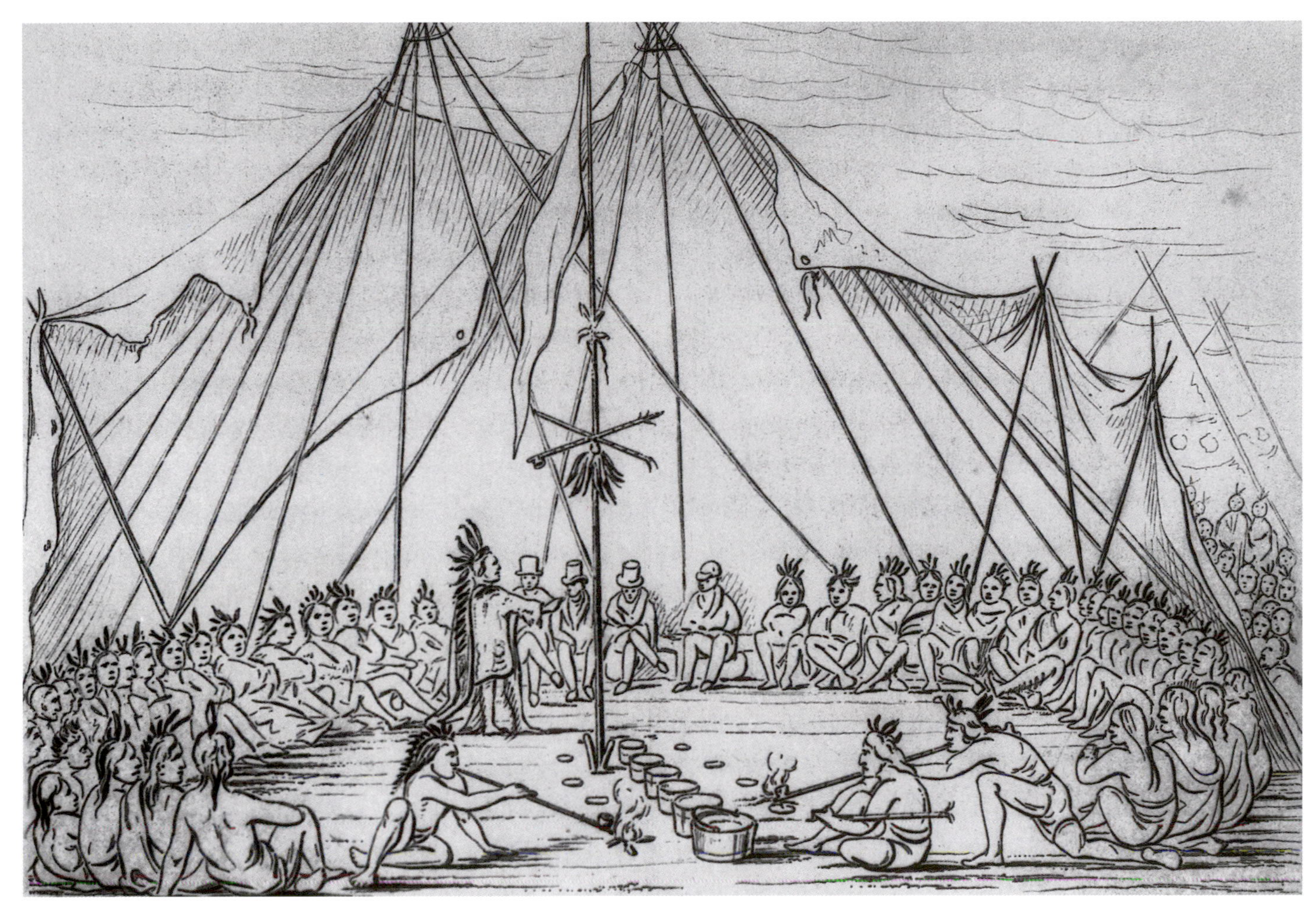

A feast given for American traders and U.S. officials by the Lakota chiefs Ha-wan-je-tah and Tchan-dee at the mouth of the Teton River near Fort Pierre in 1832, as depicted by participant George Catlin. Calumets hang suspended above the gathering as other pipes are prepared for smoking.

and dancing, collected an abbreviated "vocabulary," and described Yankton clothing, lodges, and camp life in some detail (Moulton 1987, 3:21–38; Ronda 1984:23–26). They named their meeting site with the Yanktons "Calumet Bluff," perhaps memorializing the gift of a calumet.

The diplomatic protocol that Lewis and Clark pursued with the Yanktons followed the working script they enacted throughout their travels. Yankton diplomacy, too—the product of generations of managing intertribal relations as well as decades of contact with European traders—was culturally patterned and formally structured. The meeting between the Corps and the Yanktons, with its two central diplomatic rituals of gift exchange and ceremonial smoking, provides a good illustration of how such intercultural diplomatic encounters were ordered.

Both sides in the transaction understood that the exchange of gifts was the indispensable oil that lubricated social relationships in Native North America. To announce their presence, Lewis and Clark first sent gifts to the Yankton camp: "Tobacco, Corn, & a few Kittles for them to Cook in" (Moulton 1987, 3:22). These items were not simply utilitarian offerings but symbolically communicated the Corps's (and the government's) consideration and concern for the well-being of the Yankton people (the corn and kettles) and the captains' desire to meet with them in council (the tobacco).

The social symbolism of tobacco and smoking permeated both intertribal relations and Indian-white relations throughout the Woodlands and the Plains. In Native North America, tobacco has many meanings. It is used as an offering to the spirits, a purifying agent, and a sacramental substance, capable of transcending boundaries to unify peoples and spiritual powers. Indians have smoked tobacco, often mixed with the bark and leaves of other plants, since ancient times (see Hall 1983; King 1977), and ritual smoking is believed to evoke cosmic forces that sanctify places, events, and social relationships.

Strong forms of Euro-American tobacco were introduced to Native America by the eighteenth century and generally were distributed by traders in bound twists and layered "carrots" (Hanson 1988; Ray 1980). Trade tobacco developed a secondary symbolic meaning in the context of intercultural relations, serving as a calling card for Europeans and Americans seeking entree to Indian societies. Meriwether Lewis, aware of the importance of tobacco to his enterprise, purchased at least 130 rolls of "pigtail" (twist) tobacco, totaling 63 pounds, before leaving the East (Jackson 1978, 1:94), and tobacco was the most widely distributed of the Corp's "Indian presents." Formally dispensed to each chief and to each Indian nation at large, the Corps presented tobacco to announce its arrival in new territory, invite men to councils, ease tense interactions, and express gratitude. Tobacco was also honored in exchange for food and sundries.

Lewis collected two forms of Arikara tobacco seed and leaf, which he sent to Thomas Jefferson. He noted that one variety of seed was given to him by Lepoy, an Arikara chief. The Arikaras, Mandans, Crows, and Hidatsas grew tobacco, which was valued in the intertribal trade network. Adopting traders' jargon, Lewis described the Arikaras' method of rolling tobacco into "carrots" to dry (Moulton 1987, 3:460–61, 466, 472, n. 1).

The Yankton band that the Corps met responded to the party's initial gift presentation with a formal greeting ceremony, meeting Sergeant Nathan Pryor's advance party with bison robes on which they intended to carry the party into their camp. (Pryor declined the honor, demurring that his men were not "the owners of the boats.")

Offering food was a universal gesture of Indian hospitality, and the Yanktons served the Pryor party dog stew, a dish reserved for ceremonial occasions and honored guests.

The following morning, as the night's dew began to dry on the grass, five Yankton chiefs entered the American camp, preceded by four painted warriors singing and shaking their rattles. A delegation of attendants accompanied the group. With flags flying, Lewis and Clark fired a two-gun salute and stepped forward to take the hands of the Yankton dignitaries. Each member of the Corps presented the Yankton "musicians" with a piece of tobacco, and the proceedings got under way in the shade of an oak tree.

Lewis delivered a formal speech lasting several hours, which was translated by Pierre Dorian, a trader with a Yankton wife and family. The contents of Lewis's messages varied little as the Corps traveled west (see, for example, the "Speech for Yellowstone Indians" in Moulton 1993, 8:213–15). It announced the transfer of land and power from the Spanish and English to a new, American "Great Father" and his intent to deal fairly with Indian peoples. Benefits to the tribes would include increased trade at better prices, protection from enemies, and instruction in agriculture and other "useful" arts. But this could happen, Lewis emphasized, only if the tribes stopped fighting one another and severed their allegiances to European traders. Finally, the Great Father desired to meet personally with Indian leaders and invited them to travel to Washington, D.C.

At the conclusion of their remarks, Lewis and Clark "made," or recognized, five Yankton chiefs by presenting them with specially minted peace medals, clothing, and tobacco. Since Euro-Americans imagined leadership to be hierarchical, as it was in their own society, well-established diplomatic conventions dictated that chiefs be recognized in degrees of authority. Lewis and Clark made chiefs in three degrees: a single grand chief, a second chief, and three third chiefs. Three sizes of peace medals were distributed accordingly. The older Shake Hand, whom the captains designated the grand chief of the Yanktons, was given a red-laced coat, an American flag, a shirt, and some wampum. Similar gifts went to the other chiefs, and all received carrots of tobacco. (This list is according to Clark; Sergeant Ordway gave a slightly different list; see Moulton 1995, 9:47.) To formalize the investiture and close the meeting, the chiefs and the captains "Smoked out of the pipe of peace" (Moulton 1987, 3:24). Each party then retired separately to eat and discuss the day's events; Lewis and Clark probably committed a faux pas by not inviting the Yanktons to dine with them. As was expected of Indian leaders, the Yankton chiefs immediately redistributed some of their gifts to their followers.

That night, the Yanktons invited the Americans to their camp, where their most esteemed warriors performed testimonial dances to the beating of drums. One by one, the men rose, brandishing their weapons, and recited their feats in war. The following morning, it was the chiefs' turn to respond to the captains. Dressed in their finery, the Yankton leaders arranged themselves in a row, Shake Hand flanked by two warriors "in the uniform and armer of their Nation . . . with a War Club & Speer each, & Dressed in feathurs"—possibly raven bustles (Moulton 1987, 3:28). When Lewis and Clark reached the oak-shaded meeting side, the chiefs greeted them "with elligent pipes of peace all pointing to our Seets" (Moulton 1987, 3:33). Each chief, by rank, delivered a response to Lewis and Clark's requests of the day before, all affirming in principle their intent to promote intertribal harmony, recognize the United States, and send a delegation to visit Jefferson. In return, they asked for a trader and ammunition. Lewis and Clark distributed more gifts, and then, "After all[,] the chief presented the pipe to us" (Moulton 1987, 3:31). The Corps spent a second evening in the Yankton camp, where the men distributed knives, tobacco, belts, cloth tape, and binding to the dancers.

Lewis and Clark not only respected and participated in the smoking rituals of the peoples they met but also incorporated pipe rituals into their own diplomatic initiatives. Clark described the ritual of smoking as "the greatest mark of friendship and attention" (Moulton 1987, 3:311). The captains carried their own pipes (a pipe tomahawk, and probably pipes given to them by Indian leaders) into diplomatic encounters. For example, when counciling with Mandan and Hidatsa leaders shortly after the Corps' arrival at the Mandan villages in the autumn of 1804, Clark introduced the Arikara chief Arketarnashar, who had traveled upriver with the party, and presented a "pipe of peace," urging the tribal leaders to smoke and relinquish their grievances (Moulton 1987, 3:208–9). And when, on August 13, 1805, Lewis finally made contact with a band of Lemhi Shoshones, from whom he fervently hoped to obtain directions and horses, his first act was to initiate the smoking ritual. As he wrote in his journal, "I now had the pipe lit and gave them smoke; they seated themselves in a circle around us and pulled of[f] their mockersons before they would receive or smoke the pipe" (Moulton 1988, 5:79). By a remarkable turn of fate, Lewis had located Sakakawea's brother, Cameahwait. Cameahwait immediately escorted Lewis's party to his own camp, where he conducted a much more elaborate pipe ceremony (Moulton 1988, 5:80–81).

By the time Lewis and Clark began their return trip from the Pacific Ocean, they had become fluent in the language of the pipe. During the spring of 1806, the party spent a month with the Nez Perce before attempting to recross the Rocky Mountains. In order to understand and smooth over the enmity they found between two Nez Perce leaders, Clark wrote, "[W]e Sent Drewyer with a pipe to Smoke with the twisted hair and lern the Cause of the dispute between him and the Cut nose, and also to invite him to our fire to Smoke with us" (Moulton 1991, 7:232).

During that month, Lewis and Clark spent many hours with Nez Perce leaders, discussing their wars with the Shoshones, Blackfeet, and others. The previous year, Lewis and Clark had encouraged the Nez Perce to pursue peace, so that traders could be received in their country and their young men could travel unmolested. When the Corps returned to the Nez Perce the following spring, the chiefs informed them that they had attempted to reconcile with their enemies, sending out a delegation of three men carrying pipes, but the party had been ambushed and killed. Now they were reluctant to send any of their men with Lewis, who offered to broker a peace with the Blackfeet and Atsinas, or with Clark, who proposed negotiating directly with the Shoshones.

Portrait of Grizzly Bear by George Catlin, 1831. The Menominee chief, who led a delegation to Washington, is portrayed holding a pipe as a symbol of his leadership.

On June 6, 1806, Broken Arm summoned Clark to his lodge to inform him that a party of Shoshones had arrived at the allied Cayuse villages for the purpose of discussing peace with that nation and the Nez Perce. Runners had been sent to the Shoshones, inviting them to meet Broken Arm to "Smoke the pipe of peace," which would then be carried to the Shoshone nation and presented as a pledge of their agreement. After explaining this, Broken Arm "produced two pipes," wrote Clark,

> one of which he said was as a present to me the other he intended to Send to the Shoshones &c. and requested me to take one, I receved the one made in the fascion of the Country, the other which was of Stone curiously inlaid with Silver in the common form which he got from the Shoshones. I deckorated the Stem of this pipe with blue ribon and white wampom and informed the Chief this was the emblem of peace with us. (Moulton 1991, 7:341)

Clark continued using pipes throughout his post-expedition life as the federal government's chief Indian ambassador and diplomat. He customarily smoked with Indian leaders in his St. Louis council chamber, the walls of which were lined with more than forty pipes, many presented by visiting Indian delegations. John Ewers (1981:65) reported that chiefs also sent pipes to Clark through intermediaries, as tokens of respect and to foster good relations. Clark himself adopted this Indian practice. He presented a fanned calumet to Duke Paul of Württemberg that is now in the British Museum (King 1999:fig. 276). In the spring of 1811, Henry M. Brackenridge, who traveled up the Missouri with fur trade king Manuel Lisa (then a partner with Clark in the Missouri Fur Company), acted as Clark's emissary to an Osage leader. Upon arriving at Fort Osage, Brackenridge wrote,

> I went to deliver a pipe to *sans Oreille* (a warrior and a principal man of this tribe), sent him by general Clark. . . . I then presented him the pipe, which was handsomely decorated with ribbands and beads of various colours, and told him that it was given at the request of general Clark, and that it was intended as proof of the esteem and consideration in which he was held not only by the general himself, but by all the Americans. He replied "that he was pleased with this proof of general Clark's good will towards him, that he was the friend of the Americans." (Thwaites 1904b:58)

Although the 1809 Peale memorandum indicates that Lewis and Clark received pipes from at least three other Sioux groups, as well as representatives of other nations, those presentations are apparently not described in the journals. The pipe that Shake Hand gave to Lewis and Clark may have been the "Yankton" pipe later included in the Peale memorandum, but there is no record of Peale's receiving a Shoshone or Nez Perce pipe. Clark might have exchanged Broken Arm's pipe later in the expedition, or he might have kept it himself after his return to St. Louis. One pipe in the Peabody's collection, 99-12-10/53110 (see p. 237), is decorated with "blue ribbon and white wampom," materials Clark described adding to Broken Arm's pipe, but the quilled stem suggests that the pipe was made in the eastern Plains or western Great Lakes.

Origins of the Peale Pipes

Curiously, many of the fourteen pipes listed in the Peale memorandum are attributed to groups that Lewis and Clark are not known to have encountered or formally treated with during their exploring mission. These include the Winnebagos (one pipe), Foxes (two pipes), Ioways (two pipes), and Sauks (three or four pipes). Another pipe was "given to the company by White Pigeon," who may have been a Potawatomi. All of these peoples dwelt principally north and east of the Missouri, near the Mississippi River and its tributaries. They frequented the region between the two river systems, however, and were active allies (or enemies) of tribes on both sides of the Mississippi as well as of the French, English, Spanish, and United States governments. During Jefferson's presidency and throughout the following decade, the alliances and actions of Mississippi River tribes were of considerable concern to U.S. officials. Lewis and Clark interacted with those peoples in their respective post-expedition roles as territorial governor and Indian agent of Louisiana.

Men from these and other nations composed part of a large delegation of "Missouri and Mississippi chiefs," orchestrated by Lewis, that visited Washington, D.C., in 1805–6. According to General James Wilkinson, territorial governor of Louisiana, the group included twenty-six representatives from eleven nations: Oto, Missouri, Pawnee, Kansa, Osage, Sauk, Fox, Ioway, Kickapoo, Potawatomi, and Miami.[3] Wilkinson commented that "eight of these nations are strangers to us, and the last seven embrace the belligerents among whom we have been making Peace. . . . the Machinations of the British on one side and the Spaniards on the other, and the apparently increasing jealousies of the Indians, recommended to us to swell the Deputation to its present extent" (Jackson 1978, 1:265–66).

Over the winter of 1804–5, Lewis had sent forty-five leaders of Missouri River nations to St. Louis, the staging point for eastward travel. For various reasons, their numbers became significantly reduced over the course of the following summer. According to Ewers (1966:10–11), delegates from the "belligerent" Mississippi River, western Great Lakes, and lower Missouri River tribes were then selected by Wilkinson, whose primary diplomatic goal was to establish peace between "tribes of the Old Northwest [Ohio Valley] and their enemies west of the Mississippi." General Wilkinson and Indiana territorial governor William Henry Harrison convened a treaty council involving those tribes at St. Louis in October 1805, afterward choosing tribal leaders to visit Jefferson. Later that month, the delegation, now made up of

Wilkinson's delegates and the remnants of Lewis's, began wending its way to the capital. The group was led by Amos Stoddard, the military and civil commandant of Upper Louisiana Territory and Lewis's agent in St. Louis.[4]

Members of the delegation reached Washington on December 22, 1805, and spent the following winter there and in Philadelphia, New York, and Boston. Sir Augustus J. Foster recorded that Jefferson received them so warmly on New Year's Day that the British minister, who was also visiting, felt slighted and left (Wollon and Kinard 1952:192–93). The delegation captivated Washington society for months. The president formally addressed representatives from allied groups in a series of speeches delivered during the first few days of January 1806. The delegates were given tours of the city, visited Congress, and were hosted by New York senator Samuel Latham Mitchill (Janson 1935 [1807]:232–34). Later in the month, part of the group traveled to Philadelphia, where Charles Willson Peale cut silhouettes of at least thirteen Osage, Missouri, Pawnee, Sauk, and Fox tribal representatives. Ewers (1966:2) called the silhouettes, which Peale sent to Jefferson and of which ten are now at the Smithsonian, "probably . . . the earliest known likenesses of any members of the several tribes represented." The delegation also toured Boston, Providence, and New York before returning home the following spring—the Pawnee and Osage delegates escorted by Lieutenant Zebulon Pike—after a formal farewell ceremony conducted by Jefferson.

The pipes that the Peabody Museum received from the Boston Museum, which are described in the following pages, seem to be from both the Missouri River area and from the eastern Plains and western Great Lakes. This distribution is consistent with the tribal attributions listed in the 1809 Peale memorandum, which in turn corresponds to the tribal affiliations of the members of the 1805–6 delegation. Some of these pipes might have been presented to Governor Wilkinson or to President Jefferson in the context of that delegation. Jefferson might well have associated such pipes with the Lewis and Clark expedition, which he repeatedly invoked in his speeches to the assembled Indian leaders during their visit to Washington.

The Peabody Pipes

The predominance of regional rather than tribal styles, the pervasiveness of trade and exchange, and the scarcity of comparative materials make tribal identification of pre-1850 Plains and Great Lakes pipes as difficult as it is for garments and other objects. A further complication is that the nature of their use dictated that pipes were often transferred from one person to another and from group to group. It was not uncommon for tribal leaders to carry, use, and exchange pipe stems and bowls made by members of other tribes or given to them by non-Indian traders.

We can say that in form, the calumets and flat-stemmed pipes presented here are all of types that were used during the early 1830s, when Prince Maximilian and the American artist George Catlin each journeyed up the Missouri River. Both travelers collected similar pipe stems and bowls, and both Catlin and Karl Bodmer, Prince Maximilian's artist, illustrated them. Catlin had a special interest in pipes and prepared an illustrated manuscript on his collection that John Ewers published in 1979. Comparison with such early collections and illustrations affirms that the Peabody pipes are of the right vintage to have been collected by George Hutter in 1825 or to have been acquired by C. W. Peale as part of his large 1826 purchase. Most, if not all, of these pipe styles must already have been in use by the time of Lewis and Clark's travels twenty years earlier. This is confirmed by the small sample of surviving pipe stems known to have been collected during the eighteenth century. For example, several quilled pipe stems collected before 1789 and now at the Musée de l'Homme in Paris are very similar to a number of the quilled, flat-stemmed pipes described here.

Silhouette believed to be of Tahawarra, a Missouri Indian man. These silhouettes of members of the delegation of "Missouri and Mississippi chiefs" who traveled to Washington, D.C., and other eastern cities in 1805–6 were made at the Peale Museum in Philadelphia.

Calumets

Nine of the seventeen pipes that were transferred to the Peabody from the Boston Museum in 1899 have stems that are circular in cross section and can therefore be considered calumets.[5] Five of those pipes also have pendant eagle feather fans and might be called "classic" calumets, the most formally elaborate and symbolically charged of all historic-era pipes. Two of the original five fanned calumets, 99-12-10/53102 and 99-12-10/53103 (pp. 216, 217), were traded by the Peabody to the Denver Art Museum in 1952.[6] As pictured here, one of those calumet stems is now fragmentary. The three most elaborate and intact examples—99-12-10/53099.2, 99-12-10/53100.2, and 99-12-10/53101.2—remain at the Peabody.

A variety of Native American pipe forms drawn by George Catlin in the 1830s.

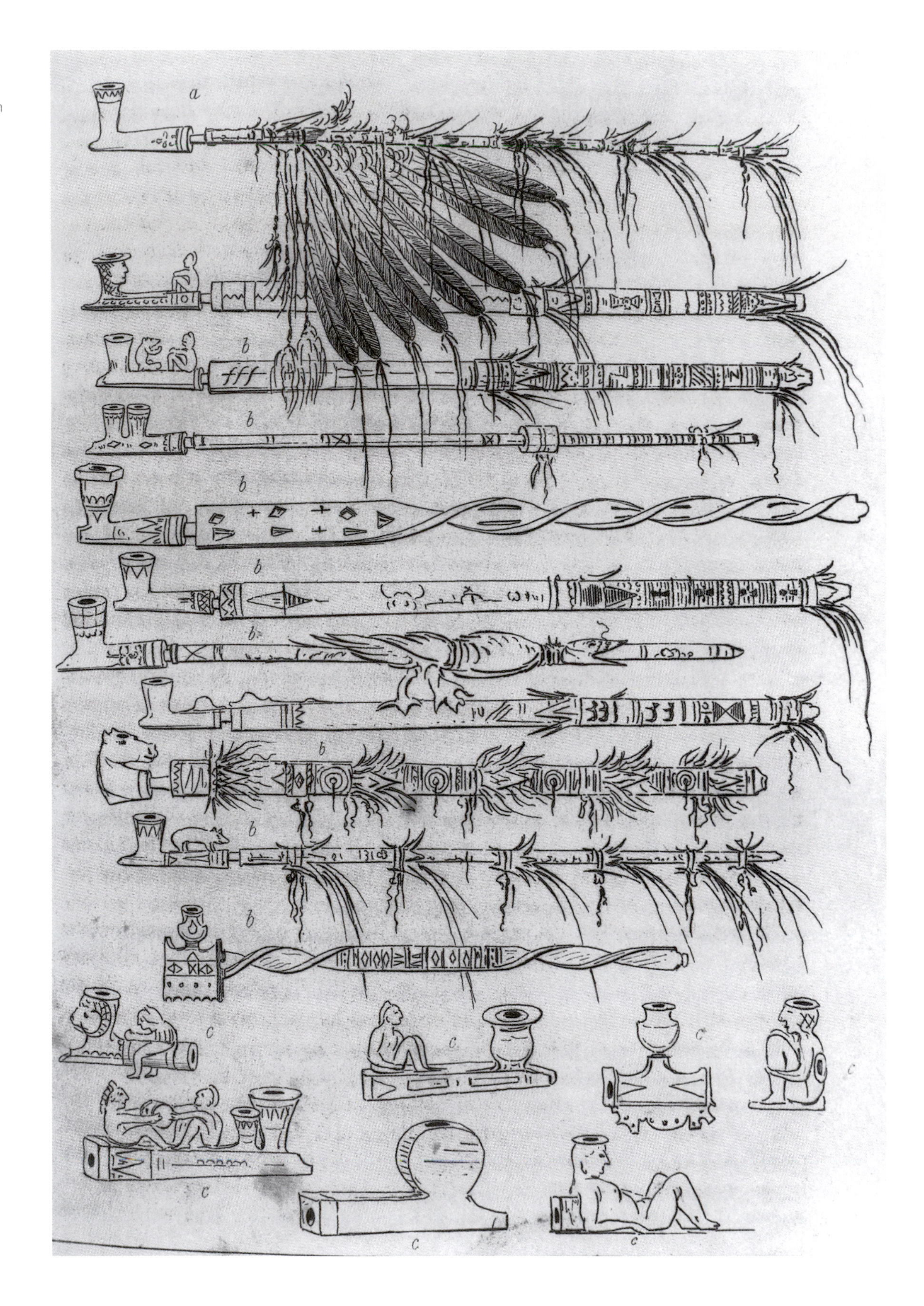

Because Indian people classify pipes by function as much as by form, there can be no hard-and-fast typologies of pipes based on their physical features (e.g., Noder 1999). But many early European explorers and contemporary scholars have reserved the term "calumet" for pipes with cylindrical stems, a form that was widely distributed and perhaps increasingly standardized during the historic period (Blakeslee 1981; Brown 1989; Hall 1983, 1987, 1997). Jonathan Carver, one of the few American-born explorers to describe the interior of North America during the eighteenth century, referred to the calumet as "the Pipe of Peace" and "the sacred badge of friendship." He traveled with one affixed to the prow of his canoe. Carver, who penned some of the earliest descriptions of Great Lakes calumets and calumet ceremonies, noted, "Every nation has a different method of decorating these pipes, and they can tell at first sight to what band it belongs. It is used as an introduction to all treaties, and great ceremony attends the use of it on these occasions" (Gelb 1993:175). Some fifty years later, George Catlin wrote, "The Calumet, much the same in all the American Indian tribes, has invariably a small bowl, and a round stem, and is ornamented in this manner, with Eagles quills. This pipe is considered sacred, and is always in the possession of the chief, and never smoked except in Treaties of peace" (Ewers 1979:32). The early French explorer Pierre d'Esprit, Sieur de Radison, famously called calumets "pipes of peace and of the wars" (cited in Hall 1997:2).

Given the widespread diplomatic use of calumets, it is likely that Indian leaders presented some of the calumets in the Peabody collection, or others like them, to Lewis and Clark. Peale accessioned his Lewis and Clark pipes as "Pipes, or calmets," and because his wax model of Lewis depicted the explorer holding a "feathered peace pipe" that he gave to Peale, we know that the captains returned with at least one classic calumet.

Determining which of these calumets are most likely to have come from the expedition and which tribes might have presented them is more difficult. Before this research project, tribal attributions had been ventured for only three of the Peabody pipes in the 99-12 accession, the three remaining fanned calumets (99-12-10/53099.2, 99-12-10/53100.2, and 99-12-10/53101.2). Decades ago, these were identified as "Mandan" by Eric Douglas of the Denver Art Museum. Both Bodmer and Catlin painted Mandan people carrying fanned calumets, and the style is often associated with upper Missouri River horticultural peoples and with the northern Plains more generally. However, Prince Maximilian and Catlin each described seeing similar calumets in use from the lower Missouri River to the mouth of the Yellowstone, and

Calumet fragments
Denver Art Museum 1952.408
(Formerly PM 99-12-10/53102)
Missouri River(?), circa 1780–1830
Wood, eagle feathers, woodpecker mandibles and scalps, unidentified hair, probable porcupine quills, wool cloth, silk ribbons?, sinew, glass beads, pigment
(Materials identified from photograph only.)

this observation seems to be supported by historic paintings and existing museum collections. Calumets now in museums testify that diverse tribal groups used common forms throughout the nineteenth century. Travelers and collectors, including Catlin (Ewers 1979:32–33) and Prince Maximilian (Bolz and Sanner 1999:fig. 56), acquired similar calumets from the Sioux, Potawatomis (Feder 1964:fig. 10), Ioways (West 1934:pl. 185), Blackfeet (West 1934:pl. 182), Crows (Conn 1982, cover), Ojibwas (Brasser 1976:fig. 55), and others.

Despite the close relationships that developed between members of the Lewis and Clark expedition and Mandan people, none of the calumets itemized in the 1809 Peale memorandum was identified as Mandan. If any of the Peabody calumets are Mandan, they may not have been acquired during the Lewis and Clark expedition. According to the Peale memorandum, most of the upper Missouri pipes donated by Lewis and Clark were presented by Sioux peoples. Four divisions of the "nation Saux" are specified as having presented "Pipes, or calmets," to Lewis and Clark: the Yanktons, "Sharone's,"

Calumet
Denver Art Museum 1952.409
(Formerly PM 99-12-10/53103)
Missouri River(?), circa 1780–1830
Wood, mammal hide, eagle feathers, downy feathers, woodpecker mandibles and scalps, unidentified hair, weasel skin, probable porcupine quills, wool cloth, glass beads
(Materials identified from photograph only.)

Tetons, and "Dacoto's or Sue's."[7] The "Saux" attribution of one pipe may also be interpreted to mean "Sioux," because three pipes are described as "from the Sauke's" [Sauk], implying that in this instance "Saux" meant Sioux, not Sauk. Considering the generality of these terms and the range of Missouri and Mississippi River tribes who used fanned calumets, the tribal identities of the Peabody examples remain uncertain.

Calumet ceremonialism among Missouri River tribes may have been inspired by or distilled in the Pawnee ceremony that Alice Fletcher (1996 [1904]) called the Hako, a complex rite performed as a prayer for the gift of children, to consummate adoption, and to unify dissimilar or antagonistic social units such as clans and tribes. Most early calumets acquired along the Missouri from above St. Louis to the Mandan villages seem to share a number of features related to the two sacred pipe stems or wands used in the Hako, as Tahirussawichi, the last custodian of the ritual, described them to Fletcher. The common patterns in calumet form and meaning doubtless resulted from the fact that the Hako was often performed as an intertribal ritual, and the host group was required to transfer the calumets used in the ceremony to its guests. Early European descriptions of cylindrical calumets carried by Indians in the Great Lakes, Woodlands, and Southeast indicate that both their general form and general function were widely shared as early as the seventeenth century. The use and presentation of calumets in diplomatic encounters with non-Indians must have derived from this existing indigenous complex of ideas and practices. In turn, intercultural diplomacy further institutionalized calumet ceremonialism.

Omaha Sacred Pipes of Fellowship. This photograph was probably taken for Alice Fletcher and Francis La Flesche's study of Omaha culture, which was published by the Bureau of American Ethnology in 1911. Fletcher, a Peabody Museum anthropologist, and La Flesche, an Omaha anthropologist, recorded valuable information about Pawnee and Omaha calumet ceremonialism.

Examples of dance wands used in Hako-like calumet ceremonies are pictured in publications by Fletcher (1996 [1904]:39, 41), Fletcher and La Flesche (1992 [1911]:fig. 87), Hartmann (1973:pl. 15), and Skinner (1926:pl. 37). Although Hako wands were often not smoked (or even bored) and are defined in part by an arrowlike feather fletching on the stem, they share many characteristics with early calumets, including the Peabody examples. These features include the division of the stem into distinct design segments demarcated by bundled wrappings of materials such as bird heads, bird beaks, and sweet grass. For the Hako priest, both the elements and their placement on the stem had specific meaning; together, they achieved a unity of opposing powers. Feathers were widely traded, and most calumet accoutrements combined the scalps, bodies, and/or feathers of waterfowl (particularly mallard ducks), raptors (especially owl and eagle feathers), and, often, pileated or ivory-billed woodpeckers. Each of these birds had cosmological significance within the context of the Hako. Woodpeckers, for example, have ancient associations with warfare. On calumet stems, woodpecker beaks are often bent over backward, flattening the characteristic red crest and exposing the inside of the upper mandible. The Pawnee religious leader Tahirussawichi explained that this was to curb the bird's warlike tendencies (Fletcher 1996 [1904]:174), an idea that seems to have been widespread (e.g., Skinner 1926:224).

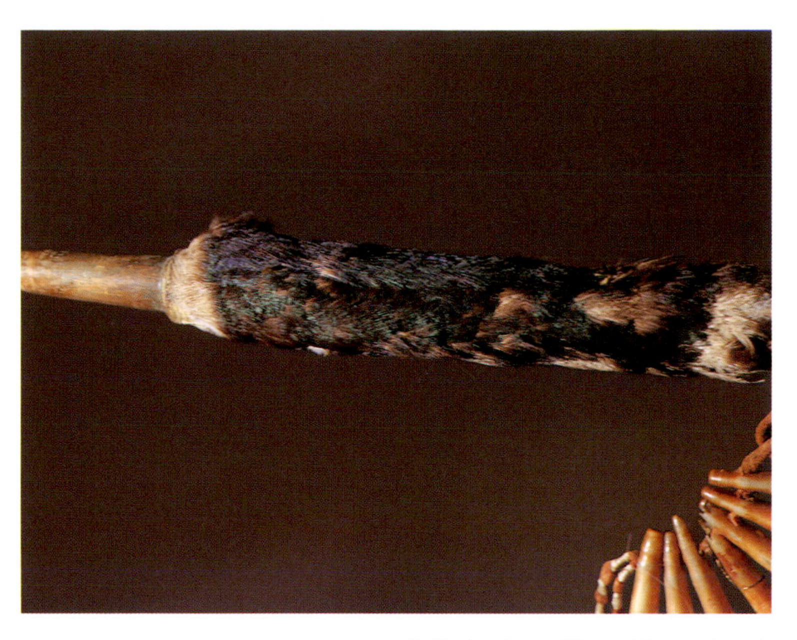

Mallard pelt on a Missouri River calumet stem. (Detail of 99-12-10/53100.2; see p. 223.)

European and American woven cloth, threads, and sewing notions were trade staples during early Indian-white interactions and soon became conventional diplomatic gifts and treaty payments. By the eighteenth century, Indian people so strongly favored British wools that the French, British, and Spanish all used them to vie for Indian allies and trading partners. Lewis and Clark distributed military coats, white and calico shirts, garments and blankets of scarlet and blue wool, bundles of gartering and ribbon, lengths of tape and braid, silk handkerchiefs, tinsel hatbands, and thread to the Indians they met (Moulton 1987, 3:492–505). The variegated patterns, textures, and colors of manufactured fabrics were readily integrated into indigenous systems of meaning and communication, and trade fabrics were often tied to pipe stems and other ritual objects. In some cases, the wool strips, silks, and other European fabrics may have been used to memorialize particular diplomatic encounters.

The three classic calumets at the Peabody are distinguished by pendant eagle feather fans, on which the feathers are supported by quill-wrapped slats. The two similar dance wands used in the Hako ritual each had a pendant fan of colored eagle feathers, symbolizing a cluster of gender-related associations. The brown feathers (from a golden eagle) represented female principles and motherhood, whereas the white feathered fan (from an immature golden eagle) symbolized male principles and the warrior (Fletcher 1996 [1904]:38–42). The feathers and bird bills attached to Hako wands are painted, and the stems are both painted and grooved.

Despite these shared design principles and the widespread use of calumets to affect social relationships and to initiate peace and war, cultural variation also shaped

IDENTIFYING FEATHERS

Carla Dove, of the National Museum of Natural History at the Smithsonian Institution, is the nation's only full-time forensic ornithologist. She is often called upon to identify a species of bird on the basis of a single feather—a task complicated by the fact that every bird has different kinds of feathers on its wings, head, and body, and the plumage of a single species differs on males and females and on older and younger birds. Her job requires a photographic memory.

Dove identified the feathers on the pipes and other objects in the Peabody Museum's Lewis and Clark collection by comparing them with bird specimens in the collection of Harvard's Museum of Natural History, both visually and through microscopic analysis. She identified nineteen species of birds, including the head of a male ivory-billed woodpecker, a species now believed to be extinct. (The Harvard museum also holds what is believed to be the only remaining bird specimen from the Lewis and Clark expedition, an example of *Melanerpes lewis,* commonly known as Lewis's woodpecker; see p. 29.)

Most of the feathers on the pipes, arrows, and other objects in the Peabody's collection came from eagles, hawks, owls, wild turkeys, and ducks. It came as a surprise to find that although golden eagles are called "calumet eagles" in the literature because their feathers were so often used on pipes, most of the eagle feathers on the Peabody pipes were from immature bald eagles. Young bald eagles have dark plumage, and their feathers are easily mistaken for those of golden eagles unless they are closely examined.

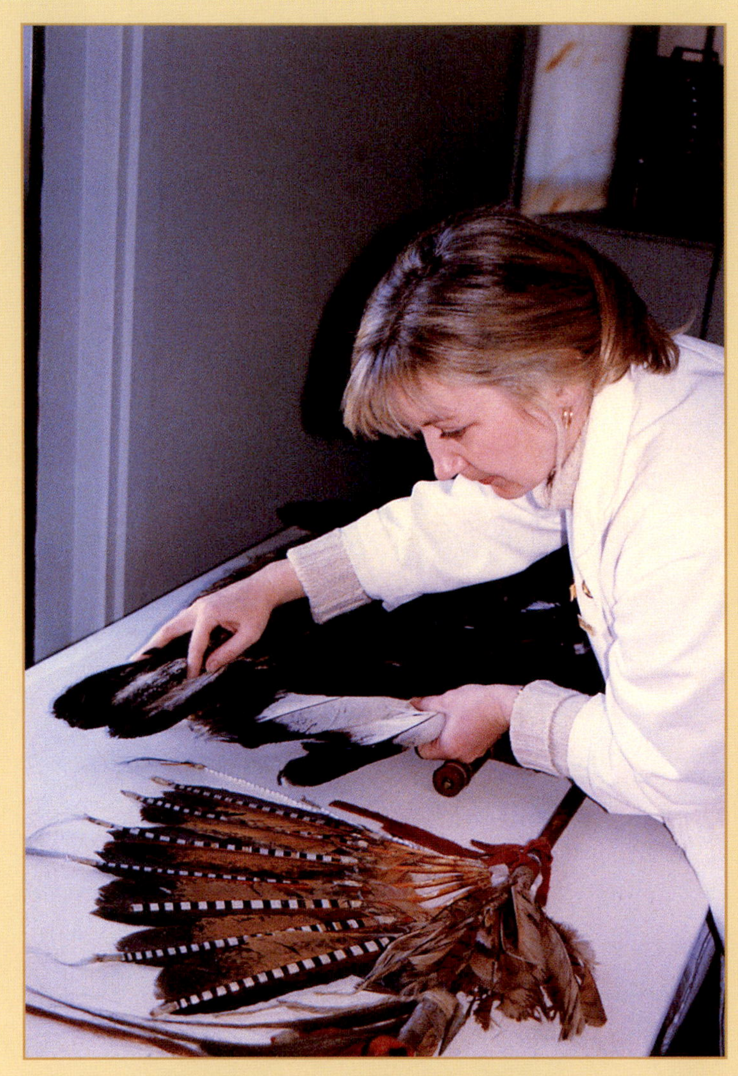

Carla Dove identifying the feathers on a Peabody calumet.

calumet ceremonialism. In some places and times, calumets were associated with diplomacy and secular authority; in others they were reserved for sacred contexts. The Blackfeet and other northern Plains groups considered calumets to be "medicine pipes" and included them in sacred bundles (e.g., Bodmer 1984:pl. 249; West 1934, 2:pl. 182). The Hako-inspired tradition of calumet ceremonialism, involving pairs of highly embellished dance wands or pipe stems, was concentrated along the lower Missouri River among peoples such as the Pawnees, Omahas, Otos, and Ioways, but variations were practiced all along the Missouri and farther east. The Ioways used Hako-style wands in the calumet ceremony along with similar calumet-style "peace pipes," one of which belonged to each of seven clans (Skinner 1926). Likewise, the Omahas used calumets both as tribal pipes and in calumet ceremonies (Fletcher and La Flesche 1992 [1911], 1:105). Calumet use probably varied among civil, religious, and corporate group leaders within many Missouri River tribes.

The decorative features of calumets, too, had overlapping, shared meanings as well as culturally and context-specific significance. For example, both Catlin and Father Marquette recorded that calumets were painted blue for peace or red for war. But the Pawnees and Omahas painted one of the wand stems blue to evoke the sky powers, and the other green, symbolizing the earth (Fletcher and La Flesche 1992 [1911], 2:380; Fletcher 1996 [1904]:38–40). The horsehair, yarn, and trade cloth that were typically wrapped around and suspended from calumet stems also had diverse, culturally specific meanings.

Calumet 99-12-10/53099.2 (see p. 222) is one of the most famous objects in the Peabody's Lewis and Clark collection. Since Charles Coleman Sellers (1980) identified it as an expedition artifact, it has become a symbol of that enterprise. Sellers, however, had no basis for attributing this particular calumet to Lewis and Clark. It is similar to a calumet that Prince Maximilian acquired near Fort Pierre in 1833, which is now in the collections of the Ethnological Museum Berlin (Bolz and Sanner 1999:fig. 56). But the presence of two separate eagle feather fans, reminiscent of those found individually on paired Hako wands, is quite unusual. The front fan, made of the dark, mottled feathers of an immature bald eagle, seems to have been added after the pipe was made, perhaps in a ceremonial or diplomatic context. The quills and vanes of the dark feathers are painted red, whereas the original white fan, made from immature golden eagle feathers, is painted orange. Brown-black, white, and orange porcupine quills are wrapped around the carved wooden slats attached to the white fan. In the center of each slat, white quills are woven vertically through the orange field, a technique also seen on quilled loops attached to the Peabody's raven

Calumet stem
99-12-10/53099.2
Missouri River, circa 1780–1830
Wood (probably semi-ring porous), eight immature bald eagle feathers (mottled), six immature golden eagle feathers, unidentified bird scalps, downy feathers, dyed and undyed horsehair, dyed artiodactyl (deer?) hair, unidentified hair, dyed and undyed porcupine quills, sinew, bast fiber cord, hide thongs, silk ribbons, cotton yarn, pigments
Length 107 cm, width 59 cm (with fan)

bustles. Blue-green paint once covered the delicate stem of this calumet, which is wrapped with bird scalps, silk ribbons, quill-wrapped hide thongs, cotton yarn, and dyed horsehair.

Burnt orange pigment covers the fan, the downy feathers, and the pendant horsehair on calumet 99-12-10/53100.2 (facing page). The stem is painted blue-green and is intermittently wrapped with the iridescent blue-green scalps of mallard ducks. Five clusters of delicate, quill-wrapped thongs hang from the stem, alternating with orange horsehair. The fan is made from seven immature bald eagle tail feathers that are supported by wooden slats wrapped with brown-black and white porcupine quills. To the tip of each feather is attached white weasel fur, orange-dyed downy feathers, and a pendant of orange horsehair. The impression of this calumet is softened by the dark brown downy feathers strung along the center of the feather fan and by the orange, quill-wrapped hide pendants suspended from the fan, which terminate in white and orange downy feathers.

In contrast, 99-12-10/53101.2 (p. 224), a calumet with classic Hako features, presents a strong, powerful character. Its formal features that correspond to those

Calumet stem
99-12-10/53100.2
Missouri River, circa 1780–1830
Wood (semi-ring porous), seven immature bald eagle tail feathers, mallard duck scalp, downy feathers, dyed horsehair, dyed artiodactyl (deer?) hair, unidentified (human?) hair, possible winter weasel fur, dyed and undyed porcupine quills, sinew, bast fiber cord, hide thongs, silk ribbons, pigments
Length 113 cm, width 58.5 cm (with fan)

seen on Hako wands include a (blue-green) painted and grooved stem, deliberately combined red and blue trade cloth wrappings, a center placement of great horned owl feathers, and seven attached woodpecker scalps and beaks. The blue-green paint has been identified as an earth pigment (glauconite or celadonite with quartz). The dark brown, black, and white porcupine quills wrapped around the wooden slats of the immature bald eagle feather fan create a bold impression when the fan is spread. The woodpecker beaks are painted blue-green and are bent over backward and tied down over bundles of sweetgrass and pendant horsehair. Red wool trade cloth is tied between five of the bundles, with blue fabric added near the feather fan. Six of the beaks are from pileated woodpeckers, and the seventh, obscured by the owl feather bundle, is that of a now-extinct ivory-billed woodpecker.

A notable feature of this calumet is the long, shimmering, copper-colored pendant suspended from the center, where it is tied on with strips of blue and red wool trade cloth. This pendant is made not from hair but from thin strands of wild hemp, twisted together to hold their shape (the same material is attached to pipe stem 99-12-10/53117.2; see page 233). In form, this calumet is consistent with other examples from

Calumet stem

99-12-10/53101.2

Missouri River, circa 1780–1830

Wood (probably semi-ring porous), seven immature bald eagle tail feathers, dyed and undyed great horned owl tail feathers, six pileated woodpecker mandibles and scalps, one ivory-billed woodpecker mandible and scalp, downy (mallard?) feathers, dyed horsehair, unidentified hair, dyed and undyed porcupine quills, hide thong, hemp (*Sesbania exaltata*), braided plant material (sweetgrass?), bast fiber cord, wool cloth, silk ribbons, cotton yarn, sinew, pigment, faux shell glass beads

Length 133 cm, width 72 cm (with fan)

IDENTIFYING WOOD

Because many Indians consider pipes to be sacred objects, the Peabody Museum staff did not want to remove pieces of wood from them for microscopic identification. Although we were able to sample a few loose splinters, Bruce Hoadley of the University of Massachusetts at Amherst based most of his identifications of the types of wood used to create the pipe stems and calumets on visual inspection alone.

Hoadley characterized the wood used for the calumets as "semi-ring porous" wood, a subdivision of the hardwoods that includes members of the poplar family such as cottonwood—an important tree along the Missouri River and in other riverine environments. Other Peabody pipe stems, such as 99-12-10/53108.2 and 99-12-10/53111.2 (p. 227), are probably made from ring porous hardwoods, which include shrubby woods. Buffalo Bird Woman, a Hidatsa, told the ethnographer Gilbert Wilson in 1916 that the men in her family made pipe stems from rose-bush stems, because they were hollow. The stems were split, the pith was removed, and the two sides were lashed together with sinew. All of the Peabody pipe stems, however, were made from single pieces of wood.

Hoadley was also able to confirm that several of the flat pipe stems, such as 99-12-10/53117.2 (p. 233), were made from ring porous hardwoods, including ash. Ash has a soft pith that was easily removed by boring the wood with a hot wire, and it was a widely favored material for pipes. It can be recognized by the alternating branching scars that remain on the wood surface.

Bruce Hoadley at the Peabody Museum.

Decorative woodworking on a Plains pipe stem. (Detail of 99-12-10/53114.2; see p. 238, top.)

the Missouri River area. It is similar to a Dakota (eastern Sioux) or Blackfeet example collected by Prince Maximilian (Schulze-Thulin 1987:pl. 90) and also to the Ioway clan pipes used to negotiate peace and restore intertribal social relations (Skinner 1926:pl. 32). An old label affixed to the stem and later painted over identifies this calumet as Cherokee. This label cannot be wholly disregarded, because delegations of Southeastern peoples visited Philadelphia and Washington, D.C., during the eighteenth and early nineteenth centuries.

Pipes 99-12-10/53107.2, 99-12-10/53108.2, 99-12-10/53111.2, and 99-12-10/53116.2 are also calumet-style stems, but they lack pendant fans. Stem 99-12-10/53107.2 (facing page), made from a lightweight hardwood, may be unfinished, because bowl supports are carved at each end. On one side of the stem, the branching scars in the wood have been burned, creating dark spots; on the other side, they have been incised and burned in a banded pattern. Stains on the wood suggest that the stem was once wrapped with silk that has been lost. A similar calumet stem, decorated only with a banded pattern, was identified by George Dorsey (Ewers 1986:fig. 119) as a Pawnee "sacred pipe of peace or war." Similarly banded stems, often executed by being marked with metal files, are also found on eastern Plains and Woodlands pipe-tomahawks. The body of pipe 99-12-10/53108.2 (facing page) has been carved into a spiral, and the shoulder has been carved down, leaving four intermittent, burl-like "knobs" considerably wider than the stem proper. Three lengths of red-dyed deer hair are wrapped to the shoulder with silk ribbon, which also once covered the intervening spaces between the hair attachments.

Calumets 99-12-10/53111.2 and 99-12-10/53116.2 are probably from the eastern Plains or western Great Lakes rather than from the upper Missouri or central Plains. The shoulder of 99-12-10/53111.2 (facing page) is wrapped with red wool trade cloth, silk ribbon, and warp-faced, plain-woven wool tape edged with white glass beads. A bird scalp and lengths of hair are attached to the center of the stem, and the body of the pipe is wound with more silk ribbon. The hair bundles attached to this calumet were dyed red and blue with a resist dye (masking) technique. The narrow beaded band is wrapped in a spiral pattern around the red wool trade cloth. Fabric weaving, especially incorporating beads, was a Woodlands and Great Lakes technology that also became popular among eastern Plains women during the historic era. The beaded wool tape on this pipe was probably produced on a carved wooden heddle, a device Indian peoples began utilizing around 1800.

Calumet 99-12-10/53116.2 (facing page) looks like a dance wand, but the stem is bored. Midway along the stem are circular, incised bands that appear to have been

Top to bottom:

Calumet stem
99-12-10/53107.2
Eastern Plains/western Great Lakes, circa 1800–1850
Wood (semi-ring porous)
Length 104 cm, width 1.8 cm

Calumet stem
99-12-10/53108.2
Eastern Plains, circa 1800–1850
Wood (ring porous), dyed artiodactyl (deer?) hair, dyed horsehair, bast fiber cord, wool cloth, silk ribbons
Length 86 cm, width 2.5 cm

Calumet stem
99-12-10/53111.2
Eastern Plains/western Great Lakes, circa 1800–1850
Wood (ring porous), unidentified bird scalp, dyed horsehair, bast fiber cord, wool cloth, warp-faced plain-woven wool tape, silk ribbons, glass beads
Length 96 cm, width 2.8 cm

Calumet stem
99-12-10/53116.2
Eastern Plains/western Great Lakes, circa 1800–1850
Wood (semi-ring porous), unidentified (horse?) hair, unidentified (deer?) hide, dyed and undyed porcupine quills, dyed bird quills, sinew, hide strips, bast fiber cord, wool cloth, silk ribbon tapes (one with metallic-wrapped silk yarns and silk tassels with metal alloy sequins), glass beads
Length 74.5 cm, width 1.7 cm

Below: Detail of 53116.2

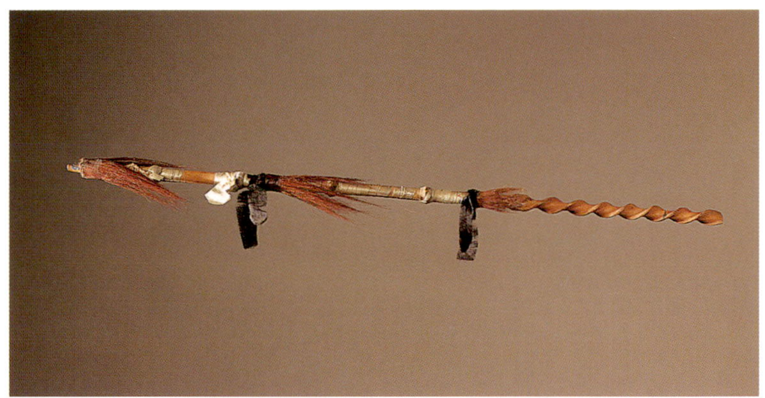

burned or darkly stained. The stem is wrapped at four intervals with red wool trade cloth faced with delicately beaded silk fabric and with a fancy, multicolored silk ribbon with fringed, gold metallic tassels. This delicate ribbon, perhaps made by a colonial American or European weaver who specialized in trim, was probably intended as a garment ornament. The red cloth streamers were once also faced with silk and then beaded. A rectangular hide strip quilled in simple bands of white and orange is affixed to the mouth of the pipe, to which are attached six wrapped bundles of unidentified (possibly horse) hair. This attachment may have been recycled from a former use. The use of European fabrics embroidered with white glass beads again suggests a Great Lakes or eastern Plains origin for this calumet.

Thunderbird design on a flat-stemmed pipe. Skilled artists could create figurative forms by wrapping together individual strands of braided quillwork. (Detail of 99-12-10/53117.2; see p. 233.)

Flat-Stemmed and Miscellaneous Pipes

The other pipes in the Peabody collection are of various styles, including wide flat-stemmed pipes, ovoid pipes, twisted pipes, and a tube pipe. Like the calumets, many of them have grooved and painted sides and incorporate materials such as bird quills, hand-spun vegetable fiber (bast) threads, unidentified plant materials, and early trade fabrics. These characteristics support pre-1850 dates of manufacture and use. Like the calumets, these pipes could have been acquired by Lewis and Clark in 1804, by Hutter in 1825, or by Peale in 1826.

Most of the flat-stemmed pipes in this collection appear to have originated in the eastern Plains–western Great Lakes region. They are the types of pipes that might have been carried by members of the Mississippi River tribes listed in the 1809 Peale memorandum as having given pipes to Lewis and Clark—Ioways, Winnebagos, Sauks, Foxes, Potawatomis, and Dakotas. They might also have been carried by members of the combined Missouri and Mississippi River delegation that Lewis sent to visit Jefferson in 1805–6 or by representatives of regional tribes that Clark and Lewis dealt with after the expedition in their capacities as U.S. government officials stationed in St. Louis. Both Lewis and Clark presented pipes to Jefferson and to Peale in the years immediately after their expedition.

George Catlin (1973, 1:235) referred to flat-stemmed, quill-wrapped pipes as "ordinary pipes, made and used for the *luxury* only of smoking; and for this purpose,

SILK RIBBONS

Silk ribbons like those that decorate many of the pipe stems in this collection were popular in Europe during the eighteenth and nineteenth centuries and were readily incorporated into colonial American and Native American clothing and accessories. Ribbons and fancy woven trims and braids served essential functional purposes and ornamented both interior furnishings and garments. Ribbons were widely favored as cockades on military and civilian hats and as adornments for fashionable wigs and hairstyles. Indians used silk ribbons as garters for leggings, trimmings for bags and sashes, and ornaments for calumets and other ceremonial items. Lists of the trade goods procured by Lewis and Clark for their Northwest journey show that they packed rolls of narrow ribbon, small bundles of ribbon, and pieces of striped silk ribbon as gifts for the wives of chiefs.

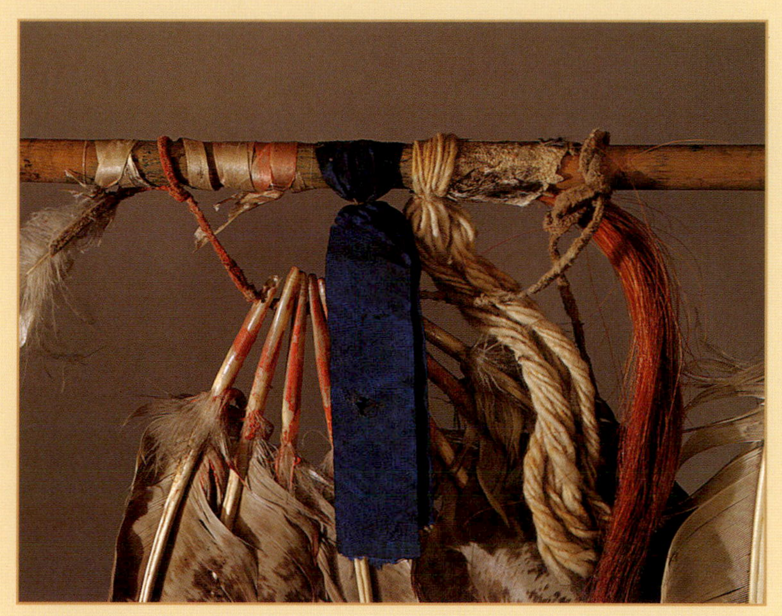

Ivory, pink, and blue silk ribbons on a calumet stem (99-12-10/53099.2; see p. 222). Light exposure has faded the ribbons on many of the Peabody pipes.

Most of the remaining ribbons on the pipe stems in the Peabody collection are of solid colors. Most prevalent are natural ivory-colored silk ribbons of varying widths (15.5 to 27.0 mm) and weave structures (satin, plain, or a fancy weft-float pattern with picot edges). Silk ribbons were generally colored with insect or plant dyestuffs. An unusually narrow (7.5 mm) satin ribbon, originally dyed a bright reddish pink, appears on eleven of the pipe stems. These ribbons may have been dyed with cochineal or perhaps with safflower, which produces a brilliant hue but is extremely sensitive to light, and the ribbons have now faded. Many once brilliantly colored ribbons now appear to be a cream color.

In European society, blue ribbons and cloth were symbolically important and played a role in diplomatic affairs. Many Indians, too, expressed a strong preference for blue. Three of the pipe stems have bluish green or navy blue ribbons still attached—colors probably derived from indigo or woad, the best blue dyestuffs. There are also three pipe stems with 24- to 25-millimeter-wide green ribbons, a color possibly created from a combination of weld (a yellow dyestuff) and indigo or logwood. Plain woven ribbons (24–25 mm wide) dyed dark brown to black were also popular and are seen on six of the pipes. One of the wider flat pipe stems has two types of silk ribbons stitched or appliquéd to narrow strips of red wool fabric, and one calumet is wrapped with a fancy, multicolored inkle or ferret trim with metal-wrapped threads.

The variety of innovative ribbon wrapping techniques found on these pipe stems is notable. Because silk is extremely fragile, in some cases only soiled and stained patterns on the wooden stems mark places where original wrappings have been lost. Some ribbons were carefully arranged, spiraling in one direction or another along the length of the stem; others spiraled in both directions, creating diamond motifs in negative space. In one location there are multiple wrappings of three or more different colors of ribbons. Many of these wrappings were symbolic rather than purely decorative—for example, the single blue-dyed silk ribbon carefully positioned close to the blue feathered bird scalp on calumet 53101.2 (see p. 224).

T. Rose Holdcraft

every Indian designs and constructs his own pipe." Many tribal, clan, and "treaty" pipes, however, are of this form and were used ritually, as were circular-stemmed calumets (e.g., Hall 1997; West 1934, 2:pls. 183, 186). Quilled, flat-stemmed pipes were seen among and collected from peoples of the central Plains, western Great Lakes, and eastern Plains-Prairie edge. Catlin illustrated and acquired many such pipes. He observed that the stem design was standardized, with the quillwork never extending to the shoulder, or front of the stem (Ewers 1979:30). As he found to be the case with calumets, Catlin noted that "the great similarity in the mode of making the [flat] stems in the various tribes, and the constant exchange among them, render it difficult to distinguish them from each other" (Ewers 1979:40).

There are four flat-stemmed pipes in the Peabody collection, all similar in design, having long, flat, tapering stems, braided quillwork, center hair attachments, and dyed horse and deer hair "caps" covering the smoking ends.[8] On each of these pipes, strips of brown (varying from red-brown to black-brown), orange, and white braided quillwork cover what Catlin referred to as the "body" of the stem. This tricolor palette, produced from organic dyes, is characteristic of early Woodlands quillwork. Braiding, the quillwork technique used on most of these stems, was time consuming and complex. After dying and flattening her quills, a woman braided one or two of them around two parallel anchoring threads of bast (vegetable stem) fiber or sinew, which were often spaced less than an eighth of an inch apart. By continually adding quills in a figure-eight pattern, she could produce a long, delicate braid that was then wrapped around the pipe stem (Orchard 1916:32–33). Most of the foundation threads used in the bands of braided quillwork on the pipes in this accession are made of bast fiber rather than sinew. Although women from the Woodlands to the Plains decorated pipe stems with braided quillwork, the spinning of bast fibers and the decoration of the stem in banded blocks of color seem to have been favored in the eastern Plains and western Great Lakes (e.g., King 1977; Skinner 1926).

Of the four flat-stemmed pipes in this collection, pipe 99-12-10/53104.2 (facing page) may be the most "western" in character, although the stem seems to have originated at the Plains-Prairie edge. Bold blocks of natural white, orange, and brown quillwork are formed by braided strips wrapped around the stem, with some variation in secondary design elements on the dorsal and ventral sides. The center of the stem is wrapped with hide loops covered in plant fiber (cornhusk?), white porcupine quills, and red-orange bird quills. Large, handmade metal (tin-plated sheet steel) cones filled with tufts of hair dangle from quill-wrapped hide thongs. The sides of the stem

are channeled and filled with blue-green paint. Horsehair wrapped to the stem with sinew seems to have been dyed with a resist-dye technique, creating an effect of banded colors. Both resist-dyed hair and "cornhusk" wrapped loops are also seen on the raven bustles illustrated in chapter 5. Whereas the loops and grooved stem seem to be "western" traits, the geometric blocks of quillwork and the placement of the hair on this pipe are reminiscent of early Dakota, Ioway, and Missouri pipes (e.g., Flint Institute of Arts 1973:fig. 56; Schulze-Thulin 1987:fig. 83; Skinner 1926). The Yankton leader Padanin Apapi (Strikes the Ree) presented a pipe of this style to the Jesuit missionary Father De Smet in 1857 (Peterson and Peers 1993:133).

Pipe 99-12-10/53104.2 arrived at the Peabody Museum with an unusual hide container (99-12-10/53104.3; at right) tied to the center of the stem. The body of the container is made from a mammal (probably bison) bladder, to which is attached a conical piece of leather covered with simple bands of orange and white porcupine quills and brown bird quills; it is finished with an orange quilled border. The interior of the cone is lined with bison hair, to which three

Top: **Pipe stem**
99-12-10/53104.2
Eastern Plains; Yankton(?), circa 1800–1850
Wood (ring porous, possibly ash), dyed horsehair, dyed artiodactyl (deer?) hair, dyed and undyed porcupine quills, dyed bird quills, unidentified plant material (cornhusk?), bast fiber cord, sinew, hide thongs, pigment, copper alloy metal cones, silk ribbons
Length 104 cm, width 3.8 cm

Bottom: **Bladder bag**
99-12-10/53104.3
Eastern Plains, Yankton(?), circa 1800–1850
Mammal (bison?) bladder, mammal (bison?) hide, bison hair, horse(?) hair, sinew, metal alloy cones, dyed and undyed porcupine quills, dyed bird quills, unidentified plant material, pigment
Length 85 cm, width 14 cm

quill-wrapped leather strips are sewn. Handmade metal (tin-plated sheet steel) cones and horsehair are attached to the pendant strips midway down their length, and the pendants are wrapped with red, yellow, and brown bird quills and white porcupine quills. Unlike most other bladder bags, the quilled neck of this container has no aperture, and it does not appear to open. Hunkpapa Lakota artist Butch Thunder Hawk examined this bag and thought that it could be a tobacco container or symbolic of a tobacco container, noting that the treatment of the neck and the dangles reminded him of a pipe tamper. George Horse Capture, who has seen only photographs of the bag, suggested that it might be a rattle.

Pipe bag
Denver Art Museum 1952.283
(Formerly PM 99-12-10/53003)
Plains, circa 1800–1840
Mammal hide, glass beads, pigment
(Materials identified from photograph only.)

According to the Peale memorandum, Lewis and Clark donated five tobacco bags to the Peale Museum, in addition to fourteen pipes. Two of them, described as "very handsomely ornamented . . . with Porcupine Quills, and Tin, &c.," were accessioned as "Saux," which could be interpreted as either Sioux or Sauk. Another was "from the Ioway's," and a fourth "from the Raneird's or Foxes." The fifth bag was said to have been "sent by the Sacks," a probable reference to the Sauk. If this bladder bag and the related pipe are indeed from the expedition, then they are likely to have been presented by Sioux (Yanktons?) or Ioways. Another pipe or tobacco bag (99-12-10/53003), formerly part of this collection, was traded to the Denver Art Museum with the two calumets in 1952. That bag, illustrated here, is of an early Plains or Prairie style, boldly and simply decorated with blue and white pony beads.[9]

Another flat-stemmed pipe, 99-12-10/53117.2 (facing page, top), has a markedly tapering stem wound with braided quillwork bearing thunderbird designs (see p. 228). Considered by some scholars to be diagnostic of Dakota (eastern Sioux) art, thunderbirds were important throughout the entire eastern Plains–Great Lakes region. The quillwork on this pipe has become loosened, and there is no hair cap at the smoking end. As on calumet 99-12-10/53101.2, a long, copper-colored twist of fine hemp fiber is pendant from this pipe. An old museum label reading "pipe of peace" and an old fourth-class postage label remain attached to the stem. The sides of this pipe are also grooved and are filled with red pigment.

A third pipe of this style, 99-12-10/53105.2 (facing page, bottom), is robust in appearance, with a thick, tapering stem braided with large, shiny porcupine quills.

Pipe stem
99-12-10/53117.2
Eastern Plains/western Great Lakes, circa 1800–1850
Wood (ring porous, probably ash), unidentified dyed (horse?) hair, hemp (*Sesbania exaltata*), dyed and undyed porcupine quills, bast fiber cord, sinew, silk ribbons
Length 116 cm, width 5 cm

Pipe stem
99-12-10/53105.2
Eastern Plains; Ioway or Missouri (?), circa 1800–1850
Unidentified wood, dyed and undyed horsehair, unidentified hair, dyed and undyed porcupine quills, bast fiber cord, sinew, wool cloth (stroud), twill-woven wool tape, silk ribbons
Length 109 cm, width 4 cm

The shoulder of the pipe is wrapped with red wool trade cloth, possibly stroud, and with black and white silk ribbons. At the middle of the pipe, where the quillwork meets the cloth, a bundle of unidentified wavy hair is attached. On one side of the stem, the braided quillwork forms orange triangular shapes, perhaps representing woodpecker beaks. Double-barred crosses (dragonflies?) appear on the reverse side. Both the quillwork on the stem and the overall design of the pipe suggest that it might have been made by an Ioway or a Missouri artisan. It is similar to a number of Ioway or Missouri pipes pictured by Skinner (1926) and to a pipe sent to President Monroe by the Ioway chief White Cloud in 1824 (Ewers 1981:fig. 2; see also the Ioway pipe in Penney 1992:fig. 198). The Peale memorandum lists two "Ioway" pipes donated by Lewis and Clark, and there were Ioway and Missouri members of the 1805–6 delegation to Washington, D.C.

The fourth flat-stemmed pipe, 99-12-10/53106.2 (facing page), shares the same basic design features as the other three. The body of the stem, which tapers markedly, is wrapped with white, orange, and brown braided quillwork strips that form color bands and geometric shapes. These designs continue around both the dorsal and ventral surfaces. A hair bundle is attached at the center, and a dyed red hair cap covers the smoking end. Traces of blue-green pigment adhere to the stem. On the shoulder is wrapped a heddle-woven wool and glass bead garter of the type popular in the eastern Plains and western Great Lakes during the early nineteenth century (e.g., Penney 1992:figs. 17, 18). This example conforms to what Dennis Lessard (1986) called "regular heddle work," which may have been the earliest heddle technique adopted by Indian women. The warp threads alternate between red and olive green, and the weft threads are light brown. White beads are woven into the garter as a base color, and sets of four translucent, cut "garnet" beads, a type Lewis and Clark distributed, create small diamond or square designs.

Pipe 99-12-10/53109.2 (p. 236) is distinctly different from the four flat-stemmed pipes. Its stem is ovoid in cross section, and different principles were used to design the stem. Whereas the flat-stemmed pipes are quilled only on the body of the stem, on this pipe the shoulder is also elaborately quilled. The color palette is the same, but a very different kind of quillwork was used, combining wrapping and weaving. The technique of wrapping flattened quills around pipe stems is associated predominantly with the Woodlands and the Great Lakes and may be quite ancient (e.g., Benndorf and Speyer 1968:cat. no. 170; King 1982:fig. 76; Orchard 1916:pl. 16). Here, some of the quills have also been woven into geometric patterns, another technique seen on eigh-

teenth-century objects from the Woodlands and Great Lakes. During the nineteenth century, the Dakotas and Ojibwas used wrapping and weaving on pipe stems and other objects (e.g., Casagrande and Ringheim 1980:nos. 28, 32; King 1977:pl. 16, 1982:fig. 76). Woven quillwork of this type is often seen on quill-wrapped feather slats, including those on the "white" feather fan attached to calumet 99-12-10/53099.2 (see p. 222). Woven quillwork also appears on the feather slats of the compound raven belt ornament 99-12-10/53049 (see p. 109) and on a quill-wrapped loop attached to 99-12-10/53051 (see p. 111), one of the two single raven belt ornaments. The warp-faced, plain-woven wool tape edged with beads that is tied to the stem of this pipe was made on a heddle and also indicates a Woodlands, Great Lakes, or eastern Plains origin.

Pipe stem 99-12-10/53110.2 (facing page) is also slightly ovoid in cross section. This stem is elaborately covered with nineteen component materials, and the

Pipe stem
99-12-10/53106.2
Eastern Plains/western Great Lakes, circa 1800–1850
Wood (ring porous), dyed horsehair, dyed and undyed porcupine quills, bast fiber cord, pigment, wool cloth, silk ribbons, wool and glass bead garter
Length 119 cm, width 5 cm

Garnet beads. Lewis and Clark distributed beads like the ones woven into the garter wrapped around this pipe stem. (Detail of 99-12-10/53106.2.)

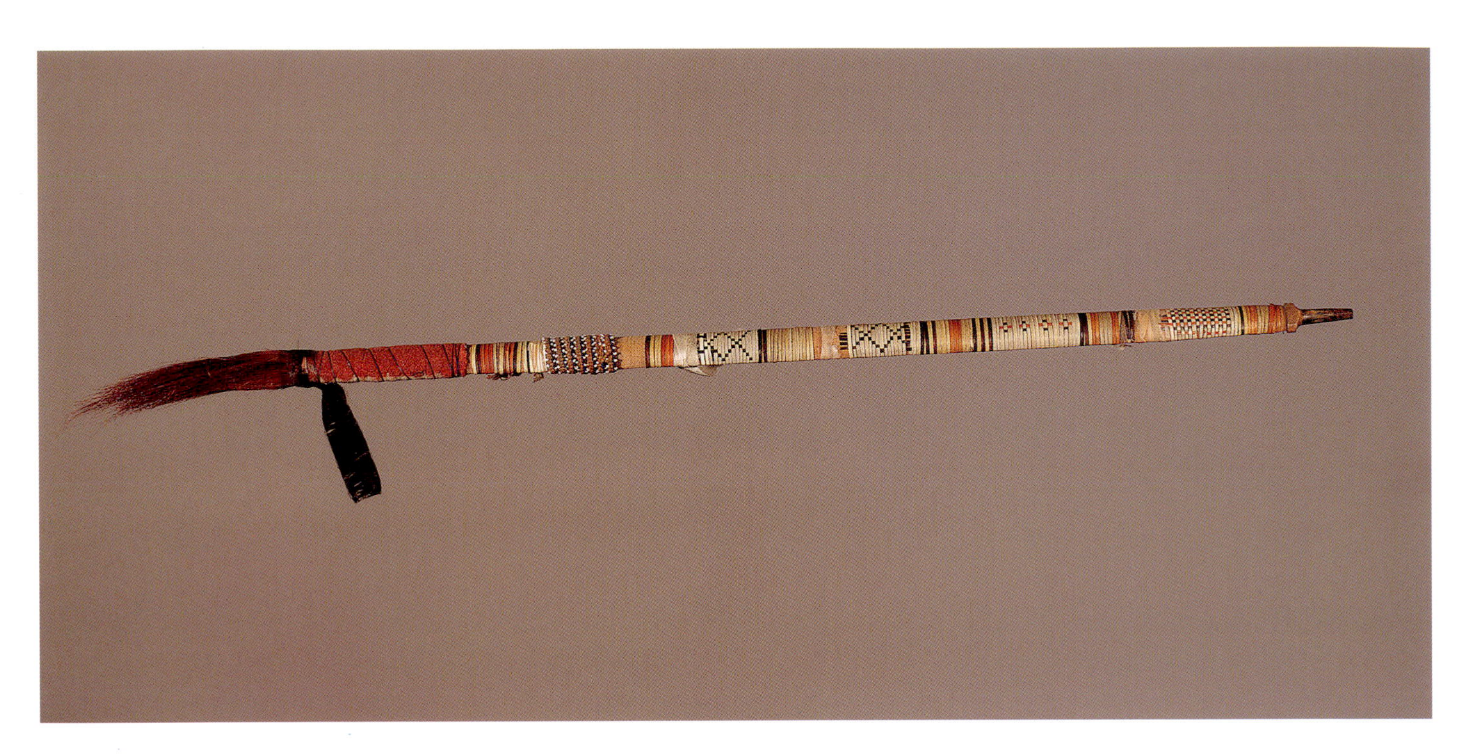

Pipe stem
99-12-10/53109.2
Eastern Plains/western Great Lakes, circa 1800–1850
Unidentified wood, dyed horsehair, dyed and undyed porcupine quills, sinew, bast fiber cord, wool cloth, warp-faced plain-woven wool tape, silk ribbons, glass beads
Length 89.5 cm, width 2.8 cm

Detail of quillwork and beaded fabric on pipe stem 99-12-10/53109.2.

braided quillwork is quite complex. Six large, primary design elements—hourglass shapes (thunderbirds?) on one side and diamonds on the other—appear in bold color blocks of yellow and dark orange. These primary motifs are all "outlined" by dark and light quills, creating a pattern of alternating positive and negative images. Smaller secondary design units, such as stripes, occur between the primary elements. Both the design elements and the color blocks differ on the ventral and dorsal faces of the pipe stem.

The blue fabric and the five strings of white wampum beads, probably quahog shell, on this pipe are materials that Clark described adding to a pipe given to him by the Nez Perce chief Broken Arm. Peabody conservator T. Rose Holdcraft observed that the blue glazed fabric wound around this stem appears indeed to have been a late addition; it is wrapped far up the stem and tied closely to the bowl, which might have made the pipe difficult to use.

Pipe stem
99-12-10/53110.2
Eastern Plains/western Great Lakes, circa 1800–1850
Unidentified wood, ivory-billed woodpecker head and scalp, wood duck (*Aix sponsa*) face patch, dyed downy feathers, dyed horsehair, dyed artiodactyl (deer?) hair, dyed and undyed porcupine quills, sinew, bast fiber cord (two types), glazed cotton fabric, twill-woven wool tapes, silk ribbons, shell beads
Length 91.2 cm, width 3.4 cm

More than a dozen materials were used on this decorated pipe stem, all of which had symbolic meaning for the maker. (Detail of 99-12-10/53110.2.)

Top: **Pipe stem**
99-12-10/53114.2
Eastern Plains, circa 1800–1850
Wood (ring-porous), pigment
Length 105.8, width 3.5 cm

Bottom: **Pipe stem**
99-12-10/53115.2
Eastern Plains, circa 1800–1850
Wood (soft maple group), pigment
Length 113 cm, width 4.8 cm

However, the quillwork and other design features, such as the presence of an ivory-billed woodpecker scalp and the face patch of a wood duck, point to an eastern Plains–western Great Lakes origin rather than a Columbia Plateau (Nez Perce) source—an argument against this pipe's having been the one received by Clark.[10] If it was given to Clark by Broken Arm, then the Nez Perce must have acquired it through trade.

Pipes 99-12-10/53114.2 and 99-12-10/53115.2 (facing page) are both beautifully carved into long, twisting spirals. This technique was already a favorite among the Dakota people by the time Catlin visited the West, although it was also practiced by other eastern Plains and western Great Lakes groups such as the Ojibwas. By Catlin's time, Plains and Great Lakes Indians were skillfully using metal tools to carve and decorate woodwork. The edge of a metal tool such as a nail was probably used to create the arched shapes impressed into the surface of 99-12-10/53114.2, which has developed a distinctive, warm patina. The surface of 99-12-10/53115.2 was scored with a metal file. The sides of both of these twisted stems are channeled and filled with red pigment.

Pipe 99-12-10/53009.2 (see p. 240) is the singular example of a tube pipe in this collection. Tube pipes are believed to have great antiquity in North America and were used in ritual contexts and by hunting parties during the historic era (e.g., Kroeber 1983:21; Thwaites 1905d:391). This tube pipe was carved from the left tibia of a small to medium-sized mammal, such as a deer. The triangular aperture is lined with sheet copper. Concentric circles carved around the center suggest an affinity to some of the pipe bowls in this collection, which are discussed later.

Among the interesting features of these pipe stems are the very long hair caps placed over the smoking ends. It is unclear whether these were conventional or whether their placement might be associated with presentation pipes. Butch Thunder Hawk notes that these caps would have been removed for smoking sessions. The caps on the pipes in this collection are made of bundles of horse and artiodactyl (probably deer) hair, dyed red. They were made by sewing multiple smaller bundles of hair together, similar to the construction of a hair roach. The cords that secure the hairs are made from hand-spun vegetable bast fibers. Bast stem cords were also used to anchor the braided quillwork on many of these pipes.

Tube pipe

99-12-10/53009.2

Eastern Plains/western Great Lakes, circa 1800–1850

Left tibia of a mammal (deer?), copper

Length 13.5 cm, width 2.5 cm

Pipe Bowls

The Peabody catalogue numbers suggest certain pairings between the pipe bowls and stems in this collection. Although my colleagues and I feel confident that these stems and bowls do represent pairs, we cannot be certain that each bowl has retained its original association with a particular stem—a caveat first offered by Charles Willoughby when he accessioned the pipes for the Peabody Museum. Unfortunately, several pipe bowls that were part of the 99-12 accession have since been lost. The formal features of many of these bowls suggest that they, like the stems, originated in the western Great Lakes

Top to bottom:

Pipe bowl
99-12-10/53099.1
Eastern Plains/western Great Lakes, circa 1800–1850
Red pipestone (probably catlinite)
Elbow form; incised
Length 12 cm, height 6 cm

Pipe bowl
99-12-10/53100.1
Plains/western Great Lakes, circa 1800–1850
Red pipestone (probably catlinite)
Elbow form with prow
Length 14 cm, height 7 cm

Pipe bowl
99-12-10/53101.1
Eastern Plains/western Great Lakes, circa 1800–1850
Red pipestone (probably catlinite), speckled
Elbow form; incised
Length 13.5 cm, height 7.8 cm

Top: **Pipe bowl**
99-12-10/53105.1
Eastern Plains/western Great Lakes, circa 1800–1850
Red pipestone (probably catlinite), speckled; silk ribbon
Elbow form with slight prow and crest, beveled and incised
Length 14 cm, height 8 cm

Bottom: **Pipe bowl**
99-12-10/53107.1
Eastern Plains/western Great Lakes, circa 1800–1850
Unidentified stone (fine grain)
Elbow form with slight prow; beveled, flat base
Length 12.6 cm, height 8 cm

or eastern Plains. As with the pipe stems, our understanding of historic bowl forms used west of the Mississippi River is based primarily on examples that were painted by frontier artists or collected after about 1830. The jutting prows, crests, carved rings, and flat bases on many of the Peabody pipe bowls are considered characteristic of pipe bowls made during the first half of the nineteenth century (Ewers 1986). Many of these features are also found on protohistoric pipe bowls from the Woodlands and western Great Lakes.

Ten of the Peabody pipe bowls are carved from a red pipestone that is probably catlinite, a malleable metamorphic clay. This beautiful material was first described to the non-Indian world by the frontier artist George Catlin, in a letter read to the Boston Society of Natural History in 1839. It was named in his honor by Charles Thomas Jackson, a Boston mineralogist (Ewers 1979:11–12, 69–77). In 1836, Catlin traveled from the East to the major catlinite quarry, located in southwestern Minnesota, which he also memorialized in a painting. During the late prehistoric and early historic era, Native Americans treated the quarry as both a sacred site and a neutral zone where people from many tribes met peacefully while excavating pipestone in a ritual manner (Hall 1991:21). This resource sharing highlights the alliance-building and peace-making function of calumet ceremonialism.

Raw catlinite and catlinite "disk" bowls were key trade items for peoples who participated in the late prehistoric Oneota tradition, which has been identified archaeologically from Minnesota to Missouri throughout the period from 900 to 1650 C.E. (e.g., Hollinger and Benn 1998). The Dakotas asserted a claim over the Minnesota quarry during the historic era and were famous in Catlin's day as producers of catlinite pipe bowls. The Ojibwas and Pawnees were also noted carvers of catlinite pipes. During the nineteenth century, catlinite pipe bowls carved in "elbow" (99-12-10/53099.1; see p. 241), prowed elbow (99-12-10/53100.1; see p. 241), and effigy (99-12-10/53106.1; see p. 247) forms were widely circulated in intertribal trade networks. Fletcher and La Flesche (1992 [1911], 2:378) listed catlinite pipe bowls as appropriate gifts to be given by the ceremonial "fathers" in an Omaha *wa'wa*$_n$ ceremony. The Mandan leader Crow's Heart, however, specified that his people used catlinite bowls only in adoption ceremonies (Bowers 1991 [1950]: 329), because Mandan women made bowls from local clays. Catlinite pipe bowls were also highly prized by non-Indians, some of whom (such as Catlin) began building pipe collections during the fur trade era. Toussaint Charbonneau, Sakakawea's husband and a member of the Corps of Northwest Discovery, collected carved pipe bowls depicting erotic themes (Ewers 1979:18).

Although other sources of catlinite have been identified throughout the upper Midwest, particularly in Wisconsin, the Minnesota quarry has remained the most significant to Native Americans. The site was designated Pipestone National Monument in 1937 and is now administered by the National Park Service. Indian people still quarry catlinite there, by permit. Although there is some disagreement among Native Americans over whether it is appropriate to sell objects made of pipestone, Native artists continue to carve pipe bowls for both Indian use and the non-Indian market.

Top: **Pipe bowl**
99-12-10/53110.1
Eastern Plains/western Great Lakes, circa 1800–1850
Red pipestone (probably catlinite), lead inlay
Elbow form; incised
Length 7 cm, height 8.5 cm

Bottom: **Pipe bowl**
99-12-10/53104.1
Eastern Plains/western Great Lakes, circa 1800–1850
Red pipestone (probably catlinite)
Elbow form; beveled
Length 11.3 cm, height 6 cm

Top: **Pipe bowl**
99-12-10/53111.1
Eastern Plains/western
Great Lakes, circa 1800–1850
Unidentified stone
Elbow form with slight prow;
perforated crest, incised, flat base
Length 11.4 cm, height 7.9 cm

Bottom: **Pipe bowl**
99-12-10/53113.1
Eastern Plains/western
Great Lakes, circa 1800–1850
Red pipestone (probably catlinite)
Elbow form with prow, low crest
Length 9.3 cm, height 5.5 cm

Several of the pipe bowls in this collection (99-12-10/53099.1, 99-12-10/53100.1, 99-12-10/53101.1, 99-12-10/53105.1, 99-12-10/53107.1, and 99-12-10/53110.1; pp. 241–43) have distinctive carved rings around both the smoking and bowl apertures. This feature may have originated in the Woodlands, but it was adopted by nineteenth-century Plains pipe bowl carvers. Contemporary Plains Indian carvers execute a similar banded design, sometimes called the "four winds." George A. West (1934, 2:pls. 174, 192) pictured a number of ringed bowls, all from Wisconsin, some apparently recovered from archaeological contexts. Another, smaller example was found at the Utz site in Missouri (Bray 1991:fig. 20e), a former Oneota village occupied into the historic period by ancestors of the Missouri tribe. Iroquois speakers carved rings around pipe bowls; a seventeenth-century Iroquois example is illustrated by J. C. H. King (1977:pl. 10). The wedge-shaped stem of bowl 99-12-10/53107.1 (p. 242) and the flat bases and faceted stems of a number of other bowls, such as 99-12-10/53104.1 (p. 243), also seem to be early-nineteenth-century features that may have derived from late prehistoric conventions. The raised "crests" on bowls 99-12-10/53105.1 (p. 242), 99-12-10/53111.1, 99-12-10/53113.1 (at left), and 99-12-10/53138.1 (facing page) are typical of pre-1850 bowls from the western Great Lakes and have been found in archaeological contexts dating a century earlier.

Bowl 99-12-10/53138.1 (facing page) is a vertical disk pipe, a type related to the horizontal disk pipes diagnostic of the late prehistoric Oneota tradition and intimately associated with the development of calumet ceremonialism. Both Catlin (Ewers 1979:24)

Pipe bowl

99-12-10/53138.1

Eastern Plains/western Great Lakes, circa 1800–1850

Red pipestone (probably catlinite)

Vertical disk form; beveled, crest, flat base

Length 18 cm, height 11.2 cm

Pipe bowl
99-12-10/53114.1
Plains/western Great Lakes,
circa 1800–1850
Red pipestone (probably catlinite),
lead inlay
Elbow form with prow
Length 8.4 cm, height 6 cm

and Prince Maximilian (Schulze-Thulin 1987:fig. 81) collected vertical disk pipes during the 1830s. Catlin identified his as Ojibwa, whereas the disk pipe collected by Prince Maximilian has been attributed to the Dakotas.

Two of the pipe bowls, 99-12-10/53110.1 (p. 243, top) and 99-12-10/53114.1 (above), have lead inlay bands around the rims. Hot lead, sometimes obtained by melting bullets or bar lead, a trade item, was poured into channels carved in the pipestone or into a secondary stone mold. Indian pipe-bowl carvers were apparently executing lead inlay by 1650. The technique flourished in the western Great Lakes and eastern Plains during the nineteenth century, when Catlin and Prince Maximilian collected some fine examples. Contemporary Indian pipe-bowl carvers still use metal inlay techniques, often employing silver solder, that are considered traditional practices.

Human and animal effigy pipe carving was developed to a high art by the prehistoric Hopewell (300 B.C.E.–600 C.E.) and Mississippian (750–1450 C.E.) cultures of the Midwest. Great Lakes and eastern Plains tribes such as the Ottawas, Ojibwas,

Pipe bowl
99-12-10/53106.1
Eastern Plains/western Great Lakes, circa 1800–1850
Red pipestone (probably catlinite)
Effigy form; anthropomorphic and zoomorphic carved figures
Length 9 cm, height 8 cm

Pawnees, Dakotas, and Otos became noted carvers of human effigy bowls during the early nineteenth century (see Penney 1992:fig. 157). The curved zoomorphic figure (an otter or dog?) on the prow of bowl 99-12-10/53106.1 (above) testifies that it was almost certainly made by a Great Lakes artist. Although it was based on ancient traditions, this style of carving seems to have flourished in the decades following the Lewis and Clark expedition, in response to demand from collectors.

Another catlinite bowl, 99-12-10/53115.1 (p. 248), is delicately and extensively carved in low relief. A pattern of fern boughs encircles the prowed stem, and a colonial-style eagle, the U.S. national symbol, is depicted on the flaring bowl. George Washington distributed silver medals engraved with this design (Hamilton 1995:156), which also appeared on the 1840 U.S. silver dollar. The style of carving seen on this bowl is often associated with the Dakotas, or eastern Sioux (see Southwest Museum 1983:fig.43), and the realistic rendering of the motif may date this bowl later than the Lewis and Clark expedition. Eagle imagery appealed to Euro-Americans and Indians

Pipe bowl
99-12-10/53115.1
Eastern Plains/western Great Lakes;
Santee(?), circa 1800–1850
Red pipestone (probably catlinite)
Elbow form with prow; carved
fern boughs, colonial American
eagle, and shield
Length 13.8 cm, height 7.2 cm

The colonial eagle, chosen
as the symbol of the new
nation in 1782, was often
depicted in profile with
wings spread, clutching
weapons or greenery.

alike, since both groups regarded eagles as emblems of power. Eagles may also have been expressly carved on pipe bowls that were made for exchange or presentation, perhaps in diplomatic contexts.

Several of the Peabody's pipe bowls were carved from local materials other than catlinite. One of these, 99-12-10/53109.1 (at right), is defined by shape as a "Micmac" or "modified Micmac" bowl, in reference to the eastern Woodlands, Algonquian-speaking people of that name. This bowl style was widely used by Great Lakes and Woodlands peoples during the early historic era. Ewers (1986:50) suggested that the Micmac shape diffused westward with the Great Lakes fur trade, and Witthoft, Schoff, and Wray (1953) suggested that it provided the model for Euro-American pipe-tomahawks. By the first half of the nineteenth century, Micmac-style bowls were in use on the Great Plains. The Crows (Conn 1982) and the Blackfeet particularly favored bowls of this shape, one of which was collected from the Blackfeet by Prince Maximilian (Schulze-Thulin 1987:figs. 85, 86).

Bowl 99-12-10/53112.1, a polished black orb probably made of basalt, is more enigmatic and unusual than the other bowls in this collection. Round pipe bowls were occasionally made and collected, but few other known examples are this large. We know little about this type of bowl, which might have been made or used in very specific contexts.

Above: **Pipe bowl**
99-12-10/53109.1
Plains/western Great Lakes,
circa 1800–1850
Unidentified stone
Micmac form
Length 7 cm, height 8.5 cm

Left: **Pipe bowl**
99-12-10/53112.1
Plains/western Great Lakes,
circa 1800–1850
Unidentified stone, possibly basalt
Enigmatic form
Height 8 cm, depth 7 cm

Chapter 9

Grizzly Claws, Garters, and Fashionable Hats

Other Possible Expedition Objects

Many additional objects in the Peabody's Boston Museum accession might have been collected by Lewis and Clark, but they lack the level of supporting documentation that allowed my colleagues and me to identify the core components of the Lewis and Clark and Hutter collections. As a group, these objects match general categories of things listed in the Peale memorandum and described in other key documents, and they are of the right types and age to have been acquired by Lewis and Clark.

The strengths of the possible associations between individual objects and the Corps of Discovery vary. Like Charles Willoughby, who accessioned the Boston Museum materials into the Peabody in 1899, we suspect that some of them were collected by the Corps but later lost their documentation. For example, a twined fiber skirt from the Columbia River (see p. 289) might well be from a village that Clark described as being populated by "petticoat

Quiver with arrows, from the upper Missouri River. (Detail of 99-12-10/52944; see p. 259.)

women." Several twined hats correspond to styles that Lewis and Clark described seeing as they descended the river to the ocean.

Other objects in this group, such as arrows and a roach feather, represent types of things that were widely collected and might simply be similar to objects that Lewis and Clark received and presented to Jefferson and Peale. More than one potential candidate exists at the Peabody for some of the objects itemized in the Peale memorandum, including a bow, a garter, and a belt. Because they did not meet our evidentiary criteria, I have omitted from consideration several objects that have appeared in previous accounts of the Peabody's Lewis and Clark collection (e.g., Richardson, Hindle, and Miller 1982), including a feather flag or banner (99-12-10/53048) and a lined, rattlesnake-skin sash (99-12-10/53025).

Arrows of Stone and Steel

Several generations of travelers on the Plains, from government and military personnel to tourists, were intrigued by and collected all manner of Indian armaments, which they often displayed in both private and institutional spaces. Bows, arrows, shields, and lances—emblematic of Indians' alienness and "savagery"—symbolized for Euro-Americans a constellation of contrasts between Plains lifeways and those of "civilized" society. Commentators on Plains warfare and related accoutrements often mixed elements of evolutionary thought and romanticism. Catlin, for instance, wrote, "There is an appearance purely classic in the plight and equipment of these warriors and 'knights of the lance,'" comparing them to the ancient Romans, Greeks, Britons, and Arabians (1973:32–34). Collectors might also have been drawn by the arrows' aura of immediacy and authenticity, in addition to their ease of transport and often intricate design.

The arrows shown on the facing page might have been collected by Lewis and Clark, by George Hutter, or by other early donors to the Peale and Boston Museums. Lewis and Clark were interested in the material technologies of Indian peoples and in technological differences between groups, in order to gauge their military force and potential for trade. Clark, for example, drawing a contrast between the Sioux and groups who were well equipped with European "fusées" (flintlock muskets), noted that the Sioux "live by the Bow and arrow" (Moulton 1987, 3:24). The explorers acquired and donated to the Peale Museum a "great number" of arrows from "different Tribes of the Saux," or Sioux (Jackson 1978, 2:477), a nation they recognized as the most powerful and resistant on the Missouri River. They also gave the Peale Museum "Stone, Spear

Points, from the natives inhabiting the Rocky Mountains." These are now lost.

Lieutenant Hutter, too, collected arrows during his trip up the Missouri in 1825–26. Military personnel worldwide seem to share a fascination with enemy arms, and army personnel posted to the Plains became avid collectors of martial material culture. The arrows Hutter presented to Peale are recorded simply as "Arrows, bundle of various." This notation suggests that Hutter acquired arrows from a variety of peoples.

I selected the six arrows shown here from among fifty-four loose specimens included in the Boston Museum accession to the Peabody Museum. They illustrate the transition from bone and stone arrow points to commercial and Indian-made steel models. Excepting epidemic disease, steel was the most profound agent of social change introduced by European colonizers of the Americas. It greatly enhanced technological efficiency, and it triggered a host of related social transformations. Members of the Corps of Discovery both observed and contributed to the transition from bone and stone tools to those made of metal. While wintering among the Mandans, they chronicled how the intense desire for metal drove other forms of trade and how central its distribution was to their own well-being.

Arrow points
Plains, circa 1790–1850
Left to right:
99-12/10/52971. Bone.
Length 4.5 cm, width 1.4 cm
99-12-10/52967. Quartz.
Length 3.2 cm, width 1.9 cm
99-12-10/52968. Chalcedony.
Length 5.1 cm, width 1.9 cm
99-12-10/52959. Metal.
Length 6.8 cm, width 2.4 cm
99-12-10/52964. Metal.
Length 3.5 cm, width 1.9 cm
99-12-10/52962. Metal.
Length 10 cm, width 1.7 cm

The most ubiquitous metal trade item was the steel arrowhead. French and English companies imported them for the Indian trade as early as the seventeenth century (Quimby 1966:33; Russell 1967:329). By 1700, Plains technologies had begun to reflect a mixture of indigenous and introduced materials, the latter initially acquired through indirect exchange. The manufacture of arrow points from slivers of the leg bones of ungulates such as deer, elk, antelope, and bison was part of an ancient bone tool technology in North America. Bone arrow points are found in small numbers in postcontact middle Missouri archaeological assemblages but are scarce in ethnographic collections. Stone points persisted longer than did bone projectiles, but, according to Prince Maximilian, were effectively passé by 1833. "The arrows of all the Missouri nations," he wrote, "are much alike, with long, triangular, very sharp, iron heads, which they themselves make out of old iron. . . . When Charbonneau first came to the Missouri, some made of flint were still in use" (Thwaites 1906a:354–57).

By the time Lewis and Clark wintered at Fort Mandan, Plains Indian men were adept at fashioning their own arrow points, knives, lances, and other metal implements by modifying Euro-American objects such as guns, hoop iron, skillets, scissors, and scrap metal. Traders also provided sheet metal blanks and an expanding array of metal tools. Metal resources were still relatively scarce, however. David Thompson, who visited the Mandans in 1797–98, wrote, "The native Arms were much the same as those that do not know the use of Iron, spears, and Arrow headed flint; which they gladly lay aside for iron. . . . Their Guns are few in proportion. . . . They have but few Hoes of iron" (Tyrrell 1916:228, 281).

Lewis and Clark's Mandan and Hidatsa hosts were eager to obtain the products of the expedition's blacksmiths, John Shields and Alexander Willard, and to have their iron hoes and war axes repaired. Throughout the winter, the Corps of Discovery conducted a lively business exchanging metal and metal working for Indian garden produce. Indeed, Clark remarked that such services were "the only means by which we precure Corn from them" (Moulton 1987, 3:281). Lewis observed that "the Indians are extravegantly fond of sheet iron of which they form arrow-points and manufacter into instruments for scraping and dressing their buffalo robes." On one occasion, he continued, four-inch metal squares salvaged from a burnt-out stove were each traded for seven or eight gallons of corn (Moulton 1987, 3:288).

A narrative relating how the Mandans dismantled a corn mill given to them by Lewis and Clark in order to use the metal components for arrowheads has become a classic account of Native recycling and secondary manufacture and underscores the value placed on metal. The transformation of the corn mill was described by Alexander Henry (the younger) when he visited the Mandan villages in 1806:

> I saw the remains of an excellent large corn mill, which the foolish fellows had demolished on purpose to barb their arrows, and other similar uses, the largest piece of it which they could not break nor work up into any weapon, they now have fixed to a wooden handle and make use of it to pound the marrow bones to make grease. (Coues 1897, 1:329)

In February 1807, Thomas Jefferson wrote a letter to his secretary of war, Henry Dearborn, advising him of the articles that Lewis considered most desired by Indians. Six of the eight items listed were metal: knives, battle axes and tomahawks, awls, needles, iron combs, brass buttons, and camp kettles. At the bottom of the letter, Jefferson noted: "Arrow points should have been added" (Jackson 1978, 1:375).

The point on arrow 99-12-10/52971, made from a bone fragment, is a type of projectile that was becoming obsolete by the nineteenth century. The stylistic features of the shafts of arrows 99-12-10/52967, 99-12-10/52968, and 99-12-10/52959, the first two tipped with stone points and the third with a handmade metal point, suggest that they might have been part of a single set constructed during the transitional period when Lewis and Clark visited the upper Missouri. The two stone points both appear to be late prehistoric in form and are made from materials that were available across the northern Plains and Rocky Mountain regions of the United States. The quartz point on arrow 99-12-10/52967 and the larger, chalcedony point on arrow 99-12-10/52968 seem to be variants of the late prehistoric style archaeologists call "Prairie side-notched" (Kehoe 1973).

Two of the steel arrow points shown here, 99-12-10/52959 and 99-12-10/52964, were probably made by Indian men before 1850. Arrow 99-12-10/52964 is part of a set of ten that includes other handmade diamond-shaped and triangular metal points. Diamonds and other irregular shapes seem to have been made most frequently prior to 1850, when Catlin collected some on the upper Missouri (see Hanson 1975). Commercially made metal projectile points and those shaped by blacksmiths are generally thinner, flatter, and more symmetrical than those that were cold-chiseled by Indian men (Hanson 1975; Russell 1967). Arrow 99-12-10/52962, long and triangular in shape, with beveled edges and a serrated base, is typical of a nineteenth-century commercial steel arrow point. Indian-made points may also have become more uniform by the second half of the nineteenth century.

Plains arrows are generally quite similar in their formal properties, such as material, length, decorative treatment, and point shape. The stylistic features of the bodies of the pictured arrows, such as the fletching, the shape of the carved nocks, and the shaft decoration, support a central Plains, possibly Siouan, origin for these examples.

Bows

On February 8, 1805, Meriwether Lewis recorded in his journal that he had received a visit from Black Cat, principal chief of the Nuptadi Mandans, who lived in what the captains referred to as the "upper" Mandan village. "The black Cat," Lewis wrote, "presented me with a bow and apologized for not having completed the shield he had

Elk antler bow
99-12-10/52946
Upper Missouri River,
circa 1800–1840
Elk antler, sinew, hide glue
Length 78 cm, width 2.5 cm

promised" (Moulton 1987, 3:289).[1] It is unclear whether Lewis had requested the bow and shield or Black Cat had volunteered to make them. Lewis reciprocated for the bow with some shot, six fishhooks, and two yards of ribbon. He also gave Black Cat's wife a small looking glass and some needles in exchange for two pairs of moccasins.

The expedition items later transferred to the Peale Museum included "A great number of arrows from different tribes of Saux. And a bow." Several bows were given to the Peabody by the Boston Museum; here I present two, either of which could be the bow given to Lewis by Black Cat.[2] The first, 99-12-10/52946 (facing page, and right), is a sinew-backed bow made from a single elk antler. The antler was scored and then covered with a thick layer of sinew, probably taken from a bison. In August 1805, Lewis and Clark saw such bows among the Shoshones (Moulton 1988, 5:150), and they were later described by travelers such as Catlin and Prince Maximilian. In the years following the expedition, Clark exhibited a number of elk antler bows in his St. Louis council chamber (Ewers 1967:58).

Mandan and Hidatsa people both made elk antler bows and traded for them. Wolf Chief, a Hidatsa man interviewed by Gilbert Wilson early in the twentieth century, described his father's manufacture of a bow from two joined elk antlers and said that such bows were not functional but were "made for show, and to hold a man's honor marks" (Wilson 1911:10). In one of his most highly acclaimed portraits (see p. 258), Karl Bodmer depicted Péhriska-Rúhpa (Two Ravens), a Hidatsa Dog Society dancer, holding a decorated elk antler bow (Bodmer 1984:pl. 330). The British Museum has an elk antler bow collected from the Hidatsas by Duke Paul of Württemberg around 1820–25 (Gibbs 1982; Schulze-Thulin 1987:68). Few survive in museum collections.

The Boston Museum's 1899 donation to the Peabody also included a bow case and quiver made of mountain lion hide (99-12-10/52944; see pp. 250 and 259). The packing list Lewis composed and sent down the Missouri River to Jefferson in the spring of 1805 noted that the shipment included a "bow an quiver of arrows—with some Ricara's tobacco seed" (Moulton 1987, 3:330). This fringed quiver and bow case contains an elm bow and ten headless arrows, the latter feature perhaps indicating that it was a gift presentation set.[3] Fletching on the arrows is made from the feathers of a wild turkey, an immature snowy owl, a hawk, and possibly a bald eagle. The strap (made from the mountain lion's tail) is lined with red wool trade cloth beaded in blue and white "bar and pendant triangle" motifs, which are characteristic of early upper Missouri materials. The wooden rod supporting the bow case is also wrapped with

The sinew side of the bow faced away from the bowyer when the bow was strung. (Detail of 99-12-10/52946.)

Bow case and quiver with bow and arrows
99-12-10/52944
Upper Missouri River;
Mandan or Hidatsa (?),
circa 1800–1840
Mountain lion hide,
wool cloth, glass
beads, sinew, wooden
support, elm "self bow"
and ten wooden arrows
without projectile
points
Length 117 cm,
width 24.5 cm

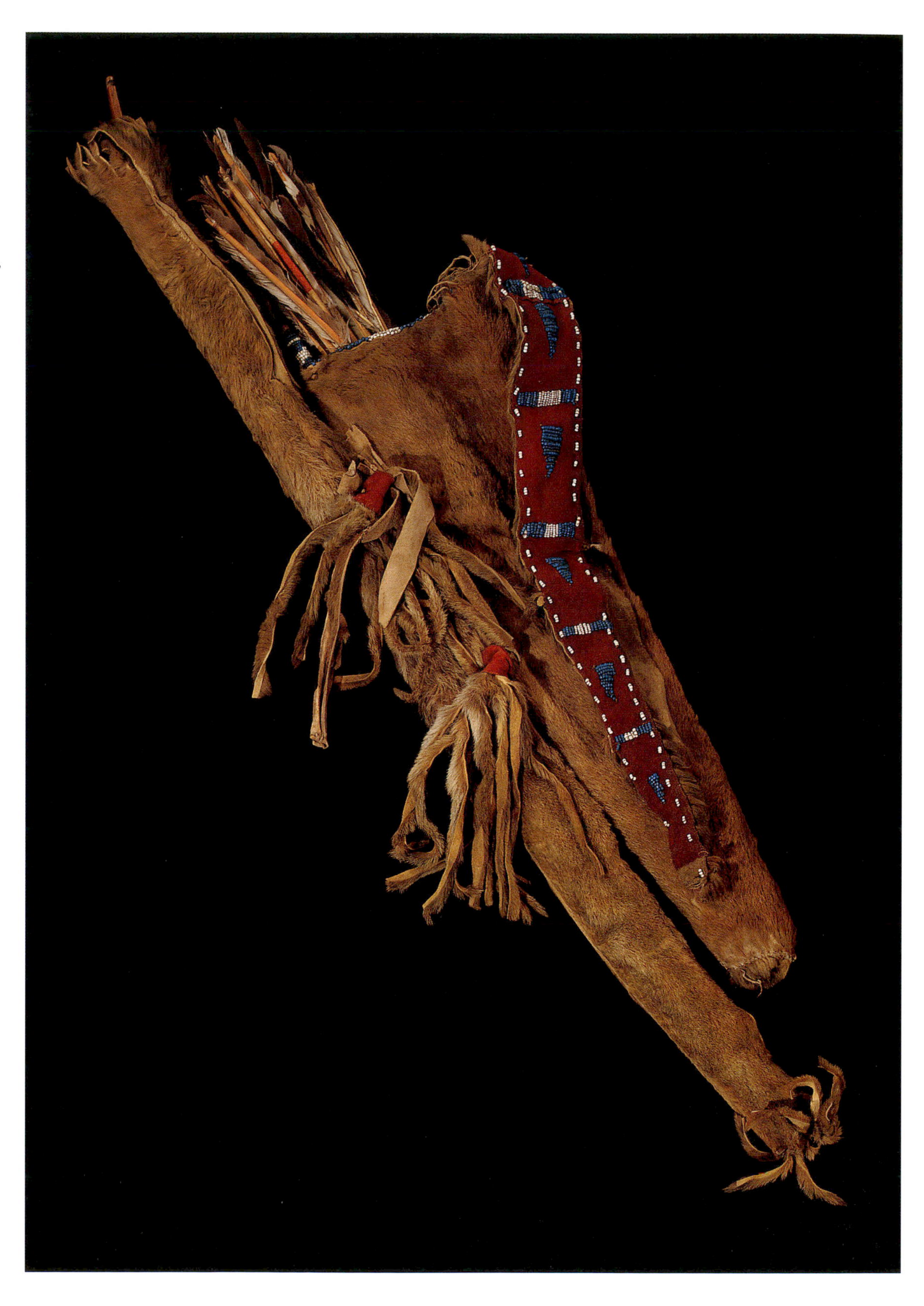

Portrait of Péhriska-Rúhpa (Two Ravens) by Karl Bodmer, 1834. In one of his most acclaimed portraits, Bodmer depicted this Hidatsa Dog Society dancer wearing a war whistle and carrying an elk antler bow.

simple, strung bands of blue and white beads. These blue beads are the same translucent, blue-green drawn beads that are seen on several other objects in this accession. Bill Billeck, of the Smithsonian's National Museum of Natural History, believes them to be diagnostic of upper Missouri River beadwork in the early nineteenth century. All of the sewing on this bow case and quiver was done with sinew.

Prince Maximilian (Thwaites 1906a:354) noted that Mandan and Hidatsa bows were made of elm or ash and described their bow case and quiver sets as similar to this one: "The quiver, to which the bow-case is fastened, is made of panther or buffalo skin; in the first case, with the hair outwards, the long tail hanging down, and, as among the Blackfeet, lined with red cloth, and embroidered in various figures with white beads." Karl Bodmer depicted the Mandan warrior Sih-Sa carrying a bow case and quiver similar to this one (Bodmer 1984:pl. 315). A later but similar design, possibly Crow, is pictured in a book by James A. Hanson (1996:fig.56).

Grizzly bear claw ornaments
99-12-10/53007a and b
Western North America, circa 1780–1825
Grizzly bear foreclaws, glass bead
Top: 53007b. Length 12.5 cm, width 2 cm
Bottom: 53007a. Length 11.5 cm, width 2 cm

Grizzly Bear Claw Ornaments

These two grizzly bear claw ornaments (99-12-10/53007a and b) came to the Peabody from the Boston Museum in 1899. They are perforated, perhaps for use in a double-stranded necklace or collar of the type worn by warriors from the lower Missouri to the Columbia Plateau. Indian men also wore grizzly claws in their hair and attached to shirts. A blue glass bead has been inlaid into one of the claws. There are three labels at the Peabody that might be associated with these bear claws, all of which are believed to be from the Peale Museum. A handwritten label, possibly in the hand of Franklin Peale, describes an "Indian Necklace, made of the claws of the Grizly Bear" and presented by Captains Lewis and Clark (see p. 5).[4] Another hand-printed label reads, "Necklace made of the claws of the Brown Bear, worn by the Osage Indians." A third printed label simply describes "Claws of the Grizly Bear." That no grizzly ornaments or necklaces appear in the Peale memorandum (Jackson 1978, 2:476–77) underscores the incomplete nature of that document.

Peale Museum label describing grizzly claws.

Powerful and bold, grizzly bears loomed large in the psyches of Indians, explorers, and the European and American publics. William Clark wrote of the Nez Perce: "This nation esteem the Killing of one of those tremendous animals (the Bear) equally great with that of an enemy in the field of action. We gave the Claws of those bear which Collins had killed to [chief] Hohâstillpelp" (Moulton 1991, 7:259). He reported that the Shoshones, too, regarded the killing of a bear "an act of . . . celebrity" equal to that of killing an enemy, commenting that "with the means they have of killing this animal it must really be a serious undertaking." Throughout their travels, from the lower Missouri to the Plateau, the explorers observed exceptional warriors and leaders wearing grizzly claw necklaces. Clark described such necklaces, some of which included the foreclaws of several animals:

Daguerreotype portrait of No-che-ninga, or "No Heart of Fear," circa 1840s. An unknown photographer made this early image of an Ioway chief wearing a bear claw necklace and what appears to be a Jefferson peace medal.

> [T]he warriors or such as esteem themselves brave men wear collars made of the claws of the brown bear which are also esteemed of great value and are preserved with great care. these claws are ornamented with beads about the thick end near which they are peirced through their sides and strung on a throng of dressed leather and tyed about the neck. (Moulton 1988, 5:135)

Although initially skeptical about the reputed dangers of traveling in grizzly country, firsthand experience taught members of the Corps of Discovery to treat the bears with caution and respect. Grizzlies were a constant presence and a formidable hazard as the Corps crossed the Great Plains and Rocky Mountains. On several occasions, members of the expedition provoked bears to attack—generally by wounding them—resulting in a series of narrow escapes. Lewis pronounced the grizzlies along the Missouri River to be exceptionally fierce. According to his museum catalogue, Clark preserved the foot of a grizzly that he shot during the expedition.

In the metropolis, even a single claw was considered a marvel—and a prized addition to a cabinet of curiosity. Enumerating the "extremely rare" items he saw in Jefferson's Indian hall at Monticello, Baron de Montlezun included "a bear's claw, from the Missouri. This species is larger and much more ferocious than the others" (Peterson 1989:67–68). In 1805, Charles Willson Peale accessioned into his museum "A claw of the Grisly bear brought from the interior of America by Alexr. McKenzie."[5]

Living grizzlies, which inspired almost universal awe and fear and were powerful tokens and symbols of wilderness, soon followed. Initially, the capture of living bears was probably motivated by both emergent scientific agendas and by a still widespread fascination with the wonders of the unknown. Zebulon Pike sent two cubs to Jefferson in 1807, which the president forwarded to Peale in Philadelphia. Peale exhibited them for a time but shot them when one escaped and trapped his family in their living quarters (Sellers 1980:206–7). Along with his Indian collection, Catlin exhibited a grizzly in Europe, where it was no doubt received as metonymic of the American West. In addition to acquiring a grizzly claw collar from the Mandans (Schulze-Thulin 1987:63), Prince Maximilian collected cubs along the Missouri to ship to Europe (Thwaites 1906b:90, 119). Eschewing science for sensation, P. T. Barnum purchased a half interest in the "California menagerie" of the original Grizzly Adams in the 1860s. Adams appeared onstage for several months, demonstrating his mastery over the "savage monsters" while slowly dying of wounds inflicted by "General Fremont," one of his star performers (Barnum 1888:262). Lewis and Clark probably regarded grizzly claw artifacts as mementos of their experiences and as hybrid natural and artificial curiosities.

Robes

We know from the packing list prepared at Fort Mandan in April 1805 that by the end of the expedition's first winter, Lewis and Clark had acquired at least seven bison robes that they considered worthy of sending to Jefferson (Jackson 1978, 1:234–36). We can assume that these were decorated robes, an assumption that seems substantiated by Secretary of War Henry Dearborn's observation that the robes were "good skins, well dressed, and highly embellished with Indian finery" (Jackson 1978, 1:254–55). Clark sent two robes to his family from Fort Mandan in 1805 and forwarded four or five more from St. Louis to Louisville in 1806 (Moulton 1993, 8:418–19).[6] The Peale memorandum specifies only two robes, suggesting that Jefferson or the explorers themselves might have kept the others.

Unfortunately, the two Peale robes were described only in terms of relative size and group of origin: "A Large Mantle, made of the Buffalow skin, worn by the Scioux, or Soue, Darcota Nation," and "A small mantle of very fine wool, worn by the Crow's nation Menetarre" (Jackson 1978, 2:476). Peale also recorded having received "A Piece of White Buffaloes skin, from the Missouri." Nothing further is known of that rare and symbolically charged item. Lewis and Clark may have received it from the Mandans, who used white robes and robe fragments in rituals and social ceremonies, often paying exorbitant prices for them in the intertribal trade network (e.g., Bowers 1991 [1950]:76–77; Thwaites 1906a:321–23).

Because descriptions of the robes in the Peale memorandum are cursory, and because that document is known to have been incomplete, I describe here all five of the decorated buffalo robes that the Boston Museum transferred to the Peabody Museum in 1899. One of the five was probably associated with a torn label that came to the Peabody along with the objects. It reads: "*Buffalo Robe* one of the 7 br . . . presented to Alexander Hamilt . . . the Peale Museum and the Old. . . ." Alexander Hamilton, the first secretary of the treasury, was killed in a duel in July 1804. He did know Peale, and it is possible that one of the Peale Museum robes was once owned by him.

One of the Peabody robes, actually a fragment of a geometrically painted robe (see p. 264), was traded to the Denver Art Museum in 1952. The robe is painted in a woman's "box-and-border" pattern, which is described later, with dark brown and red pigment and clear glue sizing. The stake holes along the edge have been painted red, and there is a leather pendant strap in one corner, beaded at the base with blue and white pony beads.[7]

Two others of the five Peabody robes were described in chapter 7: a large painted robe (99-12-10/53121; see p. 146) and a quilled robe (99-12-10/53120; see p. 189). In addition, there are two smaller decorated robes at the Peabody that were made during the first half of the nineteenth century. Both of the larger robes were probably worn by men, but the two small, geometrically painted robes would have been made and owned by women or girls.

One of these robes (99-12-10/53122; see p. 265) is an unusual, partially painted and partially quilled bison calf robe that Charles Willoughby accessioned as "Cheyenne." Calf robes were worn by children and by members of women's ceremonial societies, such as the Mandan White Buffalo Cow Society (Bowers 1991 [1950]:326). Because only the lower portion of this robe has been finished, the woman who made it

Fragmentary box-and-border robe
Denver Art Museum 1952.405
(Formerly PM 99-12-10/53123)
Central Plains, circa 1800–1840
Bison hide, pigments,
glass beads
(Materials identified
from photograph only.)

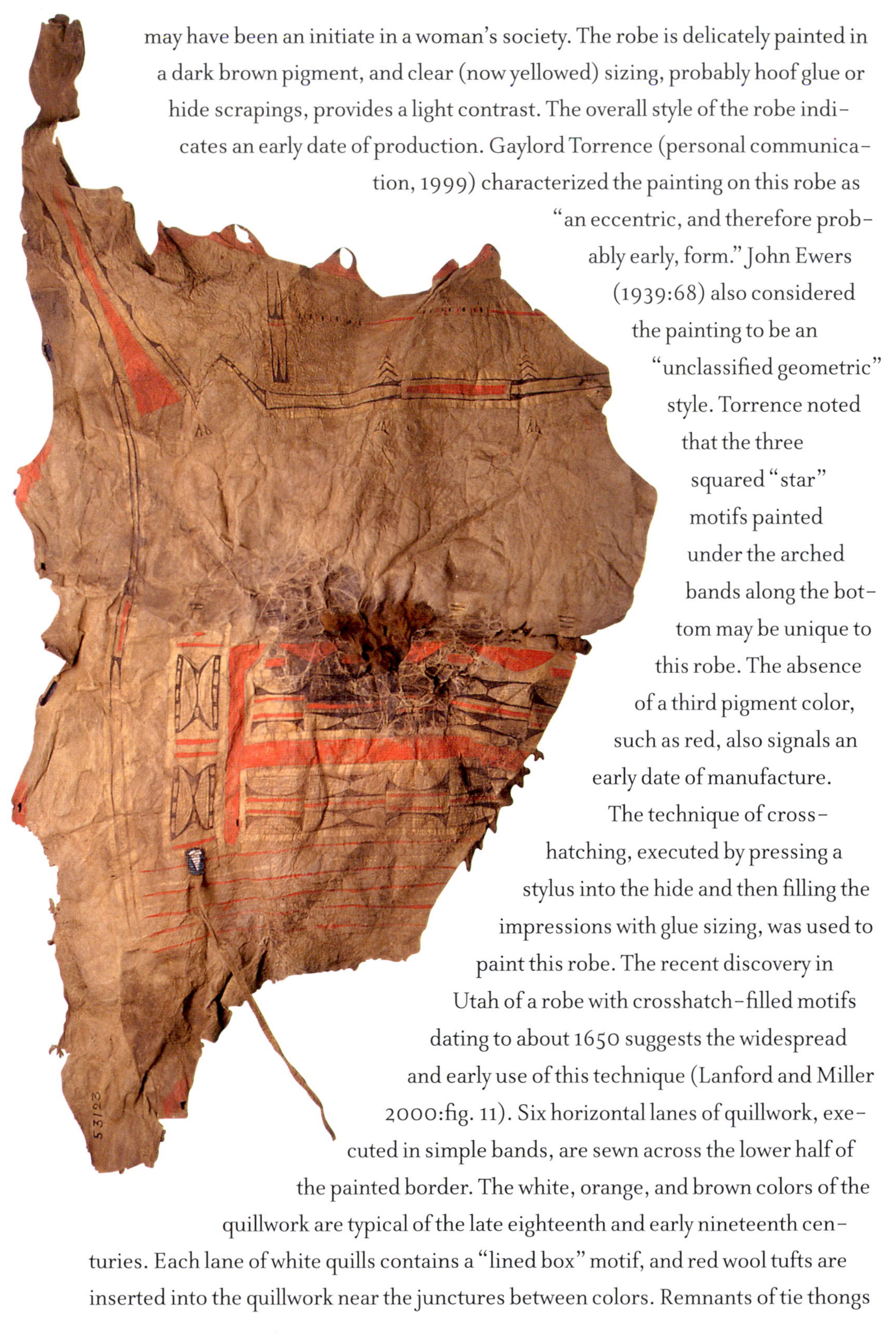

may have been an initiate in a woman's society. The robe is delicately painted in a dark brown pigment, and clear (now yellowed) sizing, probably hoof glue or hide scrapings, provides a light contrast. The overall style of the robe indicates an early date of production. Gaylord Torrence (personal communication, 1999) characterized the painting on this robe as "an eccentric, and therefore probably early, form." John Ewers (1939:68) also considered the painting to be an "unclassified geometric" style. Torrence noted that the three squared "star" motifs painted under the arched bands along the bottom may be unique to this robe. The absence of a third pigment color, such as red, also signals an early date of manufacture. The technique of crosshatching, executed by pressing a stylus into the hide and then filling the impressions with glue sizing, was used to paint this robe. The recent discovery in Utah of a robe with crosshatch-filled motifs dating to about 1650 suggests the widespread and early use of this technique (Lanford and Miller 2000:fig. 11). Six horizontal lanes of quillwork, executed in simple bands, are sewn across the lower half of the painted border. The white, orange, and brown colors of the quillwork are typical of the late eighteenth and early nineteenth centuries. Each lane of white quills contains a "lined box" motif, and red wool tufts are inserted into the quillwork near the junctures between colors. Remnants of tie thongs

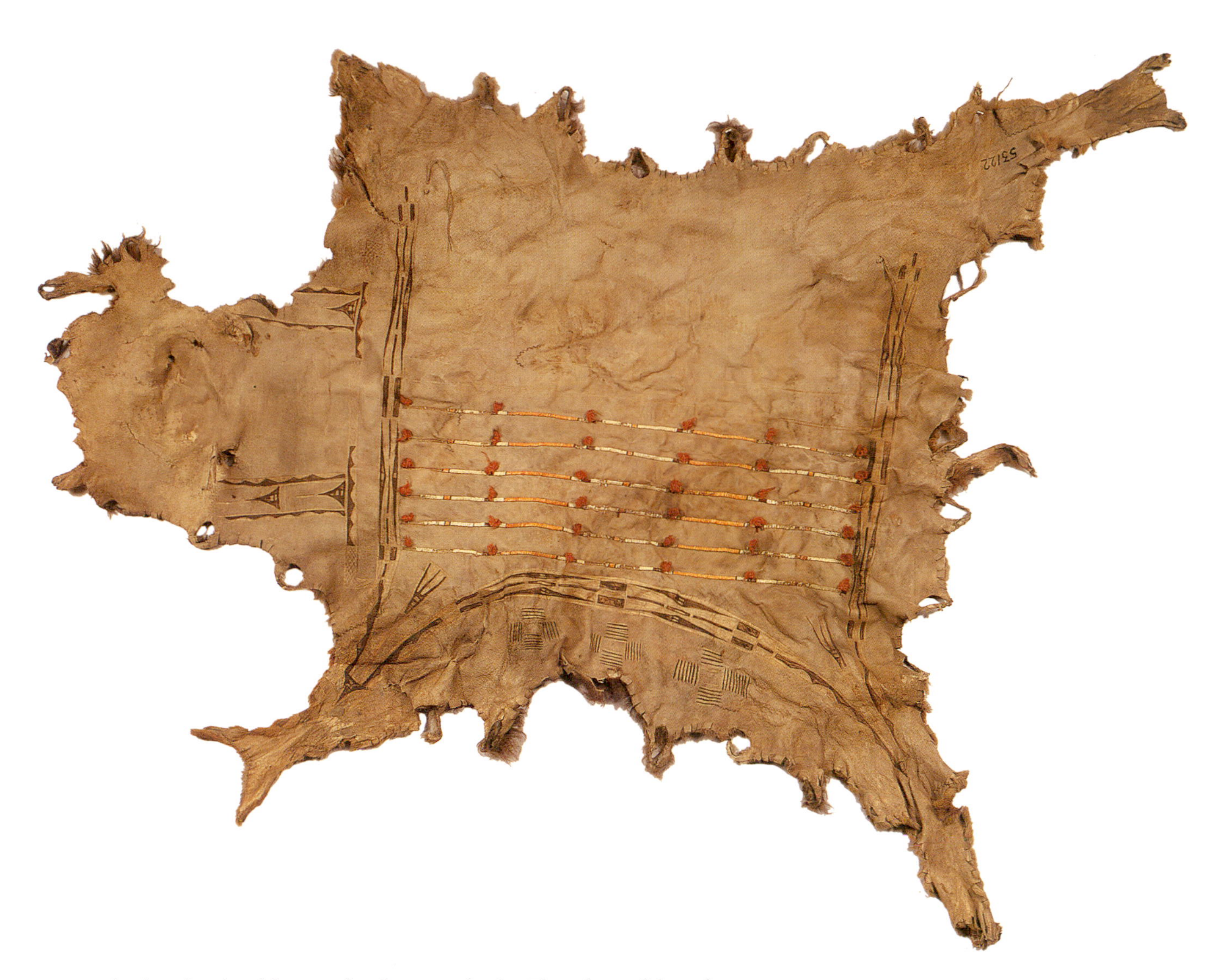

are attached to the shoulders on the decorated side. The edges of the robe are crenulated or fringed, and fragments of hooves remain on the legs.

The second of the smaller women's robes (99-12-10/53124; see p. 266) is also made from a juvenile bison skin, which has been painted in the classic Plains box-and-border design worn by women in the nineteenth century. Little is known about the antiquity of this pattern or its relationship to protohistoric painting traditions, but a second painted robe found in Utah and dated about 1550 (Lanford and Miller 2000:fig. 8) indicates that similar designs were being painted before Europeans arrived in North America. Bodmer's portrait of a Lakota woman wearing a nearly identical red and black box-and-border robe documents that the style was well developed by 1833 (Bodmer 1984:pl. 192; see p. 180).

Painted and quilled robe
99-12-10/53122
Central Plains, circa 1800–1850
Bison (juvenile) hide, dyed and undyed porcupine quills, pigments, wool cloth, glue sizing, sinew
Length 121 cm, width 100 cm

Painted box-and-border robe
99-12-10/53124
Central Plains, Sioux(?),
circa 1800–1850
Bison (juvenile) hide,
pigments, glue sizing
Length 119 cm, width 100 cm

Gaylord Torrence believes that this robe, like the one painted by Bodmer, was produced by a Sioux artist. The pigments have been identified as hematite (brown), iron oxide (yellow), and vermilion (red), mixed with a protein medium (sizing) (Miller, Moffatt, and Sirois 1990). Cross-hatching is used between the design elements, a technique that Torrence (1994:92) associates with Sioux painting traditions throughout the historic era. Some of the design elements, such as a branching, pronged motif, are simply impressed and filled with sizing. The tips of the robe's legs are fringed.

Quilled belt
99-12-10/53062
Western Great Lakes/eastern Plains, circa 1800–1840
Mammal hide, dyed and undyed porcupine quills, metal cones, unidentified mammal hair, sinew
Length 118 cm, width 10 cm

Woven wool sash
99-12-10/53002
Western Great Lakes, circa 1790–1840
Wool yarns (dyed)
Length 320 cm, width 21 cm

Sashes were worn both as belts and wound around the head as turban-type headdresses like those illustrated on pages 218 and 261.

Belts and Garters

The Peale memorandum lists "A handsomely ornamented belt, from the Winnebagou's or Puount's" and "A Handsome belt worn by the Saux as a garter." Two belts and a garter from the Peabody's Boston Museum accession are pictured here as potential matches. One (99-12-10/53062; at top) is a quilled leather belt with pendant metal cones, made in the Great Lakes region or the Woodlands.[8] Gaylord Torrence noticed its striking similarity to a mid-nineteenth-century quilled "Crow belt" in the Ethnological Museum Berlin (Bolz 1999:fig. 64) that has been identified as Dakota (Hartmann 1973:no. 49) or "Prairie region" (Torrence and Hobbs 1989:fig. 7). White, pale yellow, eggplant, and orange quills are applied in eight rows of one-quill zigzag bands. The eggplant-colored quills are large, thick, and noticeably shiny, like those on the otter bag thought to be associated with the Lewis and Clark expedition (99-12-10/53052; see p. 87), and their color seems to have been derived from the same dye.

The second potential match to Peale's description of belts and garters is the finger-woven sash illustrated above (99-12-10/53002). The sash is made from wool yarns in two shades of red, two shades of green, pale yellow, pale bluish green, and a beige or

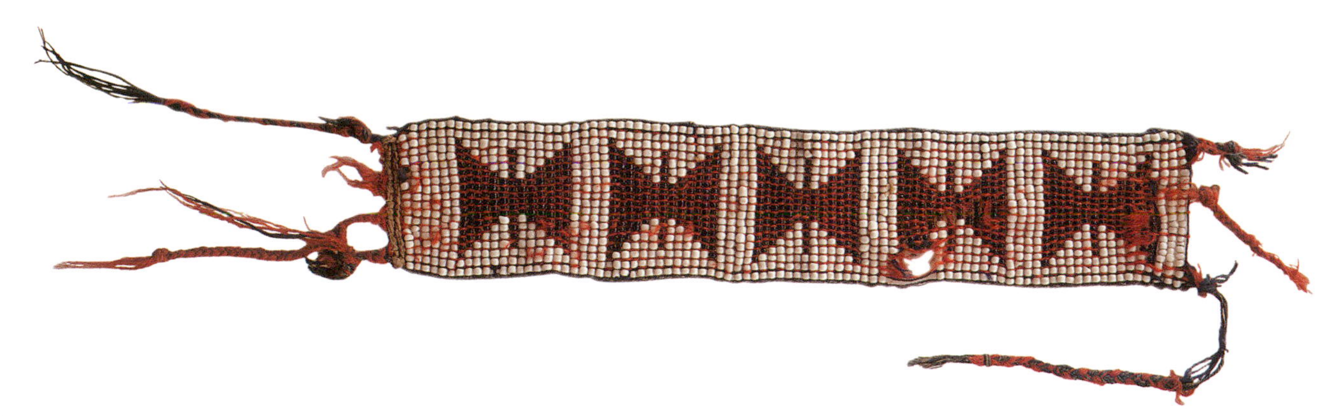

Beaded garter
99-12-10/53005
Western Great Lakes/eastern Plains; Sauk(?), circa 1800–1825
Wool yarns, cotton yarns, unidentified plant fiber (inner bark) cordage, glass beads
Length 56.5 cm, width 5.8 cm

light tan. The technique of finger-weaving predates contact with Europeans and was practiced by Indian and non-Indian peoples who participated in the Great Lakes and Woodlands fur trade. Sashes woven into "arrow" patterns are associated with French Canadian and Métis *engagés* from the region of Quebec and were widely traded. David W. Penney (1992:75, pl. 9) illustrates a similar sash, identified as Mesquakie (Fox) and dated 1800–1830; another was collected by Major Jasper Grant before 1809 (Phillips 1984:58, no. 7).

The accession also includes a single beaded garter (99-12-10/53005; above), which Gaylord Torrence (personal communication, 2000) has called "as early and as classic as you can get." Prince Maximilian collected a similar garter, identified as Sauk (Torrence 1989:fig. 4; Zepp and Ludwig 1982:20), and William Orchard (1975:fig. 101) pictured a comparable Sauk pair, with similar hourglass or thunderbird designs. On the Peabody garter, irregular white and black pony beads are patterned over wool and cotton threads. This garter was made on a heddle, a type of small hand loom introduced by Europeans and in use by Great Lakes women by about 1780 (Lessard 1986). It is an example of what Orchard (1975:119–21) called "square weave," with the warp and weft threads crossing at right angles. The warp elements are red wool and blue cotton yarns, braided at the ends; the single weft elements are of blue wool. At the beginning edge of the weaving are two rows of twining using a plant fiber (probably bast) cordage. This garter is a likely candidate for the "Saux" garter listed in the Peale memorandum.

Roach Feather

Birds and their feathers hold great importance in the Native cultures of the Americas. Feathers, symbolically encapsulating a range of cosmological meanings associated with particular birds, are widely used in ritual contexts. They also embellish objects that communicate social membership and individual status. Among Plains Indian

Roach feather
99-12-10/53057
Plains/Prairie, circa 1800–1850
Immature golden eagle (*Aquila chrysaetos*) tail feather, mammal bone socket, silk ribbon, pigment
Length 38.2 cm, width 7 cm

peoples, raptors and crows are associated with warfare. During the nineteenth century, an elaborate feather symbolism related such feathers to the martial achievements of warriors and their membership in men's societies. Not only were the types and numbers of feathers meaningful, but their arrangement, the manner in which they were worn, and the way the feathers were cut and painted served as "honor marks," signaling particular achievements, coups, or wounds suffered in war (see Wilson 1916b on Hidatsa honor marks). The art and symbolism of featherwork reached a pinnacle in the construction of men's war caps and bonnets, which became iconic of Plains culture in Euro-American society.

Lewis and Clark donated to the Peale Museum two sets of feathers relating to the war caps of warriors and leaders. The Peale ledger book lists "The Tail feathers of the Eagle, much prized by the Indians of North America, who convert them into various ornimental and war-like dresses—these being the pattern for a war-cap would be esteemed by them equal in value to two good horses" (Jackson 1978, 2:476).

According to Wolf Chief, a Hidatsa,

> If a man was one of four to strike an enemy, he had the right to wear a war eagle's tail feather. A golden, or war eagle has twelve feathers in its tail. The two middle feathers were prized most. The first four men who struck an enemy earned the right to wear a golden eagle tail feather in his hair. But the first of the four to count coup, and he only, earned the right to wear one of these two middle tail feathers. (Wilson 1916a:31)

The Peale Museum also received "Feathers which were at various times presented to Captn. Lewis and Clarke by the principal Chiefs of the nations inhabiting the Plains of

Columbia, whose custom it is to express the sincerity of their friendship by cutting feathers from the crowns of their War Caps and bestowing them on each as they esteem" (Jackson 1978, 2:477). Such feathers were quintessential chiefly gifts.

As part of the Boston Museum accession, the Peabody received a Peale Museum object label reading, "Feathers from the Tails of Eagles, intended for a War-cap by an Indian chief in the interior—esteemed equal in value to two horses." Charles Willoughby annotated a typed transcript of the Peale memorandum while trying to identify objects that might match those on the Peale list. He underlined the phrase, "The tail feathers of the eagle," probably indicating that he thought one of the newly acquired Peabody objects conformed to that description.

No loose eagle feathers, however, survive in the collection. Willoughby might have been thinking of the object illustrated on page 269 (99-12-10/53057), a single immature golden eagle feather inserted into the socket of an incised and painted bone tube.[9] Three lengths of white and red silk ribbon are attached along the vane of the feather. Feathers in bone sockets were attached to an elk antler spreader that held open a roach headdress made of erect porcupine and deer hair. The socket tube, or "plume holder," was detachable, and some people kept plume holders and feathers in specially carved wooden boxes when they were not in use. Roaches were generally worn by people of the Woodlands and the eastern Plains, but they were also associated with the Omaha (War) Dance (now known as the Grass Dance), which spread across the Plains during the nineteenth century. Richard Conn (1979:no. 149) illustrated a Pawnee roach spreader made around 1830, and Wolfgang Zepp (Zepp and Ludwig 1982:no. 13) pictured a complete Osage roach dated thirty years later. Incised antler roach spreaders and plume holders have been found in middle Missouri archaeological sites occupied during the historic period, such as Like-a-Fishhook Village, home to Hidatsa, Mandan, and Arikara people after 1850 (Smith 1972).

Wampum

Wampum is a quintessential product of the multicultural contact zone that defined colonial America. Adopting the ancient Iroquois and Algonquian Indian practice of making and exchanging marine shell beads, seventeenth-century Dutch colonists in New York rationalized their production and use in ways that were in turn integrated into Indian cultures. The chief contribution of the Dutch, necessitated by their use of shell wampum as a medium of exchange, was the standardization of beads into a cylindrical shape of consistent size (according to Ceci 1989:63, on average 5.5 mm

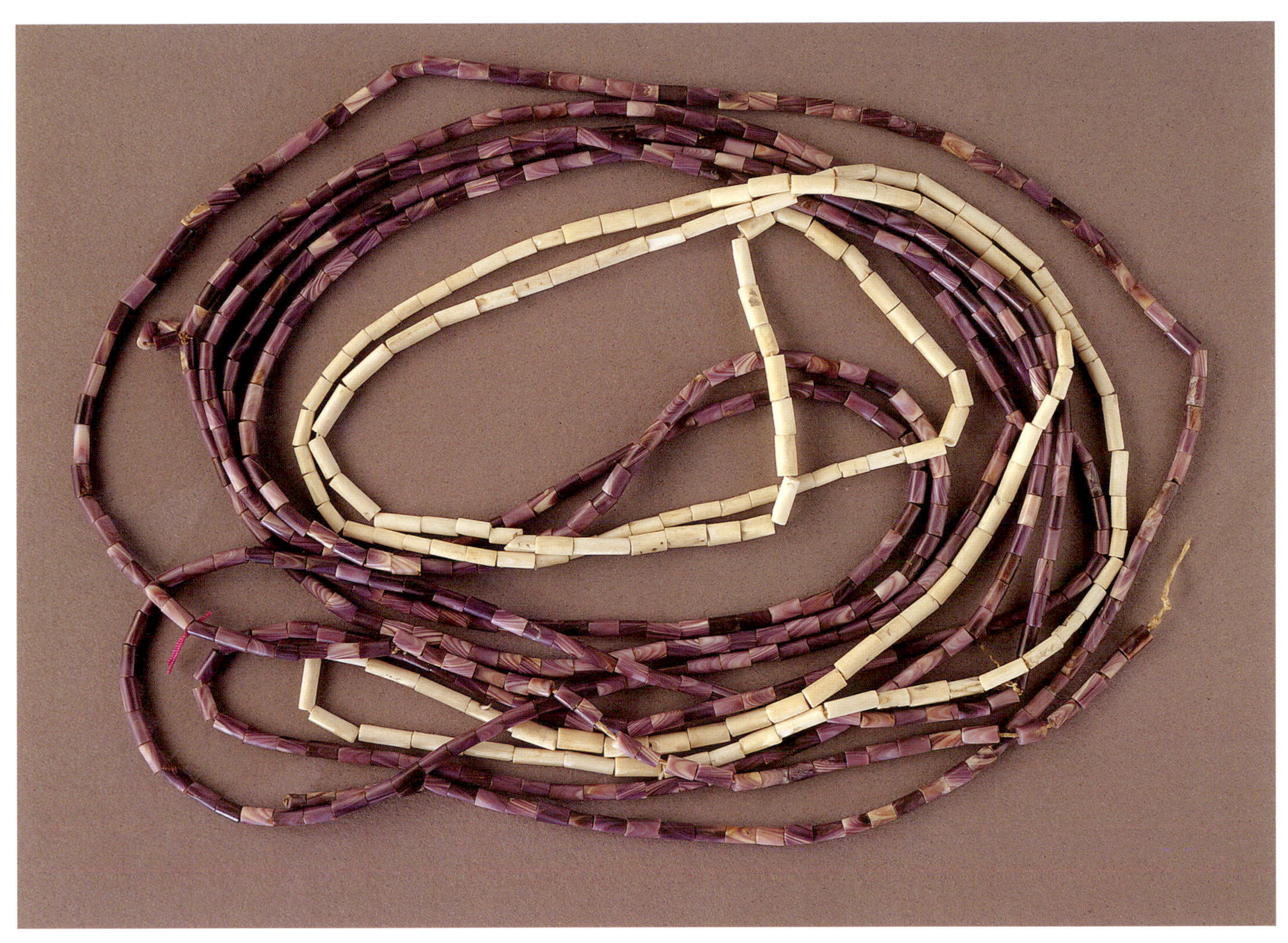

Strings of wampum
Woodlands/western Great Lakes, circa 1800–1850
Marine shell, quahog clam
99-12-10/53011 (purple). Length 232 cm
99-12-10/53012 (purple). Length 73.3 cm
99-12-10/53014 (white). Length 112 cm

in length and 4 mm in diameter) and color. This standardization was effected by the use of metal drills (soon employed by both Indian and European artisans), by the preferential use of purple and white beads made from the shell of the quahog clam (*Mercenaria mercenaria*), and by the legal establishment of values by colonial courts. During the seventeenth century, foot-long "strings" and longer "fathoms" of purple and white wampum were recognized as legal tender throughout New York and New England and were often exchanged by Indian peoples and colonial governments. Although absolute values fluctuated throughout the century, purple beads retained a consistently higher value than white beads, which were made of either quahog or northern whelk shell.

Wampum also became an integral component of formal Indian-white diplomacy and intertribal relations east of the Mississippi River, where Lewis and Clark gained their initial experience with Native Americans. The design and exchange of wampum, as of calumets, became charged with social symbolism during the contact era, particularly regarding the establishment, maintenance, and rupture of alliances. The southern Algonquians and the Iroquois developed elaborate systems of meaning and etiquette surrounding the use of wampum, and the Iroquois Confederacy extended these ideas across the Great Lakes and into the Ohio River valley. Color symbolism and the arrangement of beads were semiotic markers, so even short strands of wampum signaled particular meanings within social and ceremonial contexts such as wedding and mourning rites. The colors purple and white encoded a cluster of associated meanings for Native cultures: purple (called black when very dark) has generally been glossed as representing solemn concepts such as war and mourning, whereas white represents peace and friendship. The celebrated wampum belts of the Iroquois Confederacy and their neighbors, sometimes woven from thousands of beads into pictographic forms, memorialized groupwide initiatives and events, including declarations of war and the making of treaties.

Both Indians and non-Indians produced wampum throughout the colonial era. By the middle of the eighteenth century, the demand for wampum in the expanding fur trade was so great that its manufacture became a home industry in New York and New Jersey. Before leaving the East Coast, Meriwether Lewis purchased both wampum beads and the longer shell "hair pipes" to distribute during the western expedition as gifts to Indians (Jackson 1978, 2:69–98). Additional shell hair pipes were purchased from the fur trader August Choteau while the Corps wintered in St. Louis in 1803–4 and made preparations to embark up the Missouri River in the spring (Ewers 1968:92). The explorers presented small quantities of these shell beads, along with flags, medals, coats, armbands, and leggings, to most of the "principal" chiefs they encountered (Thwaites 1904–5, 6:270–79).

Specific references to the presentation of wampum in Lewis and Clark's journals highlight its significance as a symbol of the Corps's peaceful, diplomatic intent. For example, after the initial council with the Mandans during the fall of 1804, Clark wrote, "[W]e Sent the presents intended for the Grand Chief of the Mi-ne-tar-re or Big Belley [Minnetarees, or Hidatsas], and the presents flag and wompoms by the Old Chief" (Moulton 1987, 3:211). Later, among the Shoshones, Clark added "white wampom" to a pipe stem "and informed the Chief that this was the emblem of peace with us" (Moulton 1991, 7:341). The captains also occasionally presented wampum in less formal contexts. For example, they gave "blue wompom," powder, balls, and ribbon to two young Nez Perce men in exchange for the delivery of horses to their camp (Moulton 1991, 7:250).

The Peale Museum received from the expedition "Wampum, of various discriptions, indicating Peace, War, Choice of either, Hostilities commencing, and a disposition for them to cease &c. From different nations" (Jackson 1978, 2:477). Small quantities of marine shell wampum had no doubt reached Missouri River tribes through both intertribal trade networks and the fur trade before Lewis and Clark arrived. The trader François Larocque mentioned distributing wampum along the upper Missouri in 1804–5 (Wood and Thiessen 1985). Because many traders referred to shell beads as wampum, and because true wampum is sometimes referred to simply as shell, it is difficult to trace in the historical record. By the 1820s, however, St. Louis fur trade companies were ordering large quantities of wampum from East Coast distributors (Davis 1973:appendixes), and the western removal of eastern tribes must also have put wampum into wider circulation. During the 1830s, George Catlin found that the Mandan leader Four Bears owned a belt made "of the richest shell wampum" (Ewers 1979:56, pl. 18), and Prince Maximilian collected from the Mandans strands of purple and white wampum (Schulze-Thulin 1987:62).

Catlin, however, observed that wampum remained rare west of the Mississippi River, and it is questionable whether western Indian peoples adopted the symbolic associations between color and design that animated wampum's use in the Woodlands. Peale's allusion to these meanings in regard to the wampum he received from Lewis and Clark could be his own interpretation, based on his familiarity with cultural practices in the East. It might also indicate that Mississippi River and western Great Lakes peoples gave wampum to Lewis and Clark along with pipes and other objects. By the eighteenth century, tribes in the "Old Northwest" had integrated wampum into their cultures and often presented the shell beads during diplomatic encounters. In 1793, for example, members of an intertribal delegation to Washington, D.C., that included Kaskaskia, Wea, and Kickapoo representatives presented wampum and a pipe to President Washington at the close of a multiday council. Gomo, a Potawatomi chief, gave the president a strand of dark (black or purple) wampum beads "to honor the dead" before delivering his speech. Upon concluding his remarks, he gave the president a length of white wampum (Jackson 1981:73).

Various strands of wampum were given by the Boston Museum to the Peabody in 1899. They include two lengths of purple beads, some of which are so dark as to approach black, a short length of white beads, and two strands of mixed white and purple beads. The purple and white lengths, shown on page 271, represent the manner in which wampum was generally exchanged. Because East Coast museums and learned societies generally had wampum in their collections, the Peabody (Boston Museum)

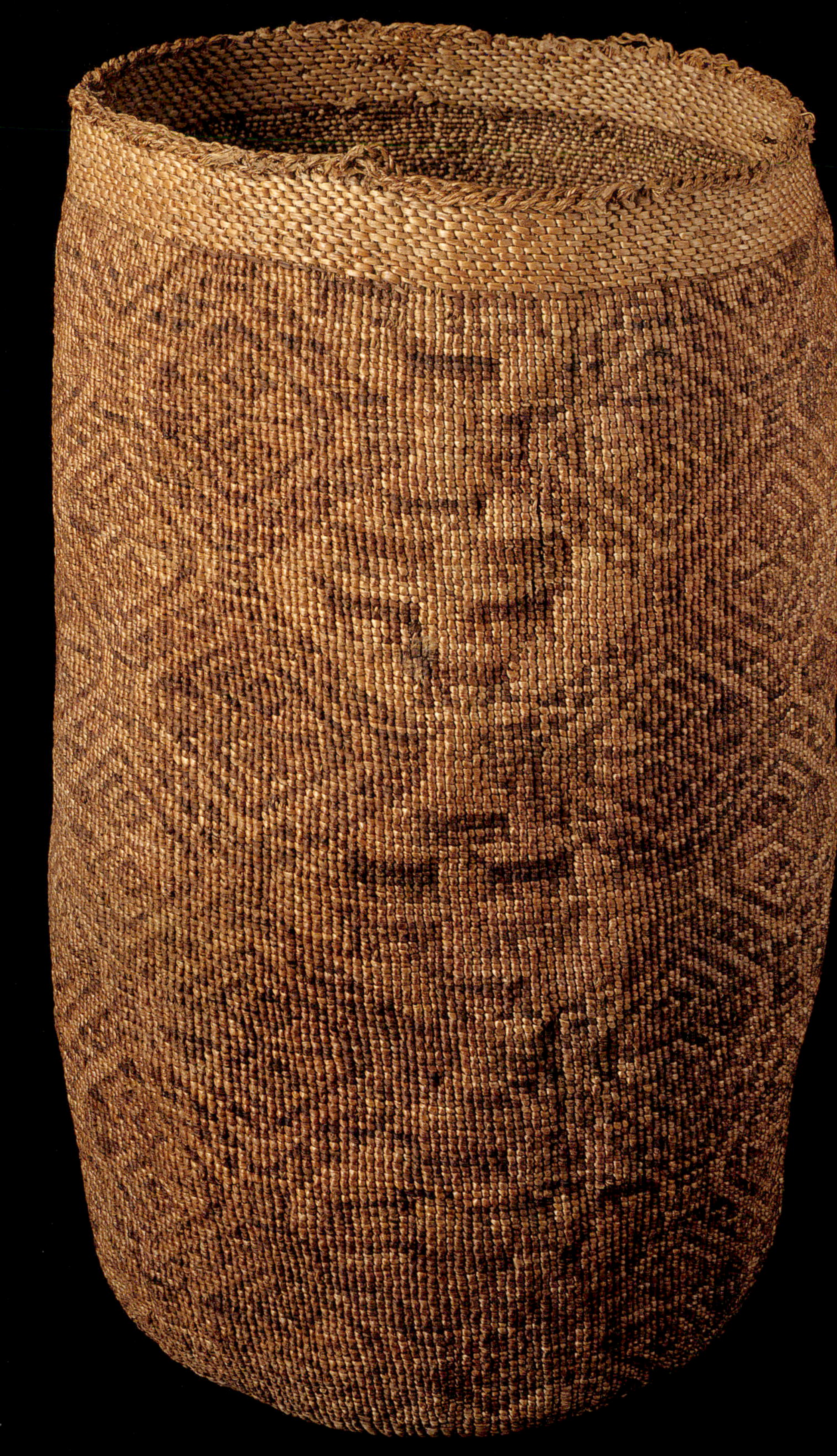

Root-gathering basket (sally bag)
99-12-10/53160
Wasco-Wishram, circa 1800–1850
Unidentified grasslike plant (sedge grass?) and fiber (Indian hemp?)
Height 30.5 cm, width 15 cm

examples might have originated with institutions other than the Peale Museum. In 1858, a Mr. Harmon A. Chambers of Carbondale, Pennsylvania, advertised for sale ten strands of wampum, each several feet long, identified as having come from the Peale Museum, but none was noted to have been associated with the Lewis and Clark expedition (Sellers 1980:321).

Root-Gathering Basket (Sally Bag)

Anne-Marie Victor-Howe

Lewis and Clark might have acquired this twined, cylindrical basketry object (99-12-10/53160) from a member of the "Pish-quit-pahs," a Sahaptin-speaking tribe they met on the north bank of the Columbia River near Umatilla Rapids.[10] Although neither expedition leader mentioned the basket in his journal, C. W. Peale's 1809 memorandum of expedition materials listed "A Bag prepared of grass by the Pishquilpahs on the Columbia River" (Jackson 1978, 2:478). Charles Willoughby did not include this bag in his 1905 article on the Peabody's Lewis and Clark collection, but he generated internal memoranda suggesting that it could have been associated with the expedition. Many years later, Mary Dodds Schlick (1979:10–13) identified this bag as an expedition object on the basis of information provided to her by Charles Coleman Sellers and a Peabody Museum cataloguer. Schlick attributed the basket to a Wasco weaver. Although no supporting documentation survives, it might well be an expedition artifact.

Lewis and Clark first encountered the Pish-quit-pahs as the Corps of Discovery made its way down the Columbia to the coast. On October 19, 1805, they observed many Pish-quit-pah lodges on the north side of the river and on several islands; the Corps later camped across from these on the southern bank (Moulton 1988, 5:306). On a map that he drew, Clark noted the presence of "44 Large Mat Lodges of the Pish quit pah Nation drying fish on large scaffolds supported by forks" (Moulton 1983, 1:74–75, maps 75, 76).

The expedition met the Pish-quit-pahs in the same area again on their way home, on April 25 of the following year. During a two-hour visit they gave two small medals to the principal chiefs of the tribe and were treated with "much rispect" (Moulton 1991, 7:165–68). Lewis counted 51 lodges and Clark 52 that day, and they estimated a population of "7 hundred souls"—significantly greater numbers of lodges and people than they had observed the previous fall, when most Pish-quit-pah men and women were presumably up in the hills, hunting for antelope, deer, and elk and gathering huckleberries, grouseberries, and low-mountain blueberries (Hunn 1990:131–32; Moulton 1991, 7:165–66). By April, the Indian men were on the river for the arrival of the first salmon run, and most of the women had come down to process and store the catch, perhaps carrying with them baskets of fresh bitterroots and lomatiums they had just harvested (Hunn 1990:121–23). Many baskets like the one in the Peabody collection would have been on hand, and it seems likely that Lewis and Clark acquired the Peabody's basket during this springtime visit. If so, they might have received it while trading for local foods, which they did throughout their journey up and down the "Great River."

What is particularly intriguing about the Peabody bag is that although it may have been obtained from the Pish-quit-pahs, it is made in the distinctive style of the Wasco-Wishrams, a branch of upper Chinookans who lived along the Columbia River from approximately the Cascades to Celilo Falls. These traditional Wasco-Wishram baskets from the mid–Columbia River region—commonly called "sally bags"—are covered with geometric patterns, human and animal motifs, and ancestral figures, often skeletal in appearance and now usually called "X-ray" figures or described as having an "exposed rib" motif (Schlick 1994:68). There are several possible explanations for the actual provenance of the bag, if it is the one given to Peale by Lewis and Clark. It might have been woven by a Sahaptin in the Wasco-Wishram style; it might have been acquired by a Sahaptin from a Wasco-Wishram in trade; or it might have been given to a Pish-quit-pah as a gift. Lewis and Clark could even have acquired the bag from a Wasco-Wishram weaver married to a Sahaptin. All four scenarios are plausible, given the close ties of trade and intermarriage between the two groups at that time.

Several explanations have also been given for the naming of these baskets. As Schlick (1994:48–49) reported: "The popular belief among collectors was that a woman named Sally was the only Wasco weaver who succeeded in making these animated subjects on her bags." Indeed, a weaver named Sally Wahkiacus, born around 1825, was known for making the baskets. It has also been suggested that early traders simply applied to the bag their own term "sally," a word for a pocket carried by English women (Schlick 1994, citing James Nason, personal communication, 1980). Schlick considered most logical the explanation that the name was an Anglicization of the local word for willow—"sallow" or "salla"—a material from which many of the early bags were made. The contemporary Wasco fiber artist Pat Courtney Gold gives another possible interpretation in her essay "A Wasco Weaver Meets Her Ancestors through Lewis and Clark," which appears as the following section of this chapter.

Mid-Columbia baskets were primarily utilitarian, and their forms and sizes were dictated by use. According to Pat Courtney Gold (personal communication, 2000), an individual basket always held the same type of food item. Soft cylindrical twined baskets such as this sally bag were typically used to carry roots.[11] Flat rectangular twined bags made of Indian hemp (*Apocynum cannabinum*), often called dogbane by Northwest basketweavers, were used to store dried roots or to carry belongings. Schlick reports that "dogbane is resistant to insects—making the twined bags ideal for dried food storage" (personal communication, 2001). Bags measuring up to two by three feet, also made from Indian hemp, were mostly used to store dried, pounded salmon and to carry belongings when families were harvesting or moving from one location to another (Miller 1991:188; Schlick 1994:14).

The Wasco-Wishrams and other Columbia tribes usually collected berries and other delicate foods in a type of basket made of strong coiled cedar root and "shaped like pails with flattened bases" to keep the fruit from being crushed (Conn 1998:48). Larger cedar root coiled baskets with rectangular bases and flat, flaring sides were used to store dried foods and household items such as clothes and ceremonial regalia. Coiled baskets of cedar root also served for cooking and for carrying water. On October 17, 1805, Clark entered a mat lodge where a man and a woman were preparing a meal using hot stones placed "into the basket of water with the fish" (Moulton 1988, 5:288).

On the Plateau, basketry objects also played important parts in ceremonies and other events. Decorated baskets and basketry clothing and hats were often given as marriage presents by the family of the bride to symbolize the bride's duties in her new household, to validate the marriage, and to consolidate the alliance of the two families (Miller 1991:182; Schlick 1994:89). Baskets were distributed along with other valued objects at events and celebrations such as those for the piercing of a child's ears, the giving of a first name, or a girl's puberty ceremony (French 1955:261, 1961:365; Schlick 1994:88). During funerals and memorials, baskets were given as payments to people who assisted the deceased's relatives (Schlick 1994:88). On October 20, 1805, near the present John Day River, William Clark saw baskets among a number of funerary objects on the ground (Moulton 1988, 5:311–12), and a few days later he reported in his journal the "sacrifice" of baskets as grave goods for the deceased (Moulton 1988, 5:347).

40
tives in diging their roots
sharpened at the lower
to a part of an Elks or
handle, standing transvers
This form A ——— B.
hart or handle. —

Top: An Indian woman from the Plateau rests during a root-digging expedition, with her basket and digging tool—called a *kapin* in Sahaptin—at her side.

Bottom: Sketch of a root-digging tool in the Lewis and Clark journals, January 24, 1806.

The sally bag in the Peabody collection is typical of the cylindrical digging bags upper Chinook women wore tied to their waists to carry their harvests of bitterroot, camas bulbs, and cous roots. To dig the roots, women used an implement called a *kapin* in Sahaptin: a gently curved hardwood stick about two feet long, sharpened at one end. At the other end a handle, generally carved from a deer antler, was fastened at a right angle. Lewis and Clark mentioned root-digging sticks several times in their journals (see Moulton 1990, 6:234). I saw similar tools in use at the Warm Springs Reservation in April 2000, but instead of wood and horn, these were made entirely of iron.

The timing of the root harvest coincided with the period when the roots reached their maximum nutritional potential and the soil was neither too muddy nor too dry to make digging difficult. When a woman secured her first root of the year, she offered a prayer to the spirit of the plant. When her small sally bag was full, she transferred her harvest to a larger bag on her back. The Root Feast, a springtime ritual of thanksgiving for roots of various species, for the return of good weather, and for the renewal of life, began in early April. Baskets such as these played a key role in the feast, which began when a group of selected elder women went out to the surrounding hillsides with their baskets to harvest roots.[12]

Dogbane could support heavy loads and was the preferred fiber for making twined root-digging bags during precontact and early contact times. The plant generally grew near rivers or sloughs, in valleys, and on the lower slopes near Wasco villages. The best dogbane, however, which had smooth, straight, thick stems, grew in more distant areas that were "kept secret by the women who know of them";

occasionally, violent conflicts erupted between rival groups of harvesters (Hunn 1990:94, 189). Harvested in summer or fall, dogbane was then "softened by burial in damp earth." To prepare the fiber, women stripped away the outer bark from the stem and then pounded the strands of inner bark until they were soft enough to be made into cordage (Hunn 1990:189).

Many sally bags were embroidered with beargrass (*Xerophyllum tenax*), the leaves of which were harvested in mid-summer and early fall and then cut to a uniform width. When dried, the grass was creamy white in color. It was used for weaving in its natural color or was dyed with vegetable pigments. Bear grass was a common item of trade in The Dalles, and on October 31 at the Cascades, Lewis and Clark observed two canoes "loaded with fish & Bear grass for trade below" (Moulton 1988, 5:363).

Dried split leaves of sedge grass (*Carex obnupta*), commonly called "basket grass" or "swamp grass," were also used as wrapping and twining material. Tall sedge grass was mostly harvested in wet places such as forest swamps, although it also grew in dry to semiarid areas. Clusters of leaves were cut at the base and tied into bundles, then processed at home and stored until the late fall and winter, when the women did most of their weaving (Turner 1998:106). Seasonal harvesting and processing activities ended in late October and early November, and families returned to their winter villages. The season ahead was a time for indoor activities such as making new songs, performing winter dances, telling stories, and manufacturing objects, including baskets, for household use or trade.

To weave the basket in the Peabody collection, the weaver started by making cordage of sedge grass for the warps. Next, she selected several warp strands about double the length of the desired bag size. Using a strand of dogbane bark, she wrapped cordage around the middle of the sedge grass to hold the spokes together. Finally, she began plain twining, interweaving her weft elements between groups of warps. The weaver went around the warps several times until she had built about one inch of the basket's base. For the weft, she then switched to sedge grass. Pat Courtney Gold (personal communication, 2000) thought she probably switched materials because producing dogbane cordage was labor intensive, and she might have wanted to use the material sparingly. The weaver continued her work for about another inch and a half and then began to add more spokes evenly to create the desired shape and size. Each time she added a spoke, she made a knot at the end of the fiber to hold it firm (Gold, personal communication, 2000). She twined around the warp strands and, as the weaving progressed, used fewer warps until she finally twined around each single warp.

Following the Wasco tradition, the weaver started her design using dyed, dark-brown sedge grass to make an "earth line"—a solid dark line at the base of the basket. Under a magnifying glass, the dark brown appears slightly purple, suggesting that its source was blackberry juice (Gold, personal communication, 2000). The color might have been obtained by mixing the juice with a decoction of willow bark or with mud from an alkaline spring.

When she began weaving designs, the maker of this basket used a full-turn twining technique with one natural and one dyed weft that she alternated to bring the design to the outside of the basket. One weft at a time remains passive, lying flat behind the group of warps, inside the basket, while the other crosses in front and then goes between the warp strands to reach behind and wrap the passive weft, finally returning to the front. The weaver ended her basket with a "sky line"—a solid dark line at the top—and the last inch of the basket before the rim was done in plain twining with natural sedge grass (Gold, personal communication, 2000). To finish the rim, "the weaver worked from right to left and involved four warps rather than three in each step of her 'braid'" (Schlick 1994:73).

X-ray faces on the sally bag. (Detail of 99-12-10/53160.)

When Pat Courtney Gold first examined the Lewis and Clark basket in May 2000, she realized that the weaver had not originally planned to use dog motifs but had intended to cover the whole bag with X-ray faces—the stylized, diamond-shaped human heads lying on their sides in pairs, crown to crown. Several scholars have pointed out a stylistic resemblance between Wasco-Wishram head motifs on basketry and prehistoric bone and stone sculptures. Others have identified diamond-shaped heads like the ones on this bag as representing the legendary leader Tsagaglalal, "She Who Watches."

As she was finishing the first row of heads at the bottom of the basket, the weaver realized that she had too little space for one more pair. Her solution was to weave one more head, starting with the crown, so that the being would face whatever the weaver placed in the remaining space. Gold ascertained that the weaver then had to weave several more rows to feature and develop the dog motif. That is why it is difficult to count and properly identify the figures represented in the first open diamond shape at the bottom of the bag. At least three dogs can be distinguished, but four additional, incomplete figures are unidentifiable. The diamond above the one woven at the bottom of the basket shows the final design adopted by the weaver: six dogs arranged in a triangle. The upper diamond also contains six dogs, whereas the beginning of the diamond closest to the rim contains one complete dog and the lower bodies of two more.

Looking at the bag as a whole, one can see a diamond-shaped head enclosed within another diamond that is connected to its adjacent diamond. The resulting general effect is one of diamond-shaped heads wrapped around a column of open diamonds containing dog figures. The pattern of horizontal and circular diamond heads contrasts with the vertical column of dogs and suggests a strong relationship between dogs and other beings. Although that may not have been the weaver's original intention, it is certainly the result she achieved.

Canids with their tails up in friendly greeting, like those on this basket, are generally identified as dogs by Columbia Plateau weavers, whereas wolves and coyotes have their tails down. It is possible that the dog figures served to warn harvesters that to avoid starvation and plague, they had to keep the world balanced. Donald M. Hines (1996:113–16) recounts a Wasco myth about a woman who marries a dog, which ends with the dog husband starving himself to death after a disastrous meeting with the woman's parents. In the Wasco-Wishram belief system, dogs are trusted helpers, situated halfway between humans and animals. A union between a dog and a human, however, was a transgression—the violation of a taboo, out of balance with the world—and so the marriage fell apart and the dog killed himself. In this basket, the column of open diamonds containing images of dogs reflects the Wasco-Wishram belief expressed in the myth—a belief in a cosmos organized on a vertical axis with dogs at the center, in an intermediate position that enables them to travel through different zones of the universe. Dogs serve as visual reminders of the importance of keeping the world in balance at all times.

A WASCO WEAVER MEETS HER ANCESTORS THROUGH LEWIS AND CLARK

Pat Courtney Gold

I first saw a photo of the Peabody Museum's beautiful Wasco basket (99-12-10/53160) in an article published by Mary Dodds Schlick in 1979 in *American Indian Basketry* magazine. The basket's intricate design attracted me, as did Schlick's statement that the basket might have been collected by Lewis and Clark on their journey some two hundred years before. My family has lived along the Columbia River for generations, and we have strong ties to the river. As a Wasco person, I could identify with this weaving. Later, when I learned to twine such baskets myself, I decided to make one similar to the basket in the Peabody's collection.

I already knew something about the unique images on the soft twined baskets known as "sally bags" that our people made for gathering and storing roots, dried salmon, medicines, and other important items. My family owned none, but I had seen them as a child during a visit to the Maryhill Museum of Art on the Columbia River near Goldendale, Washington, about one hundred miles from our home on the Warm Springs Reservation.

Pat Courtney Gold holding one of her Lewis and Clark commemorative baskets while receiving the Oregon Governor's Arts Award in 2001. The basket was inspired by the Peabody's sally bag and was made to honor the original Wasco weaver.

For thousands of years, the Wasco nation lived along the Columbia River in both Oregon and Washington. It had a language, a social structure with established standards, and a government of councils that included women. Fishing and trading were the most important activities. Historically the Columbia River was a "freeway" that connected the Native peoples who lived along it and its tributaries. The river was important for drinking, cleansing, and providing fish and plants.

The Wasco people were skilled basketweavers who had developed a unique twining technique for the beautiful soft baskets that became known as sally bags. (Anne-Marie Victor-Howe has discussed various interpretations of the term "sally bag"; I believe the most likely source is a Wasco word for basket, *schkully.*) The main design images on these bags were abstracted ancestral figures ("X-ray" human images), sturgeons, condors, mountain goats, frogs, and other familiar human, animal, and mythological forms. My ancestors also wove into the baskets abstract designs with geometric beauty: salmon gills, for the salmon so important to our lifestyle; sturgeon roe, another important food source; and symbolic images such as the step design, representing both growth and natural underwater fish ladders.

After the Treaty of 1855, my ancestors were forcibly and abruptly removed from their land along the river and relocated to a semiarid reservation in central Oregon, along with the Warm Springs and Paiute peoples—entirely separate Indian nations with their own histories, cultures, and languages. All three nations now reside at the Confederated Tribes of the Warm

Petroglyph of Tsagaglalal, or "She Who Watches," along the Columbia River at the site of Nixluidix, an old Wasco-Wishram town that the Corps of Discovery visited. Tsagaglalal is depicted on the Peabody's sally bag.

Springs Reservation of Oregon. The new environment was harsh and inhospitable, and the people's lives were governed by the War Department and the Indian Department (later the Bureau of Indian Affairs). Just surviving in this climate under army regulations was a hardship, and government policy was to acculturate the people to the lifestyle of the dominant Euro-American society. Under this policy, much of our culture was lost. My ancestors were expected to become farmers and Christians—to be like white people. Boarding schools, missions, and relocations to cities became the norm.

For three generations the people focused on survival, on how to live in this new environment. They experienced the gradual disappearance of Native cultural ways, of their language, of their arts including basketry, carving, and leatherwork, and of their legends, songs, and dances. Much of the basketry knowledge and many of the skills were lost. Throughout my childhood, visits to the Columbia River reacquainted me with my ancestral homeland. Yet what I saw were towns, trains, highways, cars, ships, and dams. Only through my grandmother's and mother's stories could I see and imagine our traditional homes, fishing sites, and lifestyle. It is now my generation, adapted to reservation life, that is helping to rediscover and revive Wasco culture. Our legends, songs, dances, and language are again being taught to the young people.

When I was growing up, I knew no one who made sally bags, although Paiute elders made willow baskets and Warm Springs elders made cylindrical root bags. I had seen Wasco sally bags with their beautiful geometric images only in museums. I thought this basket twining technique was lost and that museums were the only places where I would ever see it.

In 1992 my sister, Bernyce Courtney, another Wasco woman, Arlene Boileau, and I began to relearn Wasco sally bag twining techniques. We were committed to preserving our heritage by learning these weaving techniques to share with future generations, and the Oregon Folklife Master-Apprentice Program offered us an opportunity to work with a master weaver for one year. That weaver, Mary Schlick, is non-Indian but had lived at Warm Springs and on the Colville and Yakama Reservations for many years. Her husband worked for the Bureau of Indian Affairs, and Mary was accepted by the women weavers. Over the years she earned their respect and learned to weave from them.

During our apprenticeship we also consulted with other elders and visited museums to research designs. We worked together, sharing knowledge and skills. The twining technique was relearned, but we had to rediscover the traditional plant fibers and recreate the skills to harvest and process them.

Since completing the apprenticeship program, I have continued my studies, especially those on traditional fibers. I visit museums to examine baskets, identify designs, study the intricacies of twining techniques, identify the plant fibers used in older baskets, and study trends as the baskets evolved. I have expanded this knowledge of traditional designs into my own contemporary interpretations, and I continue to explore and experiment with plant fibers. My reward is hearing the elders express their pleasure in seeing the return of the sally bag and the continuation of this art form. Before I took up basketweaving, my parents had never visited art galleries or attended gallery openings. Now, at my exhibits, they take pride in meeting the audience and sharing their Wasco culture with other people.

My ancestors, working together, harvested plant fibers for basketweaving, as they did many food items. They sorted and processed the plants into cordage for twining. Plants used for weaving included dogbane, cattail, tule (bulrush), mountain sedge grass, and other grasses. Plant gathering was a pleasurable social activity, an opportunity to train the young people and to share and enjoy the basket stories and legends. Harvesting together created a bond. Some individuals especially enjoyed collecting and processing fibers; some enjoyed twining the baskets. The finished baskets reflected the people, the village, and their communal spirit. Although I knew no one who processed basket materials in my childhood, I remember extended family gatherings when we harvested roots and huckleberries for food, processing and drying them. We enjoyed seeing each other again and sharing many stories. It was a happy time.

When I first started weaving, I combed museums and magazines for sally bag designs. I was impressed with their intricate geometric patterns, which seemed so complicated. It was a challenge just to think of creating these designs for a basket. I thought, "When I get good, the design that I want to make is the design on the sally bag in the Peabody Museum at Harvard." This complex design is a network of horizontal faces completely covering the basket. Each face is surrounded by a geometric diamond border, and all are interwoven.

The face is that of Tsagaglalal, She Who Watches. According to legend, long ago Tsagaglalal was the honored leader of the Wasco-Wishram village of Nixluidix (Nik-loy-dik). When the trickster and teacher Coyote came to Tsagaglalal's village, he saw that all her people were happy and living well. Coyote decided that Tsagaglalal should remain the leader of Nixluidix and watch over her people forever. He turned her into stone, and to this day she watches over all of us as a well-known petroglyph above the Columbia River. Her image appears on many Wasco sally bags, including the basket thought to have been acquired by Lewis and Clark, who visited the busy trading center in October 1805 and again in April 1806.

In May 2000, the Peabody Museum invited me to study the Northwest Coast collection from the Lewis and Clark expedition. This was one of the most exciting times of my career. I spent hours studying the Lewis and Clark sally bag with curatorial associate Anne-Marie Victor-Howe. I held the basket in my hands, feeling a strong connection to the past and to my ancestors. It was an emotional and spiritual experience. As I held the basket, it told me this story: The faces are within a fishing net used in the Columbia River. My brother makes such nets. I have used them. Fishing is very important to the Wasco people, something we have done for generations. This net interweaves the people with the salmon, the river, and our heritage. The faces represent not only our ancestors but also the petroglyphs along the Columbia River—images in the rocks that were put there by generations of Wasco people. This is where my extended family lived, fished, and traded.

The maker of this basket also told her story. She was a talented weaver and skilled in creating intricate patterns. The pattern of faces within the stepped outline is complex and beautiful. It appears to have no beginning and no end.

The weaver connected twenty-eight faces, some facing different directions, as would occur in a group of people.

As a weaver, I studied the basket. I saw that the geometric pattern did not connect at the base of the design. The result was a gap, which could have thrown the pattern off. But the weaver continued and cleverly created a pattern as she advanced. She began with small geometric images, probably a study for a small face. She then experimented with four-legged images and two-legged images. These were unsatisfactory to her, so she continued experimenting. She then wove six dog images in a diamond space, while the faces continued their journey around the basket. The result was a sophisticated geometric pattern of faces, diamonds, and dogs. Even the diamond pattern was embellished with a geometric step design. I marvel at its sophistication and intricacy.

The basket itself has a beautiful structure. The weaver used very fine cordage for the warps, which appear to be fashioned from a type of sedge grass. I was excited to see that the start at the center base was identical to a start I had learned in my apprenticeship and still use in my own baskets. The weaver's bundle of starting warps was wrapped with a piece of dogbane bark and twined with dogbane bark for the first two inches. Thereafter, the weft used for twining was sedge grass cordage, very finely spun. After twining the bottom for two inches, the weaver added a series of warps to increase the size of the base. Looking down into the basket, I could see a circular design of "fuzzy knots" where the warps were added. I was learning from my ancestor, whose work was more than two hundred years old.

When I began weaving my first interpretation of the Peabody basket, I found the start of the design to be very difficult. There was no image to guide me. As I wove the first inch, I could feel my ancestors emerging from within the basket design and inspiring me. Many variations of the image visited my imagination. As a weaver, my challenge is to do variations of one theme, which is the design from the original 1805 basket. I now weave these designs into other baskets. The original basket had one column of dogs; I weave dog images into my baskets in different places—on the rim, and randomly at the bottom. I enjoy doing variations of the face designs, to reflect the diversity of people. On one of my baskets, many of the faces are different: smiling, whistling, winking, with different noses and different hairstyles.

When I received the Oregon Governor's Arts Award in 2001, I twined a sally bag to commemorate the Peabody's Wasco basket. The geometric faces on my bag are based on the petroglyph Tsagaglalal, and the geometric diamond design around the faces represents the fishing nets of our people and symbolizes the importance of fishing in our culture. The dogs are our companions and were also a food source for the Corps of Discovery. At the rim, I wove images that tell the story of the Corps' experiences along the Columbia River. There is a Chinook canoe with canoe "pullers" (paddlers) wearing cedar hats, and two Corps of Discovery dugout canoes. Lewis and Clark are each at the head of a canoe, one of which also carries Sakakawea with baby Baptiste on her back. I also wove the whale that the Corps went to see at Ecola Beach (Tillamook Head) during the winter of 1806, which Sakakawea found so fascinating, and the sturgeons Clark noted, tied along the banks of the river. These large river fish were important to the Wasco people as a food source and also as a symbol of strength and long life.

I experience many different feelings as I weave. I discover peace and calm. I imagine and feel the "old ways." I see the rapids and falls of the Columbia River. I hear the water churn and gurgle and I smell the fresh air. I feel the spirits of my ancestral weavers. The designs and the culture flow through me. I see designs from my life as geometric images. I am inspired to create my own story baskets.

As I study the old sally bags, the creativity of the designs and their sophisticated abstractness never cease to amaze me. These are integral parts of myself and influence my view of the world. My contemporary work is one of exploration, of

Pat Gold (center, wearing cap) and her students harvesting tule in eastern Oregon.

experimenting with color and texture, of adding a different dimension to the basketweaving. I enjoy using an array of commercial fibers and experimenting with colors to bring out and highlight certain design elements. I embellish the surfaces.

Testing the boundaries of tradition, I vary basket shapes and interpret designs with contemporary colors and views. While traveling in the Southwest, for example, I created a pottery-shaped basket with a mixture of ancestral Pueblo pottery designs and traditional Wasco images. To emphasize the dynamics of a changing culture, I weave images that are around me. I take pleasure in the humor of my *Indian Yuppie Couple* basket, weaving the male in a suit and hat, the woman in a flashy dress with jewelry. In another basket I attempt to raise environmental awareness by using color to focus on the endangered red frog. The goal of my design *Hanford Sturgeon* is to focus attention on the pollution of the Columbia River, the home of the now threatened sturgeon.

As a contemporary weaver, I have to be a botanist, a politician, an educator, and a computer specialist. As botanists, Native weavers have to recognize healthy plants, know which plants and which parts of plants to harvest, and learn the best times of year to harvest. We must also recognize and protect the habitats of the plants we use. Today, many of our traditional plants grow on land that is managed by federal and state agencies, and some of them are considered weeds. We have to know which plants are sprayed, which herbicides are used, and how the sprays affect our health. We also have to work with federal agencies to educate them about the "weeds" they spray.

As a politician and an educator, I was one of the organizers of the Northwest Native American Basketweavers Association, founded in 1995. With a membership now at 250 weavers, its mission is "to preserve, promote, and perpetuate Northwest Native American basketry traditions." I also try to teach federal, state, and local agencies about weavers' concerns. I mentor

From Generation to Generation, a series of linked baskets representing past, present, and future generations. The baskets are connected by a knotted cordage calendar, similar to a Plateau "time ball," on which the beads represent the gift of basketry.

new weavers and encourage them to use the resources of museums to research baskets and their culture. I work with museums to inform the public about our tribal nations' heritage, culture, and art.

As a computer specialist, I record information on basketry, basketweavers, and my museum research. I create databases of my research and data libraries on tribal culture and heritage for use in lectures. I use computer software to develop promotional biographies, create and archive slide images, and write historical commentaries on Columbia Plateau culture.

I see myself as a temporary "basket," a container with which to receive, develop, and incubate my skills, expand my weaving abilities, and share with other weavers. I feel honored to have this knowledge of my heritage and my culture to preserve and pass on to subsequent generations.

Our elders are "living museums." Their knowledge cannot be learned in institutions of higher education. It is taught in special ways and handed down to the next generation. With this in mind, I made the set of baskets called *From Generation to Generation*. The large basket represents my Wasco ancestors, the medium-sized basket represents my generation, and the small basket represents future generations. Handmade thigh-rolled dogbane cordage connects the baskets and represents the Plateau "time ball"—a calendar in which knots indicate the passage of time and beads represent special events. The trade beads on this cordage represent the gift of basketry from one generation to the next. The "end" bead is the gift that continues into the future.

The Wasco basket at the Peabody Museum has been my inspiration, and it continues to stimulate my creativity for future baskets. I weave to commemorate that special basket, and especially its weaver. I am the seventh generation since that early nineteenth-century weaver. The circle of Wasco culture continues.

Woman's Fiber Skirt

Anne-Marie Victor-Howe

Detail of twined cordage on woman's fiber skirt (99-12-10/52990).

Although in their journals Lewis and Clark mentioned seeing fiber skirts of the style illustrated here, neither man specifically recounted acquiring one from the inhabitants of the lower Columbia River. Circumstantial evidence, however, suggests that this garment (99-12-10/52990) might have been collected by the Corps of Discovery. In Clark's "Summary Statement of Rivers, Creeks, and most Remarkable Places," for example, when describing the upper village of the Wahkiacum nation, he referred to "Petticoat women at this village" (Thwaites 1904–5, 6:68). Later, C. W. Peale, in a letter to John Hawkins written on May 5, 1807, mentioned that "M[ss]rs. Lewis and Clark have actually visited the sea shore, and I have animals brought from the sea coast, also some parts of the dress &c. of the Natives of the Columbia River" (Jackson 1978, 2:410). No such skirt is described in the 1809 Peale memorandum, but that document is known to have been incomplete, and this skirt is the of right type and vintage to have been acquired by the explorers in 1804–6. There are two old, handwritten labels at the Peabody that might have been associated with this skirt. One reads, "Apron worn by the Women of the No.W Coast . . . Pres'd by Maj Horace Moses," and the other, "Indian Stroud or Apron." One or both of these labels might have originated at the Peale Museum.

We know that on November 7, 1805, in the Columbia River estuary just below Grays River, the expedition met a group of Indians in canoes and decided to follow them to their village. "Those people called themselves *War-ci-â-cum,*" Clark wrote, adding that they "Speake a language different from the nativs above with whome they trade for the *Wapato* roots of which they make great use as food" (Moulton 1990, 6:31–32).[13] The Wahkiacums provided the explorers with food such as fish, wapato roots, and dogs and traded two otter skins for some fishhooks. The expedition spent about an hour and a half in the Wahkiacum village before camping nearby for the night. (For

Woman's fiber skirt

99-12-10/52990

Lower Chinook, circa 1790–1830

Cattail leaves

Height 54 cm, width 74 cm

more on the Wahkiacums, see Boas 1901; Hodge 1910; Martin 1980; Minor 1983; Ray 1938; Rubin 1999.)

Describing the Wahkiacums in his journal that night, Clark specifically referred to the "peticoats" worn by the women:

> (Their peticoats are of the bark of the white Cedar) . . . The garment which occupies the waist and thence as low as the knee before the mid leg behind, cannot properly be called a petticoat, in the common acception of the word; it is a *Tissue* formed of white Cedar bark bruised or broken into Small Strans, which are interwoven in their center by means of Several cords of the Same materials which Serves as well for a girdle as to hold in place the Strans of bark which forms the tissue, and which Strans, Confined in the middle, hang with their ends pendulous from the waiste, the whole being of Suffcent thickness when the female Stands erect to conceal those parts useally covered from familiar view, but when she stoops or places herself in any other attitudes this battery of Venus is not altogether impervious to the penetrating eye of the amorite. This tissue is Sometims formed of little Strings of Silk grass twisted and knoted at their ends' &c. (Moulton 1990, 6:32)

On March 19, 1806, as the expedition was preparing to leave Fort Clatsop to begin the long journey home, Lewis and Clark made even more extensive notes about garments worn by men and women along the lower Columbia River, observing that "The *Kilamox, Clatsops, Chinnooks, Cath lah mahs Wau ki a cum* and *Chiltz* I—resemble each other as well in their persons and Dress as in their habits and manners" (Moulton 1990, 6:437). The women wore skirts like aprons that were made of strands of shredded cedar bark or tule rushes, also known as cattail. Lewis described a skirt much like the one in the Peabody Museum as "formed of little twisted cords of the silk grass knoted at the ends" (Moulton 1990, 6:435). He then observed that "this kind is more esteemed and last much longer than those of bark."

Hilary Stewart (1984:145) confirmed that among the peoples of the Northwest Coast, well-made clothing of fibers such as shredded cedar bark, cattail leaves, silk grass, and rushes was an extremely valuable trade commodity: "Westcoast people traded four blankets of yellow cedar bark for one good stone hammer. Among the Quileute, twenty cedar bark skirts could be traded for two bird skin blankets, one whaling canoe, or one slave. A single skirt could be worth one fathom of dentalium shells, which were used for jewelry or sewn in designs on ceremonial garments."

The skirt at the Peabody Museum is trapezoidal in shape. It is made of broadleaf cattail, also called bulrush (*Typha latifolia*), a tall perennial plant with long, swordlike, grayish green leaves that were widely used by Northwest Coast tribes for making mats, raincoats, skirts, capes, hats, baskets, screens, rugs, and bedding. To harvest this aquatic plant, women had to travel to ponds, shallow marshes and swamps, lakes, and tidal freshwater areas. They usually harvested flat cattail leaves from their canoes in late summer when the plant was sufficiently mature. After being processed to a certain point, the fibers were cut into even lengths, dried in the sun, and stored in bundles until they were needed.

The very fine twined cordage used to make this skirt was obtained by first stripping off the thin edges of the bases of the leaves. The fiber was then plied in both S-twists (to the right) and Z-twists (to the left), alternating through the skirt and ending in a knot at the bottom. In this skirt, approximately 750 "strings" averaging about 0.2 centimeters in diameter are doubled over a thin sinew belt that ties in back. The strings are gathered and secured in bundles just below the belt by two weft strands of cedar bark. On average, about fourteen strings make up each bundle; roughly half are S-twisted and half Z-twisted, so that, according to Pat Courtney Gold (personal communication, 2000), the strings would hang in a complementary fashion and the skirt would flow evenly and harmoniously around the front of the wearer's body.

Such skirts and other garments made of broadleaf cattail, split cedar bark fibers, silk grass, and rushes were extremely practical for people living along the southern part of the Northwest Coast in a temperate rainforest environment characterized by heavy rainfall, chilly winds, and frequent fog. Outer clothing had to be both water repellent and warm. Women's wraparound skirts were knee length in front and calf length in back; they weighed little and had oil worked into them to help repel rain. They were both comfortable and practical. As Stewart (1984:141) pointed out, "while loose garments of hide were certainly not unknown, they were not suitable as outerwear in this inclement weather. Clothing made from the shredded and oiled cedar bark, however, gave protection from both rain and cold. The bark's multiple layers of fiber afforded good insulation and, when wet, dried quickly by the fire."

The Peabody skirt is one of only four such skirts of fine twined cording from the Northwest Coast known to exist (Wright 1991:34). The three similar skirts, which are in the Smithsonian Institution's National Museum of Natural History, almost certainly date from the early 1800s, the same period as the Lewis and Clark expedition. They, too, are made of broadleaf cattail with similar bundles of complementary S- and Z-twisted strings.

The Smithsonian's skirts once belonged to George Catlin, who perhaps acquired them from William Clark. In 1830, Catlin went to St. Louis to meet Clark and visit his museum. The two men became friends, and Clark reportedly taught Catlin much of what he knew about Indian culture, particularly that of the Northwest Coast. According to Thomas Donaldson, one of Catlin's students, during the early 1830s Clark gave Catlin "a portion of those objects collected by Lewis and Clark, and by Governor Clark" (Donaldson 1887:389).

By 1850, despite years spent traveling and exhibiting his "Indian Gallery" in Europe (Dippie 1990:67), Catlin was financially ruined. His collection lay "in the hands and at the mercy of [his] creditors" (Dippie 1990:145). It would have been lost to the United States if not for the intervention of Joseph Harrison, Jr., a patron of the arts from Philadelphia who paid Catlin's debts, taking as security his entire original collection. From that point on, Catlin made no more additions to his "gallery"; it belonged entirely to Harrison. At some time in 1852 or 1853, Harrison shipped it back to Philadelphia and placed it in storage. In 1872, after trying in vain to persuade Congress to pass a bill providing for the purchase of the collection, Catlin died without having found a home in Washington for his life's work.

Harrison, who still owned the collection, died in 1874. In 1879, at the request of Harrison's widow, her son-in-law Thaddeus Norris gave the Catlin gallery to the Smithsonian. A large number of the Indian-manufactured objects had been badly damaged by moths, fire, and water. Objects that could not be salvaged were buried in the yard of Harrison Boiler Works, and the surviving items—including a Crow lodge and clothing, masks, pipes, moccasins, Catlin's sketchbook, his notes taken while traveling among the Indians, his palette, and some correspondence—were packed up in boxes by Thomas Donaldson and shipped at once to Washington (Donaldson 1887:6).

Among the Catlin ethnological specimens in accession number 10,638 were the three woman's fiber skirts, "probably Chinook or Salish," that are so similar to the skirt at the Peabody. It is possible that they were among the Indian artifacts Clark gave to Catlin in the early 1830s, although this does not mean that the fiber skirts were collected during the Lewis and Clark expedition. Clark might have acquired them in the years following his return. Catlin himself could not have added the skirts to his collection later, because his "rambles" along the West Coast of North America did not take place until the 1850s, by which time the collection had been seized by his creditors.

The skirts at the National Museum are in much worse condition than the skirt at the Peabody, presumably because of the damage they suffered during their years of stor-

CEDAR: THE TREE OF LIFE

Red and yellow cedar trees have always played a fundamental role in the social and religious life of the Native peoples of the Northwest Coast, providing everything from building materials to medicinal agents. Deeply interwoven in mythology and traditional beliefs (Turner et al. 1990:54), the trees could speak; they had spirits, feelings, thoughts, and life-giving properties. A woman gathering bark, for example, addressed a tree with supplications such as "I come to ask for your dress," or "I came to beg you for this, Long Life Maker." A man who wanted to cure his wife of an illness would talk to a cedar, calling it "Healing woman, Long Life Maker" (Stewart 1984:182).

The trunks of both red and yellow cedars could be split into planks for building houses, or they could be used whole for house poles. They were also carved into monuments such as memorial, mortuary, and house frontal poles. Single cedar logs, hollowed out, were used as trading canoes, war canoes, hunting and fishing canoes, transportation canoes, and river canoes. Smaller, thinner planks were steamed, bent, and polished with sandstone before being manufactured into drums, storage chests, boxes, and other receptacles.

The bark of the cedar was equally important. Smaller pieces were turned into boxes, and a large sheet of yellow cedar could be converted into a small craft that allowed a man "to ferry himself across a stream or to fish on a lake" (Stewart 1984:119). Bark could be fashioned into objects as varied as canoe bailers, rattle heads, cordage for everyday and ceremonial use, and basketry items including bags, mats, and sails. It was particularly useful for manufacturing clothing. Medicinal qualities, too, were to be found in cedar bark, the leaves of yellow cedar, and cedar branches, all of which were valuable as agents of physical cleansing and purification (Stewart 1984:180–81; Turner 1998:72).

After stripping the bark from tall, straight trees, women separated the soft inner bark from the coarse outer layer. Strips of different sizes were then tied into bundles and

Ada Markishtum from Neah Bay prepares cedar bark by separating the soft inner bark, used to make objects, from the coarse outer layer, used for fuel.

Cedar bark bundle.

Bark shredder (05-1-10/65488).

Whalebone bark beater (05-7-10/64543).

brought home, where they were untied and hung outside to dry. After a few days, the strips were refolded into bundles and stored until needed for manufacturing. Hilary Stewart (1984:117) wrote that the "bark was best worked when it had thoroughly dried out for a year."

The shredder (05-1-10/65488) and beater (05-7-10/64543) shown here were used to process bark. When a woman was ready to manufacture an object, she used a shredder made of hardwood or whalebone to work the bark strips. She would pass the strip of bark "across a block of wood while hitting it at the block edge with the bark shredder," in this way separating and softening the layers of bark (Stewart 1984:123). Referring to bark beaters made of whale ribs, Stewart wrote: "Because of the amount of oil in a fresh whale bone, a beater of this material gave the tool the necessary weight and probably imparted some of the oil to the bark also" (1984:126). Once processed, yellow shredded bark was much softer and finer than red and was therefore considered more valuable.

Anne-Marie Victor-Howe

age in Philadelphia. They also have red stripes painted on them, unlike the Peabody skirt. Several other objects in the Catlin accession at the Smithsonian are covered with the same red oil paint, and Bill Holm (1992:51) is probably correct in assuming that Catlin added the paint so that the decorated objects would catch the attention of people who visited his exhibits in the United States and abroad.

Black-Rimmed Hat

Anne-Marie Victor-Howe

The hat illustrated on pages 296 and 297 (99-12-10/53081) was among the objects acquired by the Peabody Museum from the Boston Museum. In their journals, Lewis and Clark described many kinds of clothing worn by the men and women they met west of the Rockies, including several types of hats, but no Makah or Nootka black-rimmed hats like this one. Nor were black-rimmed hats described by any other explorers or fur traders who preceded Lewis and Clark on the Northwest Coast.

Nevertheless, it seems likely that this hat dates from the early eighteenth century, which, along with its association with other Lewis and Clark material, supports the possibility that it was acquired by the Corps of Discovery. Robin K. Wright, an authority on Northwest Coast material culture, agrees that "the linear quality of the painting" and "the V-shape of the head and running row of little U-shapes on the border of the horns tend to suggest an early nineteenth-century manufacture" (personal communication, 2002). She adds that "the later painted black-rimmed hats have more band-like, curvilinear, and formline derived designs."

It is worth noting, too, that this hat is somewhat different from all other black-rimmed Makah or Nootka hats now in institutions and private collections around the world. The latter are conical in shape and thus similar to onion-dome hats, but with the knob on top replaced by a concave depression. This hat has sides that are not strictly conical, and it is somewhat broader and shorter than the others. Its design motifs of red and black, with areas of natural fiber incorporated as a third color, are also much less abstract.

Looking straight down on the hat from above, one can see that its broad black rim encloses a circular space in which are woven two V-shaped heads opposite each other, their antlers interlocked around a central oval. Wright is almost certainly correct in interpreting these antlered heads as those of deer or elk (Wright 1991:84), but

Black-rimmed hat
99-12-10/53081
Makah or Nootka (Nuu-chah-nulth), circa 1800–1840
Spruce root, cedar bark, pigment
Height 12 cm, width 38 cm

considering the relative unimportance of deer to the Makahs and Nootkas, I suggest they are most likely elk.

At the time of the Lewis and Clark expedition the Nootkas and Makahs made little use of elk for food, but they did use elk antlers to make tools and implements, most importantly whaling harpoons and lances. The relationship between elk and whales in Makah and Nootka subsistence culture is well illustrated by the Makah belief that long ago a human was changed into an elk. Its antlers were then sawed off by a man who used one of them to make a whale harpoon, which ensured that his tribe had plenty to eat all winter. Albert Irving, a former whaling "spearman," told that story to Frances Densmore at Neah Bay, on the Strait of Juan de Fuca, in 1923 or 1926, adding that "unless the human being had taken the form of an elk the people would not have had food. How thankful I am that a human being took the form of an elk, as by its horns I still have food all winter and whale hide to sell and buy groceries" (Densmore 1939:126).

The Makahs and Nootkas used elk antlers to make whaling harpoons and lances because the antlers are hard yet malleable. They also believed the antlers were imbued with the physical strength and fighting ability of the male elk and thus could help them overcome the great strength and power of whales. The Elk Dance of the Makahs, for example, demonstrated the strength and power of male elk in rut. Male dancers "walked in a dainty manner, imitating elk and holding antlers in their hands," then "held the antlers at their waist and assumed a threatening attitude" (Densmore 1939:127).

Two smaller oval shapes outside the interlocked antlers and opposite each other probably represent the eyes of either humans or animals—or both, for in the worldview of virtually all Native American groups, humans and animals were able to transform into each other. As human eyes, these would have given the wearer of the hat powerful vision for hunting. As elk eyes, they would have added the elk's particularly acute peripheral vision.

This hat is twined from spruce root, a material often used by Makah and Nootka people to make basketry items. The roots were dug up in the spring and early summer from trees growing in sandy soil (Turner 1998:88). The inner crown of the hat is twined with red cedar bark. For the Makahs and Nootkas, the yellow cedar (*Chamaecyparis nootkatensis*) and western red cedar (*Thuja plicata*) provided most of the raw material for making baskets, clothing, hats, mats, and other items.

The black paint on this hat might have been manufactured in one of several ways. James Swan noted in 1870 that the Nootkas and Makahs made black paint by "grinding a piece of bituminous coal with salmon eggs which have been chewed and spit on a stone" (Swan 1870:iii, 45). Both groups also made black paint by mixing fish oil or salmon eggs and hardwood ashes ground into a fine powder (Densmore 1939:22; Koppert 1930:54; Waterman 1920:22). Sometimes charcoal was replaced by "an evil-smelling black muck, obtained in swamps" (Waterman 1920:22).

According to Swan (1870:iii, 45), the Makahs and Nootkas made red paint like that used on this hat by mixing "vermilion [obtained from the Chinooks] and chewed salmon eggs." The Makahs also used red ochre to make red paint, mixing it with fish oil or seal oil to paint the insides of their whaling canoes (Koppert 1930:54; Waterman 1920:15; see also Densmore 1939:22). The Nootkas used red alder (*Alnus rubra*) to obtain an orange-red color (Turner et al. 1982:23, 1998:150).

Top Hat and Sailor's Cap

Anne-Marie Victor-Howe

There is no extant evidence that these European-style hats (99-12-10/53176 and 99-12-10/53083; facing page and p. 300) were part of the Peale collection, but Lewis and Clark recorded seeing and acquiring similar hats along the Columbia River. On November 21, 1805, while still on the north side of the Columbia in Chinook territory, Clark purchased a hat that he said was "made in the fashion which was common in the U States two years ago" (Moulton 1990, 6:76). On January 19, 1806, as the Corps of Discovery wintered on the south side of the "Big River" in Clatsop territory, a man and a woman visited the expedition's fort with several articles to trade. Among them was another basketry top hat that the explorers purchased in exchange for several fish-hooks. Lewis described that hat and others similar to it as follows:

> [T]hese hats are of their own manufactory and are composed of Cedar bark and bear grass interwoven with the fingers and ornimented with various colours and figures, they are nearly waterproof, light, and I am convinced are much more durable than either chip or straw. These hats form a small article of traffic with the Clatsops and Chinnooks who dispose of them to the whites. [T]he form of the hat is that which was in vogue in the Ued States and great Britain in the years 1800 & 1801 with a high crown reather larger at the top than where it joins the brim; the brim narrow or about 2 or 2 1/2 inches. (Moulton 1990, 6:221)

Peale's memorandum of expedition materials lists three hats. One is described as "manufactured by a Clatsop woman near the Pacific Ocian; from whence it was brought by Capts. Clark & Lewis" (Jackson 1978, 2:476). A second hat is described as "a Cap worn by the Natives of Columbia River, and the Pacific Ocean" (Jackson 1978, 2:478). Unfortunately, these descriptions do not record the form of the hats that the

explorers presented to Peale. In my view, they probably correspond to either the whaling hat sent by Clark in 1806 and described as "1 Hat made by the Clatsops Indians" (see the onion-dome hats 9912-10/53079 and 53080; pp. 94 and 95) or a second hat forwarded by Lewis and listed only as "Clatsop hat." The third hat, described as a "Cap, worn by the women of the Plains of Columbia" (Jackson 1978, 2:476), is almost certainly not the top hat shown here, but the description could refer to the sailor's style cap. Since we know that the Peale memorandum was incomplete and that Lewis and Clark acquired both woven top hats and caps, it is possible that these hats were obtained during the expedition.

Basketry top hat
99-12-10/53176
Tlingit or Haida, circa 1790–1820
Spruce root, unidentified grass, cedar bark
Height 16.5 cm, width 34 cm

Basketry sailor's cap
99-12-10/53083
Tlingit, circa 1790–1840
Spruce root, unidentified grass, maidenhair fern stems, cedar bark
Height 14 cm, width 28 cm

The Peabody top hat is plain twined, probably with spruce root. About this hat, Holm remarked that "the style of decoration in weft-dyed bands suggests Haida work, but the twining direction is more characteristic of Tlingit basketry" (Vaughan and Holm 1990:39). Delores Churchill, a Haida basketweaver, mentioned that the Chilkat weavers who live in the northern part of southeastern Alaska sometimes use the Haida style of decoration (personal communication, 2002). The sailor's cap is also plain twined, again probably with spruce root. The design elements are executed in false embroidery with grass, most likely bear grass (Turnbaugh and Turnbaugh 1986:32). It is therefore probable that this top hat and the sailor's cap were not manufactured by members of the lower Chinookan tribes but were acquired through trade from Tlingits or Haidas or both. Considering the volume of intertribal exchange along the Northwest Coast, it is conceivable that articles manufactured by people living in southeastern Alaska and the northern part of British Columbia could have ended up in the Chinook-Clatsop area.[14]

Winter Quarters, a 1792 watercolor by George Davidson, depicts Captain Gray's vessel at anchor in Adventure Cove. Northwest Coast people were familiar with and had adopted Euro-American clothing and hat styles—like the top hats worn here—well before the Lewis and Clark party reached the Pacific shore.

Perhaps most intriguing about these hats is the use of such obviously European styling in an article of clothing traded between Native groups so early in the nineteenth century. Trade, however, had been the preeminent occupation of the tribes of the lower and mid-Columbia River for decades before European contact. Although the environment was so rich that local peoples could easily have sustained themselves with what nature offered, the Chinook-Clatsops and the upper Chinooks were among the most commercially oriented Natives in North America. The arrival of whites along the coast was just a further opportunity for them to increase their markets and to acquire goods such as iron, copper, and colored beads that had not previously been available to them.

On September 21, 1805, near the north bank the Clearwater River, Clark was informed by a Nez Perce man that his tribe obtained white beads and clothing from white men who lived below at Celilo Falls (Moulton 1988, 5:226). James Ronda (1984:159) attributed this remark to "a misinterpretation of sign language," because "there were, in fact, no white traders in the territory so jealously guarded by Wishram and Wasco Indian middlemen." By October 16, 1805, the expedition had reached the Columbia River. Between the mouths of the tributary Umatilla and John Day Rivers, Clark saw the first piece of European clothing the explorers spotted in the region—a sailor's jacket that "was a clear indication of the visits of seagoing traders on the Pacific shore" (Moulton 1988, 5:313, n. 6). From then on, ready-made European clothing obtained from fur traders—sailor's jackets and hats, scarlet and blue cloth blankets, overalls, and shirts—were observed and recorded everywhere. While still in Wasco country among the upper Chinooks, Clark noticed a man with "his hair cued" in the style worn by nineteenth-century European and American men (Moulton 1988, 5:347). This showed that many Native people of the area were drawn to the Euro-American curiosities and fashions to which they were exposed.

By October 22, the Corps of Discovery was about to encounter the largest marketplace in the Northwest—the ten-mile stretch of river between Celilo Falls and The Dalles, which was also the most productive salmon fishery on the Columbia. The most intense activities took place at Nixluidix, the main Wishram village. That day Clark observed huge stacks of dried salmon that he estimated weighed ten thousand pounds (Moulton 1988, 5:323, 325).

Although Lewis and Clark recognized that they had entered one of the "great mart[s] of trade," they had arrived too late to observe the vast quantities and variety of Native goods exchanged there. Ronda (1984:170) tells us: "Trading took place from spring through fall during the three major salmon runs, with most activities reserved for the fall season." A great variety of European goods acquired from fur traders, such as guns, blankets, clothing, blue and white beads, copper, and kettles, made its way from the coast to The Dalles, to be exchanged for goods from the Plains such as buffalo meat, skin clothing, shields, catlinite pipes, and obsidian, along with horses from the Nez Perce, pounded salmon from the upper Chinooks, and roots and berries from the Sahaptins.

On October 29, Clark was invited to the lodge of a Chinook chief, where he noticed several pieces of European clothing: "a Scarlet & blue Cloth Sword Jacket & hat" (Moulton 1988, 5:351). Farther down the river, Natives wearing items of European clothing were a common sight. "The Indians at the last village," wrote Clark, "have more Cloth and uriopian [European] trinkets than above[.] I Saw Some Guns, a Sword, maney Powder flasks, Salers Jackets, overalls, hats & Shirts, Copper and Brass trinkets with few Beeds only" (Moulton 1990, 6:15). The same day, just as the captains and their men were ready for dinner, "Several Canoes of Indians from the village above came down dressed for the purpose as I [Clark] Supposed of Paying us a friendly visit, they had Scarlet & blue blankets Salors jackets, overalls, Shirts and Hats independent of their Usial dress" (Moulton 1990, 6:17).

There probably were many reasons Northwest Coast people eagerly adopted the costumes worn by people from afar. The new garments might have been attractive because they were made of new fibers; their styles and manufacture were also different and thus desirable. But above all else, the new clothing presented, at least at the time of first contact, a certain aura that was associated with the European wearers, whom Native people perceived, according to Mary Helms (1988:173), as "manifestations of some aspect of the unregulated 'power outside.'" Helms further remarked that if indigenous people perceived Europeans as beings with supernatural power, then this belief derived from Native recognition of several cosmological realms: the secular, or earth, realm, which was "the cultural center or heartland of socialized community life," and the faraway sacred realms, with their "extraordinary, power-filled supernatural attributes" (1988:173).

On October 19, 1805, as the Corps of Discovery proceeded down the Columbia River, Clark went ashore with a few men and shot a crane. Approaching some lodges by the river, the party observed several Indians hurrying to their homes "as fast as they could run" (Moulton 1988, 5:305). Clark entered one of the lodges to discover a crowd of "32 person men, women and a few children Setting permiscuesly on the Lodg, [Some] in great agutation, Some crying and ringing there hands, others hanging their heads." Later that day Clark learned the main reason for the agitation. As the Umatillas had watched the arrival of strangers on shore and saw the crane fall from the sky, they concluded that the newcomers "came from the clouds" and "were not men" but supernatural beings or supreme deities from the sky (Moulton 1988, 5:305). This association of foreigners with gods, evil creatures, or ancestors was widely shared among aboriginal people during the period of first contact. As Helms explained, the first Europeans "had to be ogres or demons or spirits or ancestors or gods because such were the types of animate beings 'inhabiting' distant locales. As such, Europeans by definition were embodied expressions of universal *mana* and contained supernatural powers for goods or for harm" (Helms 1988:173).

The first European goods that reached the lower and upper Chinook people were almost certainly imbued with magico-religious properties. By 1805, however, when Lewis and Clark explored the lower part of the Columbia, the Chinooks who lived there, unlike the Umatillas whom the explorers had met earlier, were already familiar with foreign goods. At the mouth of the river the captains learned from Natives of the area that at least a dozen vessels anchored off the coast twice a year, during spring and autumn. The status of goods had by now shifted slightly from their original association with "supernatural powers for good or harm," and Native people looked upon foreign objects more pragmatically. They were attracted by the newcomers' abundance of goods and the perceived superiority of Euro-American technology.

Even though Western clothing was less well suited to the northern coastal climate than were the traditional local garments made of fibers, by 1795 it had become, according to James Gibson (1992:219), "the commanding inducement" on the Northwest Coast. Gibson stressed that "the long-standing demand for textiles is explained by the fact that they were not durable" and that, "moreover, they were needed to replace the fur garments that the Indians traded right off their backs."

But the issue was more complex. New goods mostly radiated out from trading centers and from headmen or chiefs of the upper class, who monopolized trade. Powerful traders were well known and exerted influence far from their tribal territories. Wealth was of great importance, and commerce was prestigious. By adopting European clothing, Native traders expressed to the world their "association with power-filled people, . . . for such adoptions served to associate the wearer (or the acquirer) more closely with the aura of the foreign Europeans" (Helms 1988:196). When a Native trader was able to display his success in interacting and dealing with the external world, in this case Euro-Americans, his power and status increased not just among his own people but also among the foreigners. Indians' adoption of Euro-American clothing offered "material proof of the social equality" between aboriginal wearers and Westerners (Philibert and Jourdan 1996:66).

Coats and hats obtained from high-ranking government representatives were prestige items, symbolizing the government or country that presented them, and "their acceptance signified special ties of allegiance and friendship" (Prucha 1988:239). Such presents were also used "to strengthen and secure the influence of tribal chiefs, whose good will was essential for American policy" (Prucha 1988:238).

Indigenous clothing, especially hats and hair ornaments worn by Northwest Coast people, served a wide range of functions, from practical to sacred. Ornamented hats often displayed the wearer's rank and lineage, so it is no wonder the headgear worn by Euro-Americans, especially high-ranking men, appealed to the upper classes of this coastal area. By wearing such attire when dealing with Euro-Americans, Native people asserted themselves as equal in rank and status with the foreigners. Soon, Euro-American designs were being copied using traditional methods of manufacture and traditional materials.

The Peabody Museum's top hat and sailor's cap illustrate perfectly the incorporation of these types of goods into local modes of consumption. Euro-American clothing and accessories were often used in combination with traditional dance regalia, as is illustrated in several old photographs of Northwest Coast people at the turn of the last century (Bill Holm, personal communication, 2001). A photo by E. W. Merrill taken at Sitka in 1904 (now at the American Museum of Natural History, 328740), for example, shows the "Raven guests at a potlatch given by the Wolf 32 clans." In it, three women at the front of the crowd are wearing a style of hat derived from European sailors' hats (Holm, personal communication, 2001); the hats are similar to the Peabody's hat in form but are fringed around the rim with beads.

Bill Holm also referred me to a "welcome speech" by Jennie Thlunaut, a Tlingit elder, given in 1985 when she addressed apprentice weavers at the Raven House in Haines, Alaska. In the speech, she referred to *yaanwaa sháa,* or "navy women" (Dauenhauer and Dauenhauer 1990:204–5). Nora Dauenhauer explains that navy women were "an important group in the Tlingit ceremonial life." The group was composed, she wrote, "originally of the Kaagwaantaan women" who "for generations had claimed the Navy uniform as a crest, and [wore] parts of it at memorials" (Dauenhauer and Dauenhauer 1990:353, n. 169). Eventually, other Tlingit clans joined the group. Frederica de Laguna (1972:440) mentioned the Kaagwaantaan at Sitka, who "once adopted the American Navy." The women at Sitka and Yakutat, she said, "wear middy blouses or dresses with sailor collars." De Laguna also explained that at potlatches in 1904 and 1916, several Yakutat men and women wore hats with a "wide flat brim (without crowns?), trimmed all round with long beaded fringes." She said this style was copied from Russians' uniforms (1972:440).

The Northwest Coast tribes, well known for their superior trading skills, seized the opportunity of inserting these new products into their market with whites. Some Euro-Americans may have found basketry top hats and other copied items attractive as trade items because they showed how Native American cultures had been influenced and adapted to change. Lewis and Clark, however, were probably more interested in the utilitarian function than the style of the hats they acquired in trade from Northwest Coast Indians. On November 28, 1805, Clark complained: "Wind Shifted about to the S.W. and blew hard accompanied with hard rain all last night, we are all wet bedding and Stores, haveing nothing to keep our Selves or Stores dry, our Lodge nearly worn out, and the pieces of Sales & tents full of holes & rotten that they will not keep any thing dry." The garments of the Chinook and Clatsop people, as the explorers observed, were perfectly suited to the damp weather, and wearing them, the men were able to go about their daily tasks in foul weather without interruption. The explorers quickly recognized the superiority of the Native clothing, particularly the waterproof qualities of their hats. "[T]hese hats are made of the bark of Cedar and beargrass wrought with the fingers So closely that it Casts the rain most effectually," Clark declared (Moulton 1990, 6:246).

Chinook Cradles

Anne-Marie Victor-Howe

The two cradles shown here (88-51-10/50695 and 88-51-10/50696; facing page and p. 308) were acquired by the Peabody Museum on December 31, 1888, in an exchange with the Smithsonian's National Museum of Natural History. Although they were part of the George Catlin collection, Catlin may have acquired them secondhand, rather than collecting them himself. This may have been the case for all of Catlin's Northwest Coast objects.

The two cradles are in delicate condition and show signs of fire damage. In addition, according to Peabody Museum conservator Scott Fulton, the "lower edge of the cradle 88-51-10/50695 near the bottom of the handle appears to have been chewed or gnawed." These features strengthen the supposition that the two objects were among the ethnological specimens acquired from Catlin by Joseph Harrison, Jr., which were placed in storage in Philadelphia. On June 11, 1879, the Smithsonian received from Harrison's estate the first shipment of the Catlin collection, which consisted mostly of the "Paintings of Indians." On October 21, 1881, a second shipment arrived that contained mostly ethnological specimens (accession no. 10,638). These had been stored in one of Harrison's buildings in Philadelphia (Donaldson 1887:6), and many of them had suffered moth, fire, and water damage. Cradle 50695 appears to have been damaged by rodents as well.

As with the cattail skirt (99-12-10/52990; see p. 289), it seems likely that these two cradles were among the Indian objects that William Clark gave to Catlin in the early

Wooden cradle
88-51-10/50695
Chinook, circa 1800–1840
Western red cedar?, cedar bark, pigment, white glass bead
Length 67 cm, width 22.8 cm, height 16.5 cm

1830s, despite the fact that in their journals neither Lewis nor Clark mentioned acquiring any cradles. The two men took a lively, often discerning interest in Indian manners and customs, but occasionally their preconceptions and cultural biases left them baffled by the strange new languages, objects, and practices they encountered. They were particularly curious about—and amazed by—the practice of flattening children's foreheads, which they observed among the tribes of the Columbia River.

As the expedition approached the junction of the Snake and Columbia Rivers, Lewis and Clark recorded their first encounter with this custom. On October 17, 1805, Clark described the Natives of the area: "Those women are more inclined to Co[r]pulency than any we have yet Seen, with low Stature broad faces, heads flatened and the

Wooden cradle
88-51-10/50696
Chinook, circa 1800–1840
Western red cedar?, unidentified plant fiber, unidentified hide
Length 77 cm, width 23 cm, height 15.5 cm

forward [forehead] compressed so as to form a Streight line from the nose to the Crown of the head, their eyes are of a Duskey black, their hair of a corse black without orniments of any kind braded as above" (Moulton 1988, 5:289).

On November 1, 1805, writing about the "Shahala Nation," or Cascade Indians, a Chinookan people living along the Columbia just below the grand rapids, Clark said that

> all the women have flat heads pressed to almost a point at the top The[y] press the female childrens heads between 2 bords when young—untill they form the Skul as they wish it which is generally verry *flat.* This amongst those people is considered as a great mark of buty—and is practised in all the tribes we have passed on this river more or less. (Moulton 1988, 5:369)

The same day, Clark remarked that in certain places both men and women had flattened heads. He made several sketches showing how it was done. In one, a child is shown strapped to a piece of board with a pad of shredded cedar bark under its head (p. 310). Its forehead is apparently protected by another pad of the same material, on top of which is laid a shorter piece of wood that extends from the eyebrows to the top of the lower board. The boards are bound together with thongs to press the forehead. Another sketch shows the profile of the "head of Flat head Indians on the Columbia the head broad at top crop wise." A third sketch represents a man and a woman in profile, their entire frontal regions flattened (Moulton 1990, 6:252).

A few months later, on March 19, 1806, Lewis again discussed head flattening:

> [T]he most remarkable trait in their physiognomy is the peculiar flatness and width of forehead which they artificially obtain by compressing the head between two boards while in a state of infancy and from which it never afterwards perfectly recovers. [T]his is a custom among all the nations we have met with West of the Rocky mountains. I have observed the heads of many infants, after this singular bandage had been dismissed, or about the age of 10 or eleven months, that were not more than two inches thick about the upper edge of the forehead and rather thiner still higher. [F]rom the top of the head to the extremity of the nose is one streight line. [T]his is done in order to give a greater width to the forehead, which they much admire. (Moulton 1990, 6:433)

Head deformation was a privilege of the upper class, and slaves and mentally inferior people were called "round headed" or "flea faced" (Rubin 1999:84). Unlike some other indigenous operations on the body, head flattening was done when the subject was still an infant and the cranial bones were soft enough to respond to the pressure of the board on the frontal-occipital area.

The Chinooks used two types of cradles, the flat type illustrated by Lewis and Clark and by Paul Kane (1996 [1859]:141) and a dugout type like the cradles shown here. In both cases, the flattening action was created at "the front and back of the head and increased the compensatory bulges with an upward as well as sideways growth" (Dingwall 1931:5). A dugout cradle was carved from a single piece of wood, generally cedar, and had a platform that served as a headrest in the bottom of its wider part. James Swan (1870:18) described the practice of flattening the head among the Makahs, saying that they used a method similar to that of the Chinook:

Lewis and Clark were struck by the practice of cranial modification, which was accomplished by placing infants in specially designed cradle boards. During January 1806 they depicted this method and its results in their journals. The Peabody cradles are of a different type.

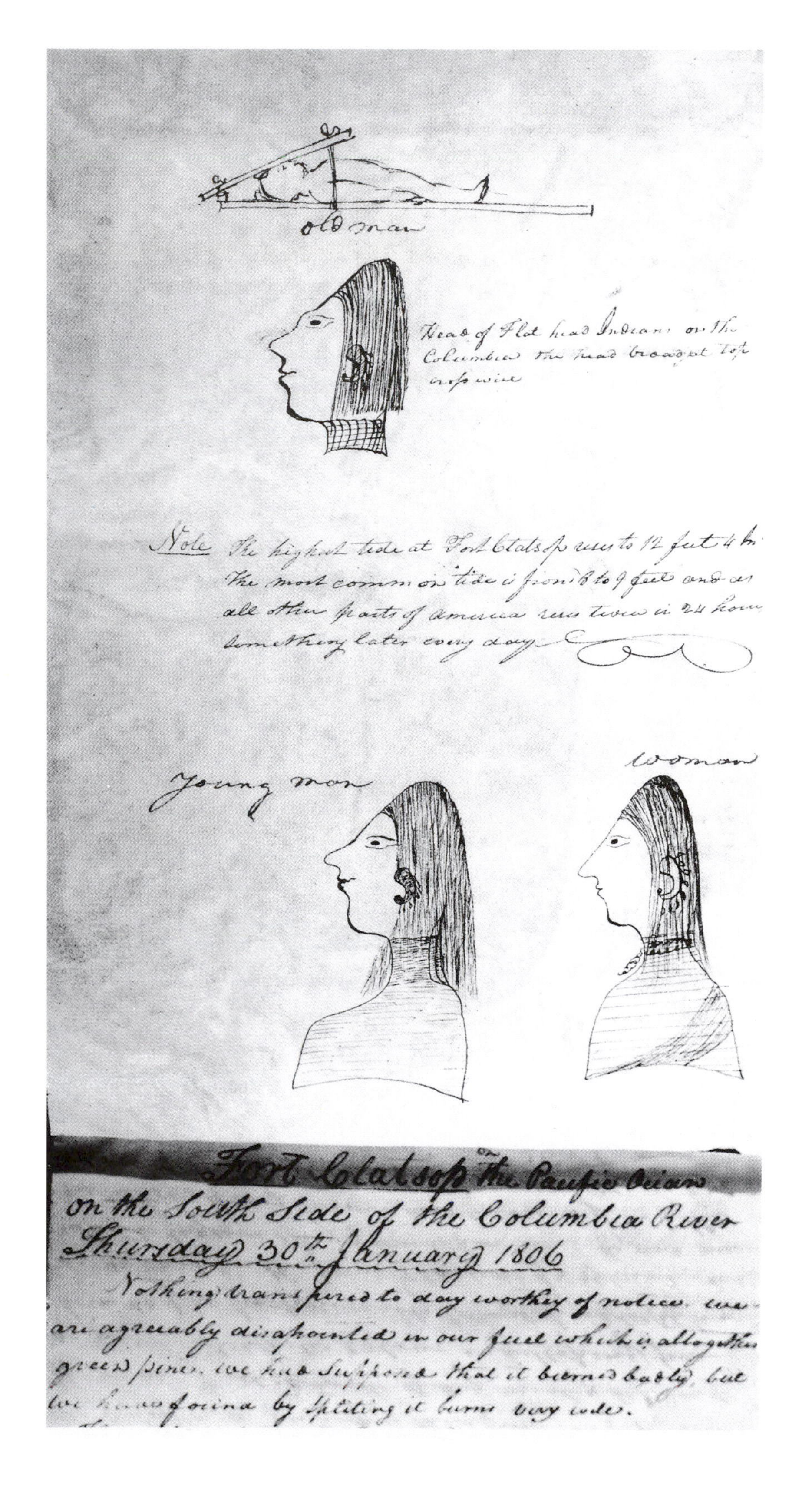

Note The highest tide at Fort Clatsop rises to 12 feet 4 In. The most common tide is from 8 to 9 feet and as all other parts of America rises twice in 24 hours something later every day

Fort Clatsop on the Pacific Ocean on the South Side of the Columbia River

Thursday 30th January 1806

Nothing transpired to day worthey of notice. we are agreeably disapointed in our fuel which is altogether green pines. we had Supposed that it burned badly, but we have found by Spliting it burns very well.

> As soon as the child is born it is wash[ed] with warm urine, and then smeared with whale oil and placed in a cradle made of bark, woven basket fashion; or of wood, either cedar or alder, hollowed out for the purpose. Into the cradle a quantity of finely separated cedar bark of the softest texture is first thrown. . . . First the child is laid on its back, its legs properly extended, its arms put close to its sides, and a covering either of bark or cloth laid over it; and then, commencing at the feet, the whole body is firmly laced up so that it has no chance to move the least.

A board bound to the back of the cradle was lowered above the child's head, creating a "powerful clamp action" as the ropes were tightly fastened (Dingwall 1931:5).

SUNDAY
MONDAY
WEDNESDAY
THURSDAY

Afterword

The Peabody–Monticello Native Arts Project

Two hundred years after the Corps of Discovery's expedition, the objects that Indian people presented to Lewis and Clark continue to create new ties between individuals and institutions. In early 2001, staff of the Peabody Museum, personnel at the historic site of Monticello, Thomas Jefferson's home and plantation, and several Native artists embarked on a joint project based in part on the Peabody's Lewis and Clark collection. The goal of the project was to create new objects for an exhibit staged at Monticello in commemoration of the Corps of Discovery's bicentennial, titled *Framing the West at Monticello: Thomas Jefferson and the Lewis and Clark Expedition.* The heart of the exhibit, which opened in January 2003, was the recreation of Jefferson's famous Indian Hall. Because Jefferson's original collection of Indian materials has been lost, Monticello commissioned contemporary Native artists to create new objects of the same types. The resulting works are not exact replicas or reproductions of vintage items. Instead, the artists studied the remaining Lewis and Clark

Contemporary Native American artists, inspired by works in the Peabody Museum's Lewis and Clark collection, created the objects displayed in this recreation of Thomas Jefferson's Indian Hall at Monticello, 2003.

Artist Dennis Fox with the Peabody's pictographic bison robe.

objects for inspiration and created new ones using historically accurate techniques and materials while incorporating their own artistic interpretations.

Butch Thunder Hawk (Hunkpapa Lakota), who teaches traditional arts at United Tribes Technical College in Bismarck, North Dakota, was a primary partner in this endeavor. Throughout the two-year project, Thunder Hawk and his students researched early Plains Indian art from every angle. They consulted with tribal elders regarding appropriate approaches to recreating objects, talked to traditional artists, and scoured the North Dakota landscape for the dye plants, woods, stones, pigments, and other materials used by nineteenth-century Indian peoples. The horsehair used on some of the objects was taken from Nokota horses, animals that had run wild in the North Dakota badlands for more than a hundred years. The Peabody Museum appointed Thunder Hawk a visiting curator and created internships for United Tribes students. With sponsorship from both the Peabody and Monticello, he and his students traveled to the Peabody to study the Lewis and Clark collection. For Thunder Hawk, seeing the old original objects was the greatest experience of the project. The opportunity to see these things firsthand, right in front of you—to see how they were put together—helps me to be accurate. It's very much a learning experience. The value for me is to have the experience and challenge of recreating history and to understand the older techniques of these art forms. I have a lot of appreciation now for art objects made back in those days, because of the limited tools and supplies they had compared to contemporary art. As some old philosopher said, "less is more." It's kind of true—these things don't have a lot of fancy beadwork and bright colors, but they are still beautiful and functional. We did a lot of spiritual work in the process—we prayed with them, and we smudged [ritually smoked] the pieces we were making every day.

Other artists participating in the project were Mary Elk (Mandan-Hidatsa), Jo Esther Parshall (Cheyenne River Lakota), Dennis Fox (Mandan-Hidatsa), and Joel Queen (Eastern Cherokee). Elk and Parshall are skilled quillworkers who created ornaments for a bison robe and for men's and women's moccasins and leggings. Queen, a stone carver, made a pair of Mississippian-style sandstone figures like those a Tennessee farmer sent to Jefferson in 1799. Fox, a painter and carver, created two painted bison hides to evoke objects that once hung prominently in Jefferson's Monticello home. One, a large robe, is based on the Peabody's famous pictographic robe, which commemorates a battle between upper Missouri River tribes. Fox also created a hide map depicting the Missouri River region between the Platte and Yellowstone Rivers. During the expedition, Jefferson received such a map from General James Wilkinson, the first governor of Louisiana Territory.

The spirit of negotiation and respect that animated many of Lewis and Clark's interactions with Indian peoples was largely forgotten in the decades following the expedition. National policies toward Indian peoples and their lands, objects, and cultures over the last two hundred years have done grave damage to Indian-white relations. The bicentennial of the Corps of Discovery has provided an opportunity both to address the problematic legacy of the American past and to create a more collaborative future. The Peabody-Monticello Native Arts project and similar bicentennial efforts can contribute to that future by fostering new relationships between Native and non-Native individuals, groups, and institutions engaged in interpreting the histories and cultures of the United States. As both a nation and as individuals, our challenge now is to sustain and strengthen these new relations.

Notes

INTRODUCTION

1. One of the raven "armbands" (now believed to be a belt) and PM 99-12-10/53047, a side-fold dress, were exhibited at the 1965 World's Fair in New York, where they represented the Lewis and Clark expedition in the Jefferson section of the "Hall of Presidents."

2. There are other surviving Native American objects believed to be associated with the expedition: an abalone hair or ear ornament and fragments of a Mandan pot, all at the University of Pennsylvania Museum of Archaeology and Anthropology (see Gilman 2003).

CHAPTER 1. THE LEWIS AND CLARK EXPEDITION

1. James Ronda (personal communication, 2002) points out that Lewis and Clark did contribute to the conceptualization of regional environments.

2. Throughout the subsequent text, all historical documents are reproduced verbatim, using original spelling, punctuation, and capitalization. Occasional interpolations, inserted to clarify particularly obscure spellings or expressions, are set in brackets.

3. As the expedition keelboat passed St. Charles the following May, this shipment was observed by a man named William Joseph Clark (no known relation to William Clark). On May 20, 1805, he recorded in his diary: "Just as we rose from Dinner news came that Capt. Lewises Barge which had bean Exploring the Mesoori was Just landed. I went on board and Saw a great many Curiosities Such as mocisons Buffaloe Robes Goat Skins Birds and a prairai Dog and the Chief of the Recarrean Indians who was a Very Large fat man very much pited with the Small Pox[.] I smookt with him

Moonrise over the Missouri River, New Town, North Dakota, summer 2000.

as a Brother." Diary of William Joseph Clark, 1804–7, entry for May 20, 1805, The Filson Historical Society.

CHAPTER 2. UP THE MISSOURI

1. James Ronda (personal communication, 2002) has observed that the imperial agenda of Jefferson's expedition expanded after the Louisiana Purchase; see Jefferson to Lewis, January 22, 1804 (Jackson 1978, 1:165–66).

2. The powerful Hidatsa chief Le Borgne, for example, in a speech to Crow visitors asking them to accept his Northwest Company trading partners, asked, "Do your Neighbors the Serpent nation [i.e., the Shoshones] enjoy the Security and happiness we enjoy? If the white men could furnish the Serpents as they furnish us with arms, we should not carry away so many of the Serpents' Scalps" (McKenzie in Wood and Thiessen 1985:247).

3. Ronda (1984:172) discusses the cultural construction of "theft" among Columbia River peoples at some length: "What Lewis and Clark saw as troublesome and potentially dangerous behavior was perceived by the river Indians rather differently. Taking axes, clothing, or rifling through the expedition's luggage probably involved two patterns of behavior not understood by the captains or their men. On one level, river people saw the things they took from the expedition as proper payment for services rendered. As the Indians viewed it, the explorers had more knives and blankets than they could possibly use. What harm could there be in taking a blanket, especially when the whites had so many and the Indians had provided valuable support in portaging around the dangerous places in the river? But the issue was more complex. The river peoples had a strong sense of private property. They left stacks of valuable dried fish standing unguarded without thought of loss. In pilfering small objects from the Americans, they sought Lewis and Clark's acknowledgment of their importance. Taking an axe was done to remind the white men of the need to offer respect and attention to the trading lords of the Columbia." Carolyn Gilman (2003) has noted that since the Corps failed to establish friendly trading relationships with the Chinooks, the exchange of property between the two groups was conducted as a form of ritual conflict.

4. The spelling "Sakakawea," which I have adopted here, is used by the Three Affiliated Tribes and the state of North Dakota. Phonetically, her name is rendered as "Tsagagawea" (Hidatsa for "Bird Woman").

CHAPTER 3. SELECTIONS

1. I have been unable to find an account of this incident in the expedition journals. Neeshneparkkeook, or Cut Nose, interacted with the party frequently during its second stay among the Nez Perce, in May and June 1806. Lewis and Clark counciled with a group of Nez Perce headmen, including Neeshneparkkeook, on May 11, 1806, but their accounts of that meeting mention no such speech.

2. Jefferson received three boxes of materials from Lewis after the expedition, as recorded in correspondence from James L. Donaldson to Thomas Jefferson on January 11, 1807 (Jackson 1978, 1:360). Clark also sent three boxes of expedition materials to his family in Louisville, Kentucky, from St. Louis in 1806 (Moulton 1993, 8:418–19). Clark evidently later sent some of this material to Jefferson; see William Clark to Thomas Jefferson, June 2, 1808, Jefferson Papers, Massachusetts Historical Society.

3. The boxes of expedition materials that Clark sent to his brother Jonathan from Fort Clark included a "Tigor Cat skin Coat" (William Clark to Jonathan Clark, April [4–7] 1805, Temple Bodley Collection, The Filson Historical Society).

4. *Poulson's Daily Advertiser,* March 10, 1810, p. 2. Transcript provided by Carolyn Gilman, Missouri Historical Society.

CHAPTER 4. INTO THE MUSEUM

1. A full account of the post-expedition history of the objects is to be published separately. For interpretations of Jefferson's Indian Hall, see Robinson (1995) and Fernandez-Sacco (n.d.). For the history of the hall and its contents, see Chew (2003).

2. Peale's tribal identifications are discussed later in passages about the specific objects.

3. I have drawn on a large body of primary and secondary material about Peale and his museum, including Sellers (1980) and Miller (1988).

4. Sellers (1980:252–53) speculated that the St. Louis traders may have been David Delaunay, "Tesson," and "Louis." According to Ewers (1966:7–8), Louis was Paul Loise, a.k.a. Paul Choteau, a mixed-blood Osage member of the prominent Choteau family. In 1805–6, Louis traveled east with the delegation of chiefs sent by Lewis and Clark and by Wilkinson, and had his silhouette made at the Peale Museum (Ewers 1966:fig. 8; see chapter 8 for more about this delegation). Louis also served as an interpreter for William Clark between 1815 and 1817. He therefore knew Clark, Peale, and Jefferson. If Louis was, indeed, a source for Peale's purchase of Indian materials from the Upper Missouri, could some of those objects have come from the collections of Clark or Jefferson?

5. "An Illustrated Catalogue and Guide Book to Barnum's American Museum," n.d., stamped "1864," was printed by Wynkoop, Hallenbeck and Thomas, New York (Special Collections, Adelphi University Library, Garden City, New York). Two other Barnum guidebooks at the New York Historical Society were examined for this project, one dated 1850 and the other 1863. Unfortunately, no proper catalogues for either the Barnum Museum or the Boston Museum are extant. A comparison of Barnum's American Museum guidebooks and those produced by the Boston Museum reveals that the two institutions not only employed the same overall design but also shared many identical illustrations, merely changing the captions to describe their separate objects.

6. Places and objects associated with the Lewis and Clark expedition have been accorded increasing historical significance in recent decades. Many objects and physical sites along the trail have been lost through benign neglect and human alteration of the landscape. Unfortunately, some of Jefferson's expedition materials may have been discarded by the University of Virginia as late as the mid-twentieth century (Jackson 1978, 2:734, n. 3; Chew 2003). During the same time period, Lewis and Clark campsites in North Dakota were flooded by the Garrison Dam project, and the lower Mandan village site visited by the expedition (near present-day Stanton, North Dakota) was destroyed by gravel and coal mining operations.

CHAPTER 5. WARRIORS AND WOMEN TRADERS

1. An otter bag similar to this one and identified as Omaha was collected by Lieutenant George Hutter in 1825–26 and presented to the Peale Museum. It was received by the Peabody Museum as part of the 99-12 accession, and the Peabody retains a paper label from the Peale Museum that identifies it as a tobacco pouch. The Hutter bag was traded to the Denver Art Museum in 1952. It is pictured in Conn 1979:155. Peale Museum records document the receipt of similar bags from other donors.

2. William Clark's "Catalogue of Indian Curiosities," William Clark Papers, Box 8, Folder 4b, Missouri Historical Society, St. Louis, Missouri.

3. Otter bags are conventionally photographed with the otter's head folded over at the neck and facing downward, but this bag has long been stored with the head extended.

4. Webber's portrait of a woman of Nootka Sound wearing an onion-dome knob-top hat presents a problem in that Nootka women were unlikely to have worn whaling hats. Jonathan King (1981:82) points out that members of Cook's expedition did not understand that whale hunting was a man's occupation and that only chiefs who also were harpooners and leaders of hunting expeditions wore decorated onion-dome hats. Webber's field sketch of the same woman (PM 41-72-10/225) shows her wearing a flat-topped painted hat, which was probably what women usually wore. His drawing of the woman wearing a knob-top hat also lacks the precise details of the artifact itself, particularly the decorative motif representing a whale hunt with the harpooner standing at the bow, his harpoon over his head, in the action of striking downward at the animal.

5. According to Christian Feest (personal communication, 2002), Sánchez Garrido (Museo de América 1991) recorded twenty eighteenth-century onion-dome knob-top hats in museum collections worldwide. Only two of them were decorated with scenes not depicting whale hunting. One of them, a hat at the Peabody Museum (67-10-10/269) collected by a Captain Magee, shows representations of four thunderbirds: two are by themselves, one hovers over a whale, and a fourth grips a whale in its claws. Thunderbirds in Makah and Nootka cosmology were giant men with huge beaks who dressed in birdlike outfits to hunt whales (Drucker 1951:153). The other hat (1949 Am 22 229) is housed at the British Museum (King 1981:38); on it are lightning and feathered serpents, who were the thunderbird's helpers.

6. William Clark to Thomas Jefferson, October 10, 1807, Jefferson Papers, Series 1, General Correspondence 1651–1827, Library of Congress, World Wide Web version.

CHAPTER 6. THE ARMY MOVES WEST

1. Biographical information on George C. Hutter was graciously provided by Christian Sixtus Hutter III of Lynchburg, Virginia, and by Gail Pond of Poplar Forest. Hutter's portrait was made available by Ann Carter Hutter. Sandusky, Hutter's home, is now listed on the National Register of Historic Places.

2. The Cobbs family, however, and later the Hutter family of Poplar Forest, apparently did "inherit" some Jefferson family articles left at Poplar Forest when Francis Eppes sold the property.

3. William Clark's "Catalogue of Curiosities," William Clark papers, Box 8, Folder 4b, Missouri Historical Society; transcript copy provided by Carolyn Gilman. Clark accumulated the bulk of his collection during his post-expedition life in St. Louis. Unfortunately, his catalogue is incomplete; it does not include objects specifically mentioned by contemporary visitors. A few entries refer to objects Clark probably acquired during the expedition, such as the foot of a grizzly bear that he killed at the "head of Missouris," a "plaited reed & sling" made of grass from the Columbia River, a Pacific sea otter pelt, and grizzly bear skins.

4. The conservator Bill Plitt suggested Blackfeet on the basis of the red pigment and a faint herbal odor that Plitt thought might be sweet balsam. Gary Spratt, a collector and dealer in North American Indian materials, feels that the baby carrier was made in the Columbia Plateau region, perhaps by a Nez Perce or Shoshone woman. Gaylord Torrence is also inclined to believe that it originated in the Plateau region.

5. Another knife sheath in the Boston Museum accession, 99-12-10/53139, could also be a candidate for the Hutter piece, but it apparently entered the Peabody tied to the shaft of an unrelated bow lance.

CHAPTER 7. ENIGMATIC ICONS

1. The "menitarras" were the Minnetarees, designated by ethnologists as the "Hidatsas proper." They were the largest of three related yet distinctive "Hidatsa" village groups living along the Knife River at the time of Lewis and Clark's visit. Lewis and Clark also referred to them as "Big Bellies." The "Ahwahharways," whom Lewis and Clark called the "Shoe Tribe," and the "Wattasoons" were the Awaxawis, the least populous and most diverse of the Hidatsa communities. A third group, the Awatixas, occupied what is now referred to as Sakakawea Village, in honor of its most famous resident. In 1804, two Mandan and three Hidatsa villages were situated along both sides of the Missouri River and a tributary, the Knife River, within about ten miles of one another (Chomko 1986; Lehmer 1971:fig. 111). Many of those village sites have been destroyed by development, but some, including Sakakawea Village and the Minnetaree village, are now protected in the Knife River Indian Villages National Historic Site near Stanton, North Dakota.

2. In a 1977 letter to Charles Coleman Sellers, a copy of which is in the Peabody Museum files, John Ewers expressed doubt that the robes visitors described at Monticello were either the Mandan robe or the Peabody robe. However, those descriptions cannot be taken at face value, since they were

provided by casual visitors who knew little about Native Americans.

3. There seem to be only two scalps represented on the Peabody robe: one hangs from the end of a stick and is circular, with projecting, rotating lines rather like a sunburst; the other dangles from the jaw of a horse and consists of a slanting vertical line with horizontal lines extending out on either side.

4. Tillie Walker, Mandan-Hidatsa, emphasized this incident and its possible meaning to me when I visited with members of the Lewis and Clark committee at Fort Berthold.

5. Carolyn Gilman at the Missouri Historical Society reports that Parks and Findlay was a mercantile company in Westport (now part of Kansas City) and that E. A. Johnson was a merchant in St. Louis (personal communication, 2000). Sutter, too, lived in Westport before finding fame in the California gold fields. Gilman also verified that Schoch did business with Pierre Choteau and Company between 1836 and 1842, so it is possible that Schoch purchased his items from these fur trade merchants rather than collecting them directly. Despite initial doubts (e.g., Bushnell 1908), however, there seems little reason not to believe Schoch's testimony that he acquired the items directly.

6. Clark to D. Madison, December 3, 1809, in Mary E. Cutts's scrapbook, Cutts-Madison Papers, Massachusetts Historical Society. This card was discovered and brought to my attention by Carolyn Gilman of the Missouri Historical Society, who found it in the course of her own research. Clark had known James Madison since childhood, and the two remained lifelong friends.

7. See Sornborger memo, April 30, 1901, Peabody Museum X Files, 99-12(c), notebook of Charles Willoughby. J. D. Sornborger, of the Boston Society of Natural History, solicited handwriting identifications from living heirs of C. W. Peale in or before 1901.

8. It is also possible that the two dresses were acquired by C. W. Peale as part of his 1826 purchase of items from the Missouri and Mississippi Rivers. Only one dress is identified as such in Peale's incomplete museum catalogue. On May 3, 1836, a Dr. Swain donated a "Sioux frock," leggings, moccasins, belt, and garters (Memoranda of the Philadelphia Museum 1804–42, Peale Papers, Microfilm copy, Roll 983.1, Historical Society of Pennsylvania, Philadelphia). Clark's museum catalogue lists two items described as "Indian squaws petticoat, Souix" (William Clark's "Catalogue of Curiosities," William Clark Papers, Box 8, Folder 4b, Missouri Historical Society; transcript copy provided by Carolyn Gilman).

9. Ted Brasser called my attention to a possible third example of a painted side-fold dress in the Malaspina collection in Madrid. That object, identified as a "hide robe" and apparently collected during the 1790s, is painted with linear and triangular black and red designs. It was illustrated by Feder (1977:fig. 15), who related it to Cree hide-painted garments.

10. John C. Ewers to Ian Brown, n.d., Ian Brown papers 993-13, Box 4, Peabody Museum Archives. For an account of Mandan women wearing strap dresses, see the narrative of David Thompson in Tyrrell (1916:233).

11. Because Plains Indian iconography in quillwork and beadwork was multivocalic, assuming different meanings in different contexts and for different people, I have made no effort to gloss designs and design elements. The possibility that these triangular motifs might represent bison heads is noted because Taylor (1997:fig. 20) identified quilled bison heads on the yoke of a "Cree type" side-fold dress from the Speyer collection (Benndorf and Speyer 1968:132; Brasser 1976:fig. 96). They might also represent bison traps. Similar motifs, with the extensions pointing upward, as horns would, appear on a tobacco bag shown in Krickeberg (1954:pl. 3c) and on an early man's shirt pictured in Feest (1968:pl. 5) and identified as Plains Ojibwa.

12. Unfortunately, Stanley Olsen (1963) used the presumed date of this dress (1804–6) as the basis for his chronological classification of brass button types.

13. Sellers (1991:121) documented that in 1795–96, General Anthony Wayne sent pipes and robes to Peale that he

had received as diplomatic gifts during an intertribal treaty negotiation in western Ohio. Among these objects was "A Large Indian Mantle, made of a Buffaloe's skin and ornament[ed] with Porcupine quills."

14. Mitchill was so closely integrated into Jefferson's social circle that one might wonder whether the Peabody's robe was originally his, had Smith not described the quillwork on the robe he drew as red and yellow. J. C. H. King (1994:60) reported that the National Museum of the American Indian has objects that once belonged to Mitchill.

CHAPTER 8. THE LANGUAGE OF PIPES

1. "Sent to Monticello March 10, [18]06." Jefferson Memorandum, March 10, 1806, Thomas Jefferson Coolidge Collection, Massachusetts Historical Society (microfilm).

2. Eleven pipe bowls, bowl fragments, or stems are listed in the Peale accession ledger between 1805 and 1822, in addition to those from Lewis and Clark, Hutter, and the 1826 purchase. Although the descriptions are minimal, four of those entries seem to refer to pipe stems. Peale included three "pipes" in a shipment he sent to an unidentified museum in France in 1833, as part of a trade.

3. Other documentation adds a "Sciouse" delegate; see Jackson (1978, 1:265–76). This delegation was months in the making, and the configuration of participants changed several times before the final group departed. The background to the delegation is provided in correspondence published by Jackson (1978, vol. 1) and summarized by Ewers (1966:10–11). Lewis and Clark intended that the delegation be twice as large, but federal officials balked at the cost. Two other delegations resulted from the expedition, an Osage group in 1804 and a small group that accompanied Lewis back to Washington in 1806, which included Sheheke.

4. See Wallace (1999:241–47) for a discussion of the Indian affairs in and around St. Louis that led up to the formation of this delegation.

5. There are seventeen pipe stems in the Peabody's 99-12 accession, two of which are treaty pipes unrelated to Lewis and Clark.

6. The Denver Art Museum has catalogued the two calumets as "possibly Cherokee."

7. Clark classified all three major divisions of the Sioux (Lakota, Dakota, and Nakota speakers) as part of the "Darcotar or Sioux Nation" (Moulton 1987, 3:419), which he also termed the "Dar-co-tans proper, the Soos or Sioux." The Teton (Lakota) groups included in his "Estimate of Eastern Indians" (Moulton 1987:3:386–447) were the Oglalas, Brules, Miniconjous, and Saones. One of the "calmets" given to the captains may have been presented by the Brules, with whom the Corps experienced its most arduous and complex encounter (see Ronda 1984:27–41).

8. Two additional flat-stemmed pipes in accession 99-12 (53112.2 and 53113.2) are not described in this chapter because they were clearly not collected by Lewis and Clark. I have, however, in a later section, included the bowls that correspond in number, because their association with the pipe stems is uncertain.

9. The Denver Art Museum gives this bag an approximate date of 1840 and attributes it to an "unknown Lakota artist."

10. Two similar pipe stems at the Musée de l'Homme (34-33/49-2 and 34-33/48-3) have been attributed to the Illinois.

CHAPTER 9. GRIZZLY CLAWS, GARTERS, AND FASHIONABLE HATS

1. The Peabody's Boston Museum accession includes an early, large shield (26 inches, or about 66 cm, in diameter), 99-12-10/52980. It is painted in simple red and blue-black geometric forms, and the blue-black pigment may contain mica.

2. Lewis also reported finding a cedar bow (Moulton 1987, 4:430), and there are several wooden bows in the Peabody's 99-12 accession.

3. A bow case collected from Omahas by Duke Paul of Württemberg around 1820 also contained headless arrows (Gibbs 1982). Sets of headless arrows have been collected in a variety of contexts, including battle sites. This no doubt reflects a stage in the cycle of warriors' tool kits. But as valu-

able as points were, it would have made sense for men to withhold them from bow sets presented to non-Indians.

4. The Peabody label book into which this label was pasted is annotated, "Reserved by Kimball heirs." If members of the Kimball family kept this necklace, however, surviving heirs who corresponded with the Peabody during the middle of the twentieth century seemed unaware of it.

5. Memoranda of the Philadelphia Museum of 1804 [1804–42], Peale Papers. Microfilm copy, Historical Society of Pennsylvania. Peale credited Mackenzie as the first to give an account of grizzlies.

6. Five robes, if a "vulture's quill with a buffalow coat" described a robe. See the description of 99-12-10/53120, the quilled bison robe, in chapter 7.

7. The Denver Art Museum attributes this robe to an "unknown Lakota artist" and dates it "about 1840."

8. The cones contain remnant mammal hair. This may be "a very handsome Indian Belt ornimented with porcupine quills &c" donated by Mr. Baltis Raser and accessioned into the Peale Museum ledger on September 3, 1814. The donation also included an ornamented otter skin tobacco pouch and "a handsome pipe, the tube ornamented with porcupine quills and hair." Possible matches to these objects are 99-12-10/53052, the otter bag, and pipe stem 99-12-10/53116.2 (Memoranda of the Philadelphia Museum 1804–42, Peale Papers, microfilm copy, roll 983.1, Historical Society of Pennsylvania, Philadelphia).

9. A second roach feather in this collection, 99-12-10/53054, serrated around the edges and at the top, is tied to a small metal roach spreader. Willoughby might also have had in mind 99-12-10/53048, a feather flag or banner made from owl and crow feathers.

10. According to Hunn and French (1998:12) and Rigsby (1965:378), the name "Pish-quit-pah" is definitely Sahaptin. It might be interpreted as "pshxu + pam," people of the place of either "sedge brush" (in the northwest Sahaptin dialects) or "rabbitbrush" (in the Columbia River Sahaptin dialects) (Eugene S. Hunn, personal communication, 2000).

11. Because the bag at the Peabody Museum has no carrying strap, it was probably used for storing only.

12. Pat Courtney Gold (personal communication, 2000) explained that being an "elder" does not necessarily refer to a person's age but rather means that he or she is recognized by the community as a wise person.

13. About fifteen hundred years ago, Chinook developed into three languages: Coastal Chinook, spoken by people living at Clatsop on the south bank of the Columbia River and at Shoalwater on the north bank; Katlamet-Chinook, spoken by Katlamet people; and Kiksht, spoken upriver by the Clackamas and people living in the Columbia River Gorge. Wishram was the most distant village, 125 miles from the Clackamas on the south bank of the Columbia (Rubin 1999:57).

14. The vast Tlingit and Haida trade networks extended throughout the Northwest, and the long distances traders traveled reflected the great demand for both manufactured and food items not produced locally. Trading voyages took place during spring and early summer, just before the major salmon harvest began. Frequently a trader from Hoonah, for example, would travel three hundred miles by canoe to trade with Haidas from Masset on Graham Island or with Tsimshians from old Metlakatla on the mainland. When the English became involved in the maritime fur trade, Tlingits, Haidas, and Tsimshians "undertook voyages of a thousand miles to Victoria and Puget Sound trading posts" (Oberg 1973:105).

References Cited

Allen, John L.

1998 "'Of This Enterprize': American Images of the Lewis and Clark Expedition." In *Voyages of Discovery: Essays on the Lewis and Clark Expedition,* edited by James P. Ronda, pp. 255–80. Helena: Montana Historical Society.

Ambrose, Stephen E.

1998 *Lewis and Clark: Voyage of Discovery.* Washington, D.C.: National Geographic Society.

Appadurai, Arjun, ed.

1986 *The Social Life of Things: Commodities in Cultural Perspective.* Cambridge: Cambridge University Press.

Appleby, Joyce

2000 *Inheriting the Revolution: The First Generation of Americans.* Cambridge, Massachusetts: Belknap Press of Harvard University Press.

Bain, James E.

1901 *Travels and Adventures in Canada and the Indian Territories between the Years 1760 and 1776 by Alexander Henry.* Boston: Little, Brown.

Barnum, Phineas Taylor

1888 *How I Made Millions: The Life of P. T. Barnum.* New York: G. W. Dillingham.

Bass, William M., David R. Evans, and Richard L. Jantz

1971 *The Leavenworth Site Cemetery: Archaeology and Physical Anthropology.* University of Kansas Publications in Anthropology 2. Lawrence, Kansas.

Interior of quilled bison robe (99-12-10/53120).

Baumgarten, Linda
2002 *What Clothes Reveal: The Language of Clothing in Colonial and Federal America.* New Haven and London: The Colonial Williamsburg Foundation in association with Yale University Press.

Beaglehole, J. G., ed.
1967 *The Voyage of the* Resolution *and the* Discovery. Cambridge: Cambridge University Press.

Bebbington, Julia
1982 *Quillwork of the Plains.* Calgary, Alberta: Glenbow Museum.

Belk, Russell W., M. Wallendorf, J. Sherry, and M. Holbrook
1991 "Collecting in a Consumer Culture." In *Highways and Buyways: Naturalistic Research from the Consumer Behavior Odyssey,* edited by Russell W. Belk. Provo, Utah: Association for Consumer Research.

Benndorf, Helga, and Arthur Speyer
1968 *Indianer Nordamerikas 1760–1860: Aus der Sammlung Speyer.* Offenbach am Main, Germany: Deutschen Ledermuseum angeschlossen Deutsches Schuhmuseum.

Berlo, Janet, and Ruth B. Phillips
1998 *Native North American Art.* Oxford: Oxford University Press.

Biddle, Nicholas D., ed.
1904 [1814] *History of the Expedition under the Command of Captains Lewis and Clark, to the Sources of the Missouri.* New York: A. S. Barnes.

Blakeslee, Donald J.
1981 "The Origin and Spread of the Calumet Ceremony." *American Antiquity* 46(4):759–68.

Boas, Franz
1901 *Kathlamet Texts.* Bureau of American Ethnology Bulletin 26. Washington, D.C.: U.S. Government Printing Office.

Bodmer, Karl
1984 *Karl Bodmer's America.* Introduction by William H. Goetzmann; annotations by David C. Hunt and Marsha V. Gallagher; artist's biography by William J. Orr. Lincoln: Joslyn Art Museum and University of Nebraska Press.

Bolz, Peter, and Hans-Ulrich Sanner
1999 *Native American Art: The Collections of the Ethnological Museum Berlin.* Seattle and Vancouver: University of Washington Press and Douglas and McIntyre.

Boorstin, Daniel J.
1993 [1948] *The Lost World of Thomas Jefferson.* Chicago: University of Chicago Press.

Bowers, Alfred W.
1991 [1950] *Mandan Social and Ceremonial Organization.* Moscow, Idaho: University of Idaho Press.

Brasser, Ted
1976 *Bojou Neejee.* Ottawa: National Museum of Man.
1984 "Backrest Banners among the Plains Cree and Plains Ojibwa." *American Indian Art* 10(1):56–63.
1987 "By the Power of Their Dreams: Artistic Traditions of the Northern Plains." In *The Spirit Sings: Artistic Traditions of Canada's First Peoples,* pp. 93–131. Toronto: McClelland and Stewart.
1999 "Notes on a Recently Discovered Indian Shirt from New France." *American Indian Art* 24(2):46–55.

Bray, Robert T.
1991 "Utz Site: An Oneota Village in Central Missouri." *Missouri Archaeologist* 52:1–146.

Brown, Ian W.
1989 "The Calumet Ceremony in the Southeast and Its Archaeological Manifestations." *American Antiquity* 54(2):311–31.

Brownstone, Arni
2001 "Seven War-Exploit Paintings: A Search for Their Origins." In *Studies in American Indian Art: A Memorial Tribute to Norman Feder,* edited by Christian F. Feest, pp. 69–85. Seattle: University of Washington Press for European Review of Native American Studies.

Bushnell, David I., Jr.
1908 "Ethnographic Material from North America in Swiss Collections." *American Anthropologist* 10(1):1–15.

Casagrande, Louis B., and Melissa M. Ringheim
1980 *Straight Tongue: Minnesota Indian Art from the Bishop Whipple Collections.* Minneapolis: Science Museum of Minnesota.
Cash, Joseph H., and Gerald W. Wolff
1974 *The Three Affiliated Tribes: Mandan, Arikara, and Hidatsa.* Phoenix, Arizona: Indian Tribal Series.
Catlin, George
1973 [1844] *Letters and Notes on the Manners, Customs, and Conditions of the North American Indians.* 2 vols. New York: Dover.
Ceci, Lynn
1989 "Tracing Wampum's Origins: Shell Bead Evidence from Archaeological Sites in Western and Coastal New York." In *Proceedings of the 1986 Shell Bead Conference: Selected Papers,* edited by Charles F. Hayes III, pp. 63–80. Research Records no. 20. Rochester, New York: Rochester Museum and Science Center.
Chew, Elizabeth
2003 "Curiosity and Enlightenment in Thomas Jefferson's Indian Hall." Unpublished manuscript.
Chomko, Stephen A.
1986 "The Ethnohistorical Setting of the Upper Knife-Heart Region." In *Papers in Northern Plains Prehistory and Ethnohistory: Ice Glider 32OL110,* edited by W. Raymond Wood, pp. 59–96. Special Publications of the South Dakota Archaeological Society, no. 10. Sioux Falls: South Dakota Archaeological Society.
Coe, Ralph T.
1976 *Sacred Circles: Two Thousand Years of North American Indian Art.* London: Arts Council of Great Britain.
Coleman, Winfield
1980 "The Cheyenne Women's Sewing Society." In *Plains Indian Design Symbology and Decoration,* edited by Gene Ball and George P. Horse Capture, pp. 50–69. Cody, Wyoming: Buffalo Bill Historical Center.
2000 "Buffalo Pounds and Sacred Robes: Hunting Magic in the Northern Plains." Unpublished manuscript.
Conn, Richard
1960 "Northern Plains Bustles." *American Indian Hobbyist* 7(1):12–17.
1979 *Native American Art in the Denver Museum.* Seattle: University of Washington Press and Denver Art Museum.
1982 *Circles of the World: Traditional Arts of the Plains Indians.* Denver: Denver Art Museum.
1986 *A Persistent Vision: Art of the Reservation Days.* Seattle: University of Washington Press and Denver Art Museum.
1998 "The Plateau Culture Area and Its Arts: Toward an Understanding of Plateau Aesthetics." In *Native Arts of the Columbia Plateau: The Doris Swayze Bounds Collection,* edited by Susan E. Harless, pp. 40–56. Seattle: University of Washington Press.
Cook, James, and James King
1784 *A Voyage to the Pacific Ocean. Undertaken, by the Command of His Majesty for making Discoveries in the Northern Hemisphere. Performed under the Direction of Captains Cook, Clerke, and Gore in His Majesty's Ships the* Resolution *and the* Discovery; *in the years 1776, 1777, 1778, 1779 and 1780.* 3 vols. London: Atlas G. Nicol and T. Cadell.
Coues, Elliott, ed.
1897 *New Light on the Early History of the Greater Northwest: The Manuscript Journals of Alexander Henry, Fur Trader of the Northwest Company, and of David Thompson, Official Geographer and Explorer of the Same Company, 1799–1814.* 3 vols. New York: Francis Harper.
1979 [1893] *The History of the Lewis and Clark Expedition.* 3 vols. New York: Dover.
Croes, Dale R.
1976 *The Excavation of Water-Saturated Archaeological Sites (Wet Sites) on the Northwest Coast of America.* Archeological Survey of Canada, Mercury Series 50. Ottawa: National Museum of Man.

1977 *Basketry from the Ozette Village Archaeology Site: A Technological, Functional and Comparative Study.* Pullman: Washington State University Press.

Croes, Dale R., and Eric Blinman

1980 *Hoko River: A 2,500-Year-Old Fishing Camp on the Northwest Coast of North America.* Reports of Investigations 58. Pullman: Laboratory of Anthropology, Washington State University.

Cutright, Paul Russell

1969 *Lewis and Clark: Pioneering Naturalists.* Lincoln: University of Nebraska Press.

Dauenhauer, Nora Marks, and Richard Dauenhauer, eds.

1990 *Haa tuwunáagu yís, for Healing Our Spirit: Tlingit Oratory.* Seattle: University of Washington Press.

Davis, Natalie Zemon

2000 *The Gift in Sixteenth-Century France.* Madison: University of Wisconsin Press.

Davis, Wayne L.

1973 "Time and Space Considerations for Diagnostic Northern Plains Glass Trade Bead Types." In *Historical Archaeology in Northwestern North America,* edited by Ronald M. Getty and Knut R. Fladmark, pp. 3–52. Calgary, Alberta: University of Calgary Archaeological Association.

de Laguna, Frederica

1972 *Under Mount St. Elias: The History and Culture of the Yakutat Tlingit.* Washington, D.C.: Smithsonian Institution Press.

Densmore, Frances

1923 *Mandan and Hidatsa Music.* Smithsonian Institution Bureau of American Ethnology, Bulletin 80. Washington, D.C.: U.S. Government Printing Office.

1939 *Nootka and Quileute Music.* Smithsonian Institution Bureau of American Ethnology, Bulletin 124. Washington, D.C.: U.S. Government Printing Office.

DeVoto, Bernard

1952 *The Course of Empire.* Boston: Houghton Mifflin.

Diamond, Beverly, M. Sam Cronk, and Franziska von Rosen

1994 *Visions of Sound: Musical Instruments of First Nations Communities in Northeastern North America.* Chicago: University of Chicago Press.

Dingwall, Eric John

1931 *Artificial Cranial Deformation: A Contribution to the Study of Ethnic Mutilations.* London: John Bale, Sons, and Danielsson, Ltd.

Dippie, Brian W.

1990 *Catlin and His Contemporaries: The Politics of Patronage.* Lincoln: University of Nebraska Press.

Donaldson, Thomas

1887 *The George Catlin Indian Gallery in the U.S. National Museum.* Annual report of the Smithsonian Institution for 1855. Washington, D.C.: U.S. Government Printing Office.

Douglas, Frederic H., and René D'Harnoncourt

1941 *Indian Art of the United States.* New York: Museum of Modern Art.

Drucker, Philip

1951 *The Northern and Central Nootkan Tribes.* Smithsonian Institution Bureau of American Ethnology, Bulletin 144. Washington, D.C.: U.S. Government Printing Office.

Dubin, Lois Sherr

1999 *North American Indian Jewelry and Adornment from Prehistory to the Present.* New York: Harry N. Abrams.

Duff, Wilson

1981 "The World Is as Sharp as a Knife: Meaning in Northern Northwest Coast Art." In *The World Is as Sharp as a Knife: An Anthology in Honor of Wilson Duff,* edited by Donald N. Abbott, pp. 209–24. Victoria: Royal British Columbia Museum.

Ewers, John C.

1939 *Plains Indian Painting: A Description of an Aboriginal American Art.* Stanford, California: Stanford University Press.

1966 "Chiefs from the Missouri and Mississippi and Peale's Silhouettes of 1806." *Smithsonian Journal of History* 1:1–26.

1967 "William Clark's Indian Museum in St. Louis, 1816–1838." In *A Cabinet of Curiosities: Five Episodes in the Evolution of American Museums,* edited by Whitfield Bell et al., pp. 50–72. Charlottesville: University Press of Virginia.

1968 *Indian Life on the Upper Missouri.* Norman: University of Oklahoma Press.

1979 *Indian Art in Pipestone: George Catlin's Portfolio in the British Museum.* Washington, D.C.: British Museum Publications and Smithsonian Institution Press.

1981 "Pipes for the Presidents." *American Indian Art* 6(3):62–70.

1986 *Plains Indian Sculpture: A Traditional Art from America's Heartland.* Washington, D.C.: Smithsonian Institution Press.

1997 "Symbols of Chiefly Authority in Spanish Louisiana." In *Plains Indian History and Culture: Essays on Continuity and Change,* by John C. Ewers, pp. 103–18. Norman: University of Oklahoma Press.

Faxon, Walter

1915 "Relics of Peale's Museum." *Bulletin of the Museum of Comparative Zoology at Harvard College* 59(3):119–48.

Feder, Norman

1964 *Art of the Eastern Plains Indians: The Nathan Sturges Jarvis Collection.* Brooklyn, New York: Brooklyn Museum.

1965 *American Indian Art before 1850.* Denver: Denver Art Museum Quarterly.

1977 "The Malaspina Collection." *American Indian Art* 2(3):40–55.

1980 "Plains Pictographic Painting and Quilled Rosettes: A Clue to Tribal Identifications." *American Indian Art* 5(2):54–62.

1984 "Side Fold Dress." *American Indian Art* 10(1):48–55.

1987 "Bird Quillwork." *American Indian Art* 12(3):46–57.

Feest, Christian F.

1968 *Indianer Nordamerikas.* Vienna: Museum für Völkerkunde.

Feest, Christian F., with contributions by Sylvia S. Kaprycki

1993 *Über Lebenskunst Nordamerikanischer Indianer.* Vienna: Museum für Völkerkunde.

Fernandez-Sacco, Ellen

n.d. "Exhibiting Cultures: Thomas Jefferson's 'Indian Hall' at Monticello." Unpublished manuscript.

Fletcher, Alice C., with assistance by James R. Murie

1996 [1904] *The Hako: Song, Pipe, and Unity in a Pawnee Calumet Ceremony.* Lincoln: University of Nebraska Press.

Fletcher, Alice C., and Francis La Flesche

1992 [1911] *The Omaha Tribe.* 2 vols. Lincoln: University of Nebraska Press. (Originally published as *Twenty-Seventh Annual Report of the Bureau of American Ethnology for the Years 1905–1906,* Washington, D.C., 1911.)

Flint Institute of Arts

1973 *Art of the Great Lakes Indians.* Flint, Michigan: Flint Institute of Arts.

Freedberg, David

1989 *The Power of Images: Studies in the History and Theory of Response.* Chicago: University of Chicago Press.

Garrett, Valery M.

1994 *Chinese Clothing: An Illustrated Guide.* Oxford: Oxford University Press.

Gelb, Norman, ed.

1993 *Jonathan Carver's Travels through America 1766–1768: An Eighteenth-Century Explorer's Account of Uncharted America.* New York: John Wiley and Sons.

Gibbs, Peter

1982 "The Duke Paul Wilhelm Collection in the British Museum." *American Indian Art* 7(3):52–61.

Gibson, James R.

1992 *Otter Skins, Boston Ships, and China Goods: The Maritime Fur Trade of the Northwest Coast, 1785–1841.* Seattle: University of Washington Press and McGill-Queen's University Press.

Gilman, Carolyn
2003 *Lewis and Clark: Across the Divide.* Washington, D.C., and St. Louis: Smithsonian Books and Missouri Historical Society Press.

Gilman, Carolyn, and Mary Jane Schneider
1987 *The Way to Independence: Memories of a Hidatsa Indian Family, 1840–1920.* St. Paul: Minnesota Historical Society Press.

Greene, John C.
1984 *American Science in the Age of Jefferson.* Ames: Iowa State University Press.

Grimes, John R., Christian F. Feest, and Mary Lou Curran
2002 *Uncommon Legacies: Native American Art from the Peabody Essex Museum.* Seattle: University of Washington Press.

Grinnell, George Bird
1923 *The Cheyenne Indians: Their History and Ways of Life,* vol. 1. New Haven, Connecticut: Yale University Press.

Gunther, Erna
1972 *Indian Life of the Northwest Coast of North America, as Seen by the Early Explorers and Fur Traders during the Last Decades of the Eighteenth Century.* Chicago: University of Chicago Press.

Hail, Barbara
1993 [1980] *Hau, Kola! The Plains Indian Collection of the Haffenreffer Museum of Anthropology.* Providence, Rhode Island: The Museum.

Hall, Robert L.
1983 "The Evolution of the Calumet Pipe." In *Prairie Archaeology: Papers in Honor of David A. Baerreis,* edited by Guy E. Gibbon, pp. 37–52. University of Minnesota Special Publications in Anthropology, no. 3. Minneapolis: University of Minnesota Press.
1987 "Calumet Ceremonialism, Mourning Ritual, and Mechanisms of Intertribal Trade." In *Mirror and Metaphor: Material and Social Constructions of Reality,* edited by D. W. Ingersoll and G. Bronitski, pp. 29–43. Lanham, Maryland: University Press of America.
1997 *An Archaeology of the Soul: North American Indian Belief and Ritual.* Urbana: University of Illinois Press.

Halvorson, Mark J.
1998 *Sacred Beauty: Quillwork of Plains Women.* Bismarck: State Historical Society of North Dakota.

Hamilton, Martha Wilson
1995 *Silver in the Fur Trade, 1680–1820.* Chelmsford, Massachusetts: Martha Hamilton Publishing.

Hanson, Charles E., Jr.
1988 "Tobacco in the Fur Trade." *Museum of the Fur Trade Quarterly* 24(4):2–11.

Hanson, James A.
1975 *Metal Weapons, Tools, and Ornaments of the Teton Dakota Indians.* Lincoln: University of Nebraska Press.
1994 *Spirits in the Art from the Plains and Southwest Indian Cultures.* Kansas City, Missouri: Lowell Press.
1996 *Little Chief's Gatherings.* Crawford, Nebraska: Fur Press.

Harrington, M. R.
1914 "Sacred Bundles of the Sac and Fox." *University Museum Anthropological Publications* 4(2):121–62. Philadelphia: University Museum, University of Pennsylvania.

Harrison, Julia
1989 "'He Heard Something Laugh': Otter Imagery in the Midewiwin." In *Great Lakes Indian Art,* edited by David W. Penney, pp. 83–92. Detroit: Wayne State University Press.

Harrod, Howard L.
2000 *The Animals Came Dancing: Native American Sacred Ecology and Animal Kinship.* Tucson: University of Arizona Press.

Hartmann, Horst
1979 *Die Plains- und Prärieindianer Nordamerikas.* Berlin: Museum für Völkerkunde.

Heinbuch, Jean
1990 *A Quillwork Companion: An Illustrated Guide to Techniques of Porcupine Quill Embroidery.* Liberty, Utah: Eagle's View Publishing.

Heiser, Charles Bixler

1979 *The Gourd Book.* Norman: University of Oklahoma Press.

Helms, Mary W.

1988 *Ulysses' Sail: An Ethnographic Odyssey of Power, Knowledge, and Geographical Distance.* Princeton, New Jersey: Princeton University Press.

Hines, Donald M.

1996 *Celilo Tales: Wasco Myths, Legends, Tales of Magic and the Marvelous.* Issaquah, Washington: Great Eagle Publishing.

Hodge, F. W., ed.

1910 *Handbook of American Indians North of Mexico.* Bureau of American Ethnology 30(2). Washington, D.C.: Smithsonian Institution.

Hogendorn, Jan, and Marion Johnson

1986 *The Shell Money of the Slave Trade.* Cambridge: Cambridge University Press.

Hollinger, R. Eric, and David Benn, eds.

1998 *Oneota Taxonomy: Papers from the Oneota Symposium of the 54th Plains Anthropological Conference, 1996.* Wisconsin Archaeologist 79(2).

Holm, Bill

1992 "Four Bears' Shirt: Some Problems with the Smithsonian Catlin Collection." In *Artifacts/Artifakes: The Proceedings of the 1984 Plains Indian Seminar,* edited by George P. Horse Capture and Suzan G. Tyler, pp. 43–59. Cody, Wyoming: Buffalo Bill Historical Center.

Holmberg, James J., ed.

2002 *Dear Brother: Letters of William Clark to Jonathan Clark.* New Haven, Connecticut: Yale University Press.

Horse Capture, George P.

1993 "A Gallery of Hides." In *Robes of Splendor: Native American Painted Buffalo Robes,* pp. 93–139. New York: New Press.

Horse Capture, Joseph D.

1998 "The Story of the Robe." Lecture presented at the Great Falls, Montana, Civic Center, July 3, 1998.

Horse Capture, Joseph D., and George P. Horse Capture

2001 *Beauty, Honor, and Tradition: The Legacy of Plains Indian Shirts.* Washington, D.C.: Minneapolis Institute of Arts and National Museum of the American Indian.

Howard, James H.

1974 "The Arikara Buffalo Society Medicine Bundle." *Plains Anthropologist* 19:241–71.

1976 *Yanktonai Ethnohistory and the John K. Bear Winter Count.* Plains Anthropologist, Memoir 11.

Howay, F. W.

1940 *The Journal of Captain James Colnet aboard the* Argonaut *from April 26, 1789, to November 3, 1791.* Toronto: Champlain Society.

Hunn, Eugene S.

1990 *Nch'i-Wána "The Big River": Mid-Columbia Indians and Their Land.* Seattle: University of Washington Press.

Hunn, Eugene S., and Katherine S. French

1998 "Wasco, Wishram, and Cascades." In *Handbook of North American Indians,* vol. 12, *Plateau,* edited by Deward E. Walker, Jr., pp. 378–94. Washington, D.C.: Smithsonian Institution Press.

Jablow, Joseph

1994 [1951] *The Cheyenne in Plains Indian Trade Relations, 1795–1840.* Lincoln: University of Nebraska Press.

Jackson, Donald, ed.

1978 *Letters of the Lewis and Clark Expedition with Related Documents, 1783–1854.* 2d ed., 2 vols. Urbana: University of Illinois Press.

1981 *Thomas Jefferson and the Stony Mountains: Exploring the West from Monticello.* Urbana: University of Illinois Press.

Jackson, J. Wilfrid

1917 *Shells as Evidence of the Migrations of Early Culture.* Manchester, U.K.: Manchester University Press.

Jacobs, Wilbur R.

1950 *Diplomacy and Indian Gifts: Anglo-French Rivalry along the Ohio and Northwest Frontiers, 1748–1763.* Stanford, California: Stanford University Press.

Janson, Charles William

1935 [1807] *The Stranger in America.* New York: Press of the Pioneers.

Jones, Dorothy V.

1988 "British Colonial Indian Treaties." In *Handbook of North American Indians,* vol. 4, *History of Indian-White Relations,* edited by Wilcomb E. Washburn, pp. 185–201. Washington, D.C.: Smithsonian Institution Press.

Kaeppler, Adrienne L., ed.

1978 *Cook Voyage Artifacts in Leningrad, Berne, and Florence Museums.* Honolulu, Hawaii: Bishop Museum Press.

Kane, Paul

1996 [1859] *Wanderings of an Artist among the Indians of North America.* Minneola, New York: Dover.

Kappler, Charles J., ed.

1975 [1904] *Indian Treaties 1773–1883.* New York: Interland Publishing.

Kehoe, Thomas F.

1973 *The Gull Lake Site: A Prehistoric Bison Drive Site in Southwestern Saskatchewan.* Publications in Anthropology and History 1. Milwaukee, Wisconsin: Milwaukee Public Museum.

Kennerly, William Clark

1948 *Persimmon Hill: A Narrative of Old St. Louis and the Far West* (as told to Elizabeth Russell). Norman: University of Oklahoma Press.

Kenseth, Joy, ed.

1991 *The Age of the Marvelous.* Hanover, New Hampshire: Hood Museum of Art.

Keyser, James D.

1987 "A Lexicon for Historic Plains Indian Rock Art: Increasing Interpretive Potential." *Plains Anthropologist* 32(115):43–71.

1996 "Painted Bison Robes: The Missing Link in the Biographical Art Style Lexicon." *Plains Anthropologist* 41(155):29–53.

2000 *The Five Crows Ledger: Biographical Warrior Art of the Flathead Indians.* Salt Lake City: University of Utah Press.

Keyser, James D., and Timothy J. Brady

1993 "A War Shirt from the Schoch Collection: Documenting Individual Artistic Expression." *Plains Anthropologist* 38(142):5–20.

King, J. C. H.

1977 *Smoking Pipes of the North American Indian.* London: British Museum.

1981 *Artificial Curiosities from the Northwest Coast of America: Native American Artefacts in the British Museum Collected on the Third Voyage of Captain James Cook and Acquired through Sir Joseph Banks.* London: British Museum.

1982 *Thunderbird and Lightning: Indian Life in Northeast America, 1600–1900.* London: British Museum.

1994 "Native Art as Depicted by Charles Hamilton Smith, 1816–1817." *American Indian Art* 19(2):58–67.

1999 *First Peoples, First Contacts: Native Peoples of North America.* Cambridge, Massachusetts: Harvard University Press.

2000 "Nuu-chah-nulth Art at the British Museum." In *Nuu-Chah-Nulth Voices, Histories, Objects and Journeys,* edited by Alan L. Hoover, pp. 257–72. Victoria: Royal British Columbia Museum.

Koppert, Vincent A.

1930 *Contributions to Clayoquot Ethnology.* Catholic University Anthropology Serial 1. Washington, D.C.: Catholic University of America.

Krickeberg, Walter

1954 *Altere Ethnographica aus Nordamerika im Berliner Museum für Völkerkunde.* Berlin: D. Reimer.

Kroeber, Alfred L.

1983 [1902] *The Arapaho.* Lincoln: University of Nebraska Press.

Lanford, Benson L., and Pamela W. Miller

2000 "Late Prehistoric Painted Rawhide and Leather Artifacts from Utah." *American Indian Art* 25(3):38–47.

Lehmer, Donald J.
1971 *Introduction to Middle Missouri Archaeology.* Anthropological Papers 1. Washington, D.C.: U.S. Department of the Interior, National Park Service.
Lehmer, Donald J., W. Raymond Wood, and C. L. Dill
1978 *The Knife River Phase.* Denver: U.S. Department of the Interior, Heritage Conservation and Recreation Service.
Lessard, Dennis F.
1986 "Great Lakes Indian Loom Beadwork." *American Indian Art* 11(3):54–61, 68–69.
Lessard, Rosemary T.
1980 "A Short Historical Survey of Lakota Women's Clothing." In *Plains Indian Design Symbology and Decoration,* edited by Gene Ball and George P. Horse Capture, pp. 70–76. Cody, Wyoming: Buffalo Bill Historical Center.
Luscomb, Sally C.
1967 *The Collector's Encyclopedia of Buttons.* New York: Bonanza Books.
Lyford, Carrie A.
1940 *Quill and Beadwork of the Western Sioux.* Edited by Willard W. Beatty. Lawrence, Kansas: U.S. Office of Indian Affairs, Education Division.
Mallery, Garrick
1972 [1893] *Picture Writing of the American Indians.* 2 vols. New York: Dover.
1987 [1886] *The Dakota and Corbusier Winter Counts.* Lincoln, Nebraska: J and L Reprint Company.
Malloy, Mary
2000 *Souvenirs of the Fur Trade: Northwest Coast Indian Art and Artifacts Collected by American Mariners, 1788–1844.* Cambridge, Massachusetts: Peabody Museum Press, Harvard University.
Marriott, Alice
1956 "Trade Guilds of the Southern Cheyenne Women." *Bulletin of the Oklahoma Anthropological Society* 4:19–27.
Martin, Irene
1980 "Ethnohistorical Notes on the Wahkiakum Indians." Appendix A in *Further Archaeological Testing at the Skamokawa Site (45-WK-5), Wahkiakum, Washington,* pp. 40–52. Reconnaissance Report 36. Seattle: Office of Public Archaeology, University of Washington.
Maurer, Evan M.
1992 *Visions of the People: A Pictorial History of Plains Indian Life.* Minneapolis: Minneapolis Institute of Arts.
Mauss, Marcel
1954 *The Gift: Forms and Functions of Exchange in Archaic Societies.* Translated by Ian Cunnison. Glencoe, Illinois: Free Press.
May, Henry F.
1976 *The Enlightenment in America.* Oxford: Oxford University Press.
McDermott, John Francis
1948 "Museums in Early Saint Louis." *Missouri Historical Society Bulletin* 4:129–38.
1954 "William Clark: Pioneer Museum Man." *Journal of the Washington Academy of Sciences* 44(11):370–73.
1960 "William Clark's Museum Once More." *Missouri Historical Society Bulletin* 16(2):130–33.
Meares, John
1967 [1790] *Voyages Made in the Years 1788 and 1789 from China to the North-West Coast of America.* Bibliotheca Australiana 22. Cambridge, Massachusetts: Da Capo Press.
Meyer, Roy A.
1977 *The Village Indians of the Upper Missouri: The Mandans, Hidatsas, and Arikaras.* Lincoln: University of Nebraska Press.
Miller, Judi, Elizabeth Moffatt, and Jane Sirois
1990 *Native Materials Project Final Report.* Ottawa: Canadian Conservation Institute.
Miller, Lillian B., ed.
1988 *The Selected Papers of Charles Willson Peale and His*

Family, vol. 2, *The Artist as Museum Keeper, 1791–1810.* New Haven, Connecticut: Yale University Press.

Miller, Lynette

1991 "Basketry Styles of the Plateau Region." In *A Time of Gathering: Native Heritage in Washington State,* edited by Robin K. Wright, pp. 178–85. Seattle: Burke Museum and University of Washington Press.

Minor, Rick

1983 "Aboriginal Settlement and Subsistence at the Mouth of the Columbia River." Ph.D. dissertation, Department of Anthropology, University of Oregon.

Mitchill, Samuel Latham

1826 *A Discourse on the Character and Services of Thomas Jefferson, More Especially as a Promoter of Natural and Physical Science, Pronounced, by Request, before the New York Lyceum of Natural History, on the 11th October, 1826.* New York: G. and C. Carvill.

Moulton, Gary E.

1998 "On Reading Lewis and Clark: The Last Twenty Years." In *Voyages of Discovery: Essays on the Lewis and Clark Expedition,* edited by James P. Ronda, pp. 281–98. Helena: Montana Historical Society.

Moulton, Gary E., ed.

1983–99 *The Journals of the Lewis and Clark Expedition.* 12 vols. Lincoln: University of Nebraska Press.

Mozino, Jose Mariana

1970 *Noticias de Nutka: An Account of Nootka Sound in 1792.* Translated by Iris Higbie Wilson. Vancouver, British Columbia: Douglas and McIntyre.

Museo de América

1991 *Indios de América del norte: Otras culturas de América.* Text by Araceli Sánchez Garrido. Madrid: Ministerio de Cultura, Dirección General de Bellas Artes y Archivos.

Nasatir, A. P., ed.

1952 *Before Lewis and Clark: Documents Illustrating the History of the Missouri, 1785–1804.* 2 vols. St. Louis, Missouri: St. Louis Historical Documents Foundation.

National Museums of Canada

1980 *The Covenant Chain: Indian Ceremonial and Trade Silver.* Ottawa: National Museums of Canada.

Norder, John

1999 "Coming to Terms with the Native American Pipe in Eastern North America." Paper presented at the annual meeting of the American Society for Ethnohistory, Mashantucket Pequot Museum and Research Center, Connecticut.

Oberg, Kalervo

1973 *The Social Economy of the Tlingit Indians.* Seattle and Vancouver: University of Washington Press and Douglas and McIntyre.

Olsen, Stanley J.

1963 "Dating Early Plains Buttons by Their Form." *American Antiquity* 28(4):551–54.

Orchard, William C.

1916 *The Technique of Porcupine Quill Decoration among the North American Indians.* Contributions from the Museum of the American Indian–Heye Foundation 4(1). New York.

1975 *Beads and Beadwork of the American Indians.* New York: Museum of the American Indian–Heye Foundation.

1982 *The Technique of Porcupine Quill Decoration.* Liberty, Utah: Eagle's View Publishing.

O'Shea, John, and John Ludwickson

1992 *Archaeology and Ethnohistory of the Omaha Indians: The Big Village Site.* Lincoln: University of Nebraska Press.

Pearce, Susan M.

1995 *On Collecting: An Investigation into Collecting in the European Tradition.* London: Routledge.

Peers, Laura L.

1987 "Rich Man, Poor Man, Beggarman, Chief: Saulteaux in the Red River Settlement." In *Papers of the Eighteenth Algonquian Conference,* edited by William Cowan, pp. 261–70. Ottawa: Carleton University Press.

Penney, David W.

1985 *Ancient Art of the American Woodland Indians.* New York: Harry N. Abrams.

1992 *Art of the American Indian Frontier.* Seattle: University of Washington Press and Detroit Institute of Arts.

Peterson, Jacqueline, and Laura Peers

1993 *Sacred Encounters: Father De Smet and the Indians of the Rocky Mountain West.* Norman: University of Oklahoma Press.

Peterson, Merrill D., ed.

1989 *Visitors to Monticello.* Charlottesville: University Press of Virginia.

Philibert, Jean-Marc, and Christine Jourdan

1996 "Perishable Goods: Modes of Consumption in the Pacific Islands." In *Cross-Cultural Consumption: Global Markets, Local Realities,* edited by David Howes, pp. 55–73. London: Routledge.

Phillips, Ruth B.

1984 *Patterns of Power: The Jasper Grant Collection and Great Lakes Indian Art of the Early Nineteenth Century.* Kleinburg, Ontario: McMichael Canadian Collection.

1987 "Like a Star I Shine: Northern Woodlands Artistic Traditions." In *The Spirit Sings: Artistic Traditions of Canada's First Peoples,* pp. 51–92. Toronto: McClelland and Stewart.

Pomian, Krystzof

1990 *Collectors and Curiosities: Paris and Venice, 1500–1800.* Translated by Elizabeth Wiles-Portier. Cambridge, U.K.: Basil Blackwell.

Pond, Gail

1984 "The Mandan Buffalo Robe." Unpublished memo, Historic Site of Thomas Jefferson's Poplar Forest, Virginia.

Prucha, Francis Paul

1971 *Indian Peace Medals in American History.* Madison: State History Society of Wisconsin.

1988 "United States Indian Policies, 1815–1860." In *Handbook of North American Indians,* vol. 4, *History of Indian-White Relations,* edited by Wilcomb E. Washburn, pp. 40–50. Washington, D.C.: Smithsonian Institution Press.

Quimby, George Irving

1948 "Culture Contact on the Northwest Coast, 1785–1795." *American Anthropologist* 50:247–54.

1966 *Indian Culture and European Trade Goods.* Madison: University of Wisconsin Press.

Ray, Arthur J.

1980 "Indians as Consumers in the Eighteenth Century." In *Old Trails and New Directions: Papers of the Third North American Fur Trade Conference,* edited by Carol M. Judd and Arthur J. Ray, pp. 255–71. Toronto: University of Toronto Press.

Ray, Verne F.

1938 *Lower Chinook Ethnographic Note.* University of Washington Publications in Anthropology 7(2). Seattle.

Reid, Russell, and Clell G. Gannon, eds.

1929 "Journal of the Atkinson-O'Fallon Expedition." *North Dakota Historical Quarterly* 4(1):5–56.

Richardson, Edgar P., Brooke Hindle, and Lillian B. Miller

1982 *Charles Willson Peale and His World.* New York: Harry N. Abrams.

Rigsby, Bruce J.

1965 "Linguistic Relations in the Southern Plateau." Ph.D. dissertation, Department of Anthropology, University of Oregon, Eugene.

Ritvo, Harriet

1997 *The Platypus and the Mermaid, and Other Figments of the Classifying Imagination.* Cambridge, Massachusetts: Harvard University Press.

Ritzenthaler, Robert E.

1978 "Southeastern Chippewa." In *Handbook of North American Indians,* vol. 15, *Southeast,* edited by Bruce G. Trigger, pp. 743–59. Washington, D.C.: Smithsonian Institution Press.

Robinson, Joyce Henri

1995 "An American Cabinet of Curiosities: Thomas

Jefferson's Indian Hall at Monticello." *Winterthur Portfolio: A Journal of American Material Culture* 30(1):41–58.

Ronda, James P.

1984 *Lewis and Clark among the Indians.* Lincoln: University of Nebraska Press.

1998a "'The Writingest Explorers': The Lewis and Clark Expedition in American Historical Literature." In *Voyages of Discovery: Essays on the Lewis and Clark Expedition,* edited by James P. Ronda, pp. 299–326. Helena: Montana Historical Society.

1998b "'A Darling Project of Mine': The Appeal of the Lewis and Clark Story." In *Voyages of Discovery: Essays on the Lewis and Clark Expedition,* edited by James P. Ronda, pp. 327–35. Helena: Montana Historical Society.

2000 *Jefferson's West: A Journey with Lewis and Clark.* Monticello Monograph Series. Charlottesville, Virginia: Thomas Jefferson Foundation.

Rubin, Rick

1999 *Naked against the Rain: The People of the Lower Columbia River, 1770–1830.* Portland, Oregon: Far Shore Press.

Russell, Carl P.

1967 *Firearms, Traps, and Tools of the Mountain Men.* New York: Alfred A. Knopf.

Schlick, Mary Dodds

1979 "A Columbia River Basket Collected by Lewis and Clark." *American Indian Basketry* 1(1):10–13.

1994 *Columbia River Basketry: Gift of the Ancestors, Gift of the Earth.* Seattle: University of Washington Press.

Schulze-Thulin, Axel, ed.

1987 *Indianer der Prärien und Plains.* Stuttgart: Linden-Museum.

Sellers, Charles Coleman

1980 *Mr. Peale's Museum: Charles Willson Peale and the First Popular Museum of Natural Science and Art.* New York: W. W. Norton.

1991 "'Good Chiefs and Wise Men': Indians as Symbols of Peace in the Art of Charles Willson Peale." In *New Perspectives on Charles Willson Peale: A 250th Anniversary Celebration,* edited by Lillian B. Miller and David C. Ward, pp. 119–29. Pittsburgh: University of Pittsburgh Press.

Skinner, Alanson

1913 "Social Life and Ceremonial Bundles of the Menomini Indians." *Anthropological Papers of the American Museum of Natural History* 1:1–165.

1921 *Material Culture of the Menomini.* Indian Notes and Monographs. New York: Museum of the American Indian–Heye Foundation.

1923–25 "Observations on the Ethnology of the Sauk Indians." *Bulletin of the Public Museum of the City of Milwaukee* 5:1–4.

1924 "The Mascoutens or Prairie Potawatomi Indians: Social Life and Ceremonies." *Bulletin of the Public Museum of the City of Milwaukee* 6(1):1–262.

1926 "Ethnology of the Ioway Indians." *Bulletin of the Public Museum of the City of Milwaukee* 5(4):181–354.

Smith, G. Hubert

1972 *Like-a-Fishhook Village and Fort Berthold Garrison Reservoir, North Dakota.* Anthropological Papers 2. Washington, D.C.: U.S. Department of Interior, National Park Service.

Southwest Museum

1983 *Akicita: Early Plains and Woodlands Indian Art from the Collection of Alexander Acevedo.* Los Angeles: Southwest Museum.

Spotted Wolf, Wilbur

1972 "Chronological List of Important Dates concerning Arikara, Hidatsa and Mandan." New Town, North Dakota: Fort Berthold A.C.D.P.

Stein, Susan

1993 *The Worlds of Thomas Jefferson at Monticello.* New York: Harry N. Abrams.

Stewart, Hilary

1982 *Indian Fishing: Early Methods on the Northwest Coast.* Seattle: University of Washington Press.

1984 *Cedar: Tree of Life to the Northwest Coast Indians.* Seattle: University of Washington Press.

Swan, James G.

1870 *The Northwest Coast, or Three Years' Residence in Washington Territory.* New York: Harper and Brothers.

Taylor, Colin F.

1990 "Reading Plains Indian Artefacts: Their Symbolism as Cultural and Historical Documents." Ph.D. dissertation, Department of Literatures, University of Essex, U.K.

1997 *Yupika: The Plains Indian Woman's Dress.* Wyk auf Foehr, Germany: Verlag für Amerikanistik.

Thomas, Davis, and Karin Ronnefeldt, eds.

1976 *People of the First Man: Life among the Plains Indians in Their Final Days of Glory.* New York: E. P. Dutton.

Thomas, Nicholas

1991 *Entangled Objects: Exchange, Material Culture, and Colonialism in the Pacific.* Cambridge, Massachusetts: Harvard University Press.

Thompson, Judy

1977 *The North American Indian Collection: A Catalogue.* Berne, Switzerland: Historical Museum.

Thornton, Russell

1987 *American Indian Holocaust and Survival: A Population History since 1492.* Norman: University of Oklahoma Press.

Thwaites, Reuben Gold, ed.

1904–5 *Original Journals of the Lewis and Clark Expedition, 1804–1806.* 8 vols. New York: Dodd, Mead.

1904a *Bradbury's Travels in the Interior of America, 1809–1811.* Early Western Travels, 1748–1846, vol. 5. Cleveland, Ohio: Arthur H. Clark.

1904b *Journal of a Voyage up the River Missouri, Performed in 1811, by H. M. Brackenridge,* part 1. Early Western Travels, 1748–1846, vol. 6. Cleveland, Ohio: Arthur H. Clark.

1905a *James's Account of S. H. Long's Expedition, 1819–1820,* part 1. Early Western Travels, 1748–1846, vol. 14. Cleveland, Ohio: Arthur H. Clark.

1905b *James's Account of S. H. Long's Expedition, 1819–1820,* part 2. Early Western Travels, 1748–1846, vol. 15. Cleveland, Ohio: Arthur H. Clark.

1905c *James's Account of S. H. Long's Expedition, 1819–1820,* part 3. Early Western Travels, 1748–1846, vol. 16. Cleveland, Ohio: Arthur H. Clark.

1905d *Travels in the Interior of North America,* vol. 1. By Maximilian, Prince of Wied. Translated from the German by Hannibal Evans Lloyd. Early Western Travels, 1748–1846, vol. 22. Cleveland, Ohio: Arthur H. Clark.

1906a *Travels in the Interior of North America,* vol. 2. By Maximilian, Prince of Wied. Translated from the German by Hannibal Evans Lloyd. Early Western Travels, 1748–1846, vol. 23. Cleveland, Ohio: Arthur H. Clark.

1906b *Travels in the Interior of North America,* vol. 3. By Maximilian, Prince of Wied. Translated from the German by Hannibal Evans Lloyd. Early Western Travels, 1748–1846, vol. 24. Cleveland, Ohio: Arthur H. Clark.

Torrence, Gaylord

1994 *The American Indian Parfleche: A Tradition of Abstract Painting.* Seattle: University of Washington Press and Des Moines Art Center.

Torrence, Gaylord, and Robert Hobbs

1989 *Art of the Red Earth People: The Mesquakie of Iowa.* Iowa City: University of Iowa Museum of Art.

Truettner, William H.

1979 *The Natural Man Observed: A Study of Catlin's Indian Gallery.* Washington, D.C.: Smithsonian Institution Press.

Turnbaugh, Sarah Peabody, and William A. Turnbaugh

1986 *Indian Baskets.* West Chester, Pennsylvania: Schiffer.

Turner, Nancy J.

1998 *Plant Technology of First Peoples in British Columbia.* Vancouver: University of British Columbia Press.

Turner, Nancy J., John Thomas, B. F. Carlson, and R. T. Ogilvie

1982 *Ethnobotany of the Nitinaht Indians of Vancouver Island.* Occasional Paper 24. Victoria: Royal British Columbia Museum.

Turner, Nancy J., et al.

1990 *Thompson Ethnobotany: Knowledge and Usage of Plants by the Thompson Indians of British Columbia.* Memoir 3. Victoria: Royal British Columbia Museum.

Tyrrell, J. B., ed.

1916 *David Thompson's Narrative of His Explorations in Western America, 1784–1812.* Toronto: Champlain Society.

Vaughan, Thomas, and Bill Holm

1990 *Soft Gold: The Fur Trade and Cultural Exchange on the Northwest Coast of America.* Portland: Oregon Historical Society Press.

Vibert, Elizabeth

1997 *Trader's Tales: Narratives of Cultural Encounters in the Columbia Plateau, 1807–1846.* Norman: University of Oklahoma Press.

Viola, Herman

1981 *Diplomats in Buckskins: A History of Indian Delegations in Washington City.* Washington, D.C.: Smithsonian Institution Press.

Wallace, Anthony F. C.

1999 *Jefferson and the Indians: The Tragic Fate of the First Americans.* Cambridge, Massachusetts: Harvard University Press.

Waterman, T. T.

1920 *The Whaling Equipment of the Makah Indians.* University of Washington Publications in Anthropology 1(1). Seattle.

Webber, Alika Podolinsky

1983 "Ceremonial Robes of the Montagnais-Naskapi." *American Indian Art* 9(1):60–77.

West, George A.

1934 *Tobacco, Pipes, and Smoking Customs of the American Indians.* 2 parts. Bulletin of the Public Museum of the City of Milwaukee 17.

Wildschut, William, and John C. Ewers

1959 *Crow Indian Beadwork: A Descriptive and Historical Study.* Contributions from the Museum of the American Indian–Heye Foundation 16. New York: Heye Foundation.

Willoughby, Charles C.

1905 "A Few Ethnological Specimens Collected by Lewis and Clark." *American Anthropologist* 7(4):633–41.

Wilson, Gilbert L.

1911 "Report for American Museum of Natural History." Typescript copy, Knife River Indian Villages National Historic Site. Original manuscript in the collections of the Minnesota Historical Society, Minneapolis.

1916a "Native Hidatsa Botany: Accounts Collected for the University of Minnesota, Summer 1916." Typescript copy, Knife River Indian Villages National Historic Site. Original manuscript in the collections of the Minnesota Historical Society, Minneapolis.

1916b "Report of Anthropological Work on Fort Berthold Indian Reservation, to the American Museum of Natural History." Typescript copy, Knife River Indian Villages National Historic Site. Original manuscript in the collections of the Minnesota Historical Society, Minneapolis.

Wissler, Clark

1915 *Costumes of the Plains Indians.* Anthropological Papers of the American Museum of Natural History 17(2). New York.

Witthoft, John, Harry Schoff, and Charles F. Wray

1953 Micmac Pipes, Vase-Shaped Pipes, and Calumets. *Pennsylvania Archaeologist* 23(3–4):89–107.

Wollon, Dorothy, and Margaret Kinard, eds.

1952 "Sir Augustus J. Foster and 'The Wild Natives of the Woods,' 1805–1807." *William and Mary Quarterly* 9(2):191–214.

Wood, W. Raymond

1980 "Plains Trade in Prehistoric and Protohistoric Intertribal Relations." In *Anthropology on the Great Plains,* edited by W. Raymond Wood and Margot Liberty, pp. 98–109. Lincoln: University of Nebraska Press.

Wood, W. Raymond, and Thomas D. Thiessen, eds.

1985 *Early Fur Trade on the Northern Plains: Canadian Traders among the Mandan and Hidatsa Indians, 1738–1818.* Norman: University of Oklahoma Press.

Woodward, Arthur

1948 "Trade Goods of 1748." *The Beaver: A Magazine of the North Outfit* 279:3–6.

1965 *Indian Trade Goods.* Oregon Historical Society Publication no. 2. Portland, Oregon: Metropolitan Press.

1970 *Denominators of the Fur Trade: An Anthology of Writings on the Material Culture of the Fur Trade.* Pasadena, California: Socio-Technical Publications.

Woody, Ken

1998 "Buffalo Robes." Unpublished manuscript.

Wright, Robin K.

1991 *A Time of Gathering: Native Heritage in Washington State.* Seattle: Burke Museum and University of Washington Press.

Württemberg, Friedrich Paul Wilhelm, Duke of

1973 *Travels in North America, 1822–1824.* Translated by W. Robert Nitske. Edited by Savoie Lottinville. Norman: University of Oklahoma Press.

Zepp, Wolfgang, and J. W. Ludwig, eds.

1982 *Reise in das Innere Nord-America.* Berlin: Amerika Haus Berlin.

Acknowledgments

This book reflects the energy, creativity, and gifts of many people: scholars, artists, consultants, mentors, friends, and teachers. It is a product of connections and collaborations, past and present.

The foundations for this and all other interpretive works that engage the Lewis and Clark story were laid by Rueben Gold Thwaites, Nicholas Biddle, Elliot Coues, Donald Jackson, and Gary Moulton, who edited and made accessible the expedition journals and related primary documents.

A great many people helped to shape our understanding of the Peabody's collection of Lewis and Clark objects. Rubie Watson, the director of the Peabody Museum of Archaeology and Ethnology, initiated our research and provided the unflagging enthusiasm and support that sustained the project's growth and development. Art historian Gaylord Torrence, Fred and Virginia Merrill Curator of American Indian Art at the Nelson-Atkins Museum of Art in Kansas City, Missouri, served as our primary objects consultant. Gaylord visited the Peabody twice to evaluate the objects, read and commented on most of the text, and contributed an important essay on raven ornaments to this book. His understanding is informed by his deep experience with early historic North American Native art and by his artist's eye. We learned a lot just watching him interact with the materials, and his unstinting enthusiasm for the project buoyed our own.

Artists Pat Courtney Gold (Wasco) and Butch Thunder Hawk (Lakota) were generous teachers, too. They visited the Peabody and shared with us the meanings that these objects hold for them. We are particularly grateful to Pat for giving us a compelling written account of her relationship with the "sally bag" in this collection and of her role in the revival of Columbia River basketry. With support from Wayne Pruse, director of art and art marketing, and from other administrators at United Tribes Technical College in Bismarck, North Dakota, Butch and his students played a central role in the

The Peabody Museum of Archaeology and Ethnology, circa 1900.

Consultant Gaylord Torrence.

Peabody-Monticello Native Arts Project. Mike Cross (Mandan-Hidatsa), a community activist on the Fort Berthold Reservation, contributed an essay on the legacy of the Lewis and Clark expedition in Indian country. We also want to thank Mike, Butch, Pat and her parents—Terry Courtney and the late Catherine Courtney—and Keith Bear and Jo Esther Parshall for their hospitality when we visited their communities and homes in North Dakota and Oregon. Anne-Marie Victor-Howe also thanks Mary Dodds Schlick and members of the Northwest Native American Basket-weavers Association, especially Colleen Ray and Karen Padnole, for sharing their knowledge and passion about basketry traditions with her when she visited their country. She thanks Haida basketweavers Lisa Telford and Delores Churchill for teaching her about weaving techniques. Castle McLaughlin thanks members of the Fort Berthold Lewis and Clark committee for their hospitality and contributions.

A number of scientific specialists helped us to identify the material components of objects in the collection. Carla Dove, the Smithsonian Institution's forensic ornithologist, spent several days at the Harvard Museum of Natural History studying the bird feathers on the objects. Alison Pirie of Harvard's Museum of Comparative Zoology, a unit of the Museum of Natural History, facilitated access to the bird collection. Bruce Hoadley (University of Massachusetts–Amherst) identified the woods from which many of the objects were made. The wampum was examined by archaeologist Kevin McBride (Mashantucket Pequot Museum Research Center). Bill Billeck (Smithsonian Institution) helped to identify glass trade beads, as did Peabody Museum volunteer Shirley Weinberg. Joyce Saler, a private consultant, analyzed the English brass buttons on one of the side-fold dresses. Richard Newman, head of scientific research, and Michele Derrick, contract research scientist, at the Museum of Fine Arts, Boston, identified pigments on the biographical warrior's shirt and on the painted side-fold dress. Archaeologist David D. Kuehn identified arrow point types.

We also relied on the expertise and generosity of several Harvard scientists who performed materials analysis. Eugene Farrell at the Straus Conservation Center, Fogg Art Museum, identified pigments on the pipes. Richard Meadow, director of the zooarchaeology lab at the Peabody Museum, supervised the identification of mammal bones, and Gustavo A. Romero, keeper of the Oakes Ames Orchid Collection, analyzed plant materials. Carl Francis, curator of geology at Harvard's Museum of Natural History, examined the pipe bowls, and associate professor of anthropology Carole Mandryke helped to identify arrow point styles. Peabody research associate Daniella Bar-Yosef identified the several species of shell that occur throughout the collection.

We solicited opinions from a number of specialists in American Indian art and ethnology about the objects

Some of the Peabody Museum staff members who worked on the Lewis and Clark project. Front row: conservator T. Rose Holdcraft, associate curator Castle McLaughlin, curatorial associate Anne-Marie Victor-Howe. Back row: conservator Scott Fulton, collections associate Nynke Dorhout, collections assistant Stuart Heebner, photographer Hillel S. Burger.

included in this book. In this regard, it is a pleasure to thank Ted J. Brasser (Canadian Museum of Civilization, retired), Thomas Buckley (Boston University), Winfield Coleman (independent scholar), Dale R. Croes (South Puget Sound Community College), Penelope Drucker (New York State Museum), Ryan Heary Head (independent scholar), Steve Henrickson (Alaska State Museum), Bill Holm (Burke Museum, University of Washington), George Horse Capture (National Museum of the American Indian), Joe Horse Capture (Minneapolis Institute of Arts), J. C. H. King (British Museum), Bill Mercer (Portland Art Museum), David Penney (Detroit Institute of Arts), Mary Dodds Schlick (independent scholar), Sarah Peabody Turnbaugh (Museum of Primitive Art and Culture), Ken Woody (National Park Service), and Robin K. Wright (Burke Museum, University of Washington). All of these people responded thoughtfully to inquiries about individual objects and categories of objects. Bill Plitt, an objects conservator from Santa Fe, and Gary Spratt, a consultant and dealer in Native American art, visited the Peabody during this project and provided insights about some of the objects.

Carolyn Gilman of the Missouri Historical Society worked in tandem with us while pursuing her own research for the national Lewis and Clark bicentennial exhibit and proved a most capable and generous colleague. So did Elizabeth Chew, associate curator of collections at Monticello, with whom we developed the Peabody-Monticello Native Arts Project. Gail Pond, at Jefferson's Poplar Forest, Ann Carter Hutter, and Christian Sixtus Hutter helped us piece together the story of Lt. George C. Hutter and his relationship to William Clark.

So many staff members at the Peabody Museum worked on various aspects of the Lewis and Clark research project that we can thank only the major contributors by name. Hillel S. Burger photographed the entire collection and produced most of the object images for this book. Stuart Heebner assisted in the photography project. Conservators T. Rose Holdcraft and Scott Fulton were responsible for conducting and contracting materials analysis and for cleaning, repairing, and handling the objects. In addition, T. Rose undertook detailed analysis of the fabrics and other materials on many of the objects in this collection, examined pipes at the Musée de l'Homme, and contributed text for several sidebars. Nynke Dorhout, collections associate, assisted the early phases of the research and conducted technical analysis of the quillwork in this collection. Senior collections manager David Schafer, collections associate Susan Haskell, and staff in the Peabody's photographic archives (especially Victoria Cranner and Jeff Carr) provided all manner of logistical assistance with the objects and the publication project. Archaeologists Neal Trubowitz and Eric Hollinger fielded questions about the North American prehistoric record. Last but certainly not least, research assistants Betty Duggan, Jenny Marsh, Nicole Newendorp, and Juliette Rogers accomplished many of the tasks that enabled us to pull together the story of this collection.

We also want to express our appreciation to the staff at the following institutions and archives, who skillfully and generously facilitated our research: the Tozzer Library, Harvard University; the library at Adelphi University, Garden City, New York; the Harvard University Theatre Collection; Monticello/The Thomas Jefferson Foundation; Thomas Jefferson's Poplar Forest; the New York Historical Society; the Missouri Historical Society; the Massachusetts Historical Society; the North Dakota Historical Society; the Harvard Museum of Natural History; the National Museum of Natural History, Smithsonian Institution; the Oregon Historical Society; the Makah Cultural and Research Center, Neah Bay, Alaska; and the Musée de l'Homme in Paris.

The Peabody Museum was particularly fortunate in benefiting from the expertise of editorial consultant Joan K. O'Donnell, who shepherded this project into print with creative diligence. The authors gratefully acknowledge Joan's intelligence and commitment to polishing the text and organization of this book. In many ways, it is her creation. Christian Feest, an authority on Native North American art, and James Ronda, the preeminent Lewis and Clark scholar of his generation, graciously took time to carefully review the manuscript, much to our benefit. Jane Kepp's editorial acumen improved the final text, and Kristina Kachele's imaginative and graceful design created a visually rich experience of the collection. At the University of Washington Press, we thank Pat Soden, John Stevenson, Marilyn Trueblood, Mary Anderson, Alice Herbig, and Gigi Lamm.

A few notes of personal appreciation are also in order. Castle McLaughlin wishes to thank George Horse Capture, Irma Bailey, Gerard Baker, Barbara Kreutz, the late Dennis Lessard, and her parents, Mr. and Mrs. M. K. O'Daniel and Mr. and Mrs. W. F. McLaughlin, for shaping and supporting the course of this endeavor in diverse but fundamental ways. Anne-Marie Victor-Howe would like to thank John R. Howe, Bruce Howe, Mary Schlick, David Maybury-Lewis, Janine Bowechop, and Maria Pascua for their support and advice. Gaylord Torrence thanks Marjorie Alexander for her generous and thoughtful support. And Pat Courtney Gold dedicates her essay in chapter 9 to Mary Schlick, in gratitude for her many hours of editorial help, her encouragement, and her years of mentoring.

THE AUTHORS
May 2003

Contributors

HILLEL S. BURGER has been head of the photographic studios of the Peabody Museum since 1971. His photographs have been featured in dozens of publications, including *The Glass Flowers at Harvard* (1982).

MIKE CROSS, Mandan-Hidatsa, grew up on the Fort Berthold Reservation in North Dakota, where his father served as tribal chairman. He is an independent grant writer and consultant and an avid gardener of traditional plants.

Wasco fiber artist PAT COURTNEY GOLD is in the forefront of the Columbia River basketry revival. In addition to creating innovative and award-winning fiber art, Gold travels and consults widely as a teacher and an advocate for indigenous arts.

T. ROSE HOLDCRAFT is an objects conservator and the head of the conservation department at the Peabody Museum of Archaeology and Ethnology.

Lead author CASTLE McLAUGHLIN is a social anthropologist whose research interests include the American West, past and present; politics and political economy; and art and visual culture. She is associate curator of Native American ethnography at the Peabody Museum of Archaeology and Ethnology.

JAMES P. RONDA is a specialist in the history of the exploration of the American West. He holds the H. G. Barnard Chair in Western History at the University of Tulsa and is a past president of the Western History Association. His books include *Lewis and Clark among the Indians* (1984), *Finding the West: Explorations with Lewis and Clark* (2000), and *Beyond Lewis and Clark: The Army Explores the West* (2003).

GAYLORD TORRENCE is the Fred and Virginia Merrill Curator of American Indian Art at the Nelson-Atkins Museum of Art, Kansas City, Missouri, and professor emeritus, Drake University. He is the author of *The American Indian Parfleche: A Tradition of Abstract Painting* (1994) and coauthor, with Robert Hobbs, of *Art of the Red Earth People: The Mesquakie of Iowa* (1989).

ANNE-MARIE VICTOR-HOWE is a cultural anthropologist and independent consultant with expertise in the Native peoples of the Pacific Northwest. She was formerly a curatorial associate at the Peabody Museum.

Social anthropologist RUBIE WATSON is the William and Muriel Seabury Howells Director of the Peabody Museum of Archaeology and Ethnology.

Picture Credits

The authors thank the individuals, institutions, and collections listed here for generously allowing us to reproduce the images used in this book. "T," "S," "P," and "N" numbers identify transparencies, slides, prints, and negatives in the collections of the Peabody Museum Photographic Archives. Unless otherwise noted, all Peabody Museum photographs are by museum photographer Hillel S. Burger and are copyright the President and Fellows of Harvard College. The map on pages 18–19 and the diagram on page 174 were drawn by Deborah Reade and are copyright the President and Fellows of Harvard College. All other images are copyright the listed photographers, collections, or institutions, unless otherwise noted.

COVER: Calumet by Hillel S. Burger, T3009.2; landscape by Aneata Hagy © Aneata Hagy; author portrait by Nynke Dorhout.

FRONT MATTER

i, T2993.1; **ii**, T2930.4.1; **vi–vii**, T3042.1.1; **xii**, N35447, Courtesy Tozzer Library, Harvard University; **xx**, N35445, Courtesy Tozzer Library, Harvard University; **xxiv–1**, T3836.1

INTRODUCTION

2, N36065; **5**, T3739.1; **7**, Courtesy Castle McLaughlin; **10**, T2926.5.1; **12**, T2513d.1; **14–15**, T4262.1; **16**, T2921.1

CHAPTER 1

18–19, Map by Deborah Reade; **23**, Courtesy Independence National Historical Park; **25**, Courtesy Independence National Historical Park; **26**, Courtesy Independence National Historical Park; **29**, Courtesy Ernst Mayr Library, Museum of Comparative Zoology, Harvard University; **31**, T3019.1

CHAPTER 2

32, T4278.1; **34**, Courtesy Library of Congress, Prints and Photographs Division, LC-USZ62; **36**, N35352, Courtesy Tozzer Library, Harvard University; **38**, T2914.1.1; **41**, T4266.1; **42 top**, T3054.1; **42 bottom**, Courtesy American Philosophical Society, neg. 965; **45**, Courtesy Jefferson National Expansion Memorial/National Park Service; **46**, T2480.1; **47**, T2481.1; **48**, Courtesy Castle McLaughlin

CHAPTER 3

52, T3001.1.1; **54**, T4276.1; **55**, N35448, Courtesy Tozzer Library, Harvard University; **56**, Courtesy American Philosophical Society; **60**, T2924.4.1; **61**, T3001.1.1; **62**, T3056.1; **65**, Courtesy American Philosophical Society, neg. 940

CHAPTER 4

68, Oil on canvas; 103¾ x 79⅞ in., acc. no. 1878.1.2. Courtesy the Pennsylvania Academy of the Fine Arts, Philadelphia. Gift of Mrs. Sarah Harrison (The Joseph Harrison, Jr. Collection); **74**, Courtesy New York Historical Society © Collection of the New York Historical Society, neg. 51322; **76**, Courtesy the Barnum Museum, Bridgeport, CT; **79**, Courtesy the Harvard Theatre Collection, Houghton Library; **80**, Courtesy the Harvard Theatre Collection, Houghton Library; **81**, N1784

PART 2

82–83, T4238.2

CHAPTER 5

84, T2928B.1; **86**, T3741.1; **87**, T2930.1.1; **90**, T2930.5.1; **91**, Courtesy North Dakota Council on the Arts, Dennis Gad photo; **92**, Courtesy North Dakota Council on the Arts, Dennis Gad photo; **94**, T3831; **95**, T3832; **99**, N36016, cat. no. 41-72-10/496; **101**, T4238.2; **106**, T3742.1; **109**, T2927.1.1; **110**, T2928A.1; **111**, T2929.1; **112**, N35447, Courtesy Tozzer Library, Harvard University; **113**, Courtesy Kunsthistorisches Museum/Museum für Völkerkunde, Vienna, photo no. 41.463; **123 top and bottom**, Courtesy State Historical Society of Iowa, Des Moines

CHAPTER 6

126, T2914.5; **128**, Portrait courtesy Ann Carter Hutter; Paul Nichols photo; **131**, N36450, Courtesy Tozzer Library, Harvard University; **135**, T2914.3.1; **137 top**, T2915.1; **137 bottom**, Courtesy Joslyn Art Museum, Omaha; **139 top**, T2913.1; **139 bottom**, T3743.1; **140**, T2911.1; **141**, Courtesy Jenna N. Schafer, Minot, ND; **143**, T2916.1; **144**, T3061.1

CHAPTER 7

146, T4279.2; **148**, T4262.1; **151**, N35440, Courtesy Tozzer Library, Harvard University; **155**, T4274; **157**, T4269.1; **159 top**, Courtesy Castle McLaughlin; **159 bottom**, T3062.1; **161**, T3745.1; **162**, T3746.1; **165**, T2924.2.1; **166**, T2924.1.1; **167**, T2924A; **169**, T2924.6.1; **171**, T2926.2.1; **172**, T3747.1; **174**, Drawing by Deborah Reade; **176**, T2925.5.1; **179**, T2925.1.1; **180**, Courtesy Joslyn Art Museum, Omaha; **181**, T2926.1; **183**, N35445, Courtesy Tozzer Library, Harvard University; **185**, T3074.1; **187**, T2926.8; **189**, T2956.1.1; **190**, T2959.1.1; **194**, T2966.1.1; **198**, T2970.2.1

CHAPTER 8

200, T2999.1.1; **205**, N35443, Courtesy Tozzer Library, Harvard University; **209**, N35444, Courtesy Tozzer Library, Harvard University; **213**, Courtesy National Anthropological Archives, Smithsonian Institution, NAA Inv. 08682600; **214**, N35449, Courtesy Tozzer Library, Harvard University; **216**, Denver Art Museum Collection: Museum Exchange, 1952.408 © Denver Art Museum 2002; **217**, Denver Art Museum Collection: Museum Exchange, 1952.409 © Denver Art Museum 2002; **218**, N36015, Peabody Museum Photo Archives, cat. no. H5859; **219**, T3007.1.1; **220**, Courtesy Castle McLaughlin; **222**, T2997.1.1; **223**, T3005.3.1; **224**, T3010.1; **225 top**, Courtesy Castle McLaughlin; **225 bottom**, T3031.1; **227 top to bottom**,

T3835, T3024.1, T3047.1.1, T3034.1.1; **227 detail**, T3035.1; **228**, S14140; **229**, T3041.1; **232 top left**, T3015.1.1; **232 top right**, T2954.1.1; **232**, Denver Art Museum Collection: Museum Exchange, 1952.283 © Denver Art Museum 2002; **233 top**, T3036.2.1; **233 bottom**, T3016.1.1; **235 top**, T3021.1.1; **235 bottom**, T3022.2.1; **236 top**, T3026.1; **236 bottom**, S14117; **237 top**, T3042.1.1; **237 bottom**, T3044.1.1; **238 top**, T3833; **238 bottom**, T3834; **240**, T2912.1; **241 top to bottom**, S14069, S14070, S14071; **242 top**, T2983.3.1; **242 bottom**, T2985.3.1; **243 top**, T2987.1.1; **243 bottom**, S14072; **244 top**, T2988.2; **244 bottom**, T2064.1.1; **245**, T2996.1; **246**, T2990.1.1; **247**, T3038.1.1; **248 top**, T2992; **248 bottom**, T2995; **249 top**, T2986.1.1; **249 bottom**, T2989.1.1

CHAPTER 9

250, T2921.2.1; **253**, T3083.1; **256**, T2696.1; **257**, T1261.1; **258**, T2921.1.1; **259**, Courtesy Josyln Art Museum, Omaha; **260 top**, T2582.1; **260 bottom**, T3748.1; **261**, T1905.1, Peabody Museum Photo Archives, cat. no. 41-72-10/53023c; **264**, Denver Art Museum Collection: Museum Exchange 1952.405 © Denver Art Museum 2002; **265**, T2973.2.1; **266**, T2980.2.1; **267 top**, T2933.1; **267 bottom**, T2909.1; **268**, T2910; **269**, T2931.1; **271**, T3055.1.1; **274**, T2512B.1; **278 top**, Courtesy Oregon Historical Society, OrHi 104293; **278 bottom**, Courtesy American Philosophical Society; **280**, T2512A.1; **282**, Photo © Jim Piper 2001; **283**, Courtesy Pat Courtney Gold; **286**, Photo by Judy Battin, courtesy Pat Courtney Gold; **287**, Photo by Bill Bachhuber, courtesy Pat Courtney Gold; **288**, T2920.1; **289**, T3836.1; **293 top**, Courtesy of the Burke Museum of Natural History and Culture, cat. no. 2.5A 489; **293 bottom**, T3749.1; **294 top**, T2901.1; **294 bottom**, T2900.1; **296**, T2514A.1; **297**, T2952; **299**, T2517A.1; **300**, T2953.1; **301**, Courtesy the Oregon Historical Society, OrHi 59298 #317; **307**, T2908.1.1; **308**, T2918.1; **310**, Courtesy the Missouri Historical Society, MHS Neg. L/A187b

AFTERWORD

312, Charles Shoffner/Monticello, Thomas Jefferson Foundation, Charlottesville, Virginia; **314**, T3661

BACKMATTER

316, Photo by Aneata Hagy © Aneata Hagy; **324**, T2957.1; **340**, N26701A; **342**, N34753; **343**, P95600004

INDEX

Boldface page numbers refer to illustrations.

Adoption ceremonies: pipes and, 202, 243
Agriculture: traditional, 48, 50–51, 56
Algonquians, 143; pipe bowls and, 249; wampum and, 270, 271, 272
Ambrose, Stephen, 160
American Fur Company, 153
American Museum, 69, **76**, 77, 319n5. *See also* Barnum
American Philosophical Society, 22, 70, 75, 196
Amulets, 66–67
Anishnabeks, 144
Anthropology, 26, 27
Antlers. *See* Bows; Elk
Appadurai, Arjun, 55, 64
Arapaho, 178; bison calling and, 195; ceremonial objects of, 113; quilled ornaments and, 194, 195; robes and, 195; rosettes and, 194
Arcawecharchi (Half Man), 41
Arikara Buffalo Society, 193
Arikaras, 37, 134, 147, 157, 178, 270; bison calling and, 192; Cheyenne and, 183; Corps of Discovery and, 152, 195; glass beads and, 38; history of, 48–51; Sioux and, 149–51, 153, 154–55; smallpox and, 49; tobacco and, 62, 206, 257; trading by, 40, 150, 151–52
Arketarnashar, 208
Armbands, 106, 114–15, 317n1. *See also* Raven belt ornaments
Arrows, 28, 72, **250**, 252–55, **253**
Artist in His Museum, The (Peale), **68**
Assiniboins, 158, 175, 178; bison calling and, 192; quillwork of, 160; war whistles and, 140
Atkinson-O'Fallon expedition, 132, 150, 204; Hutter and, 127, 128, 131, 133, 134, 136, 153
Atsinas, 42, 209; bison calling and, 192, 195; quilled ornaments and, 194; robes and, 193, 195; rosettes and, 194
Awatixa Hidatsa village, drawing of (Catlin), **151**
Awatixas, bison calling and, 192

Baby carriers, 135–36, **135**, 320n4
Backrests, 193, 194, 195
Bark beaters, 294, **294**
Bark shredders, 293, **294**
Barnum, P. T., 77, 262; catalogues of, 77–78; Kimball and, 77, 78; Lewis and Clark and, 11, 77–78; museum of, 69, **76**, 78, 319n5; Peale Museum and, 76, 77
Barter, 11, 34, 45–46, 47, 56, 60–61, 63–66; gifts and, 39; Indian-white relations and, 36
Barton, Benjamin Smith, 25
Bar-Yosef, Daniella, **7**
Baskets, 97, 283–87, 297; cedar root, 279; funeral/memorial, 279; Haida, 300; Lewis and Clark commemorative, **282**; root-gathering, **274**, 275–81; shapes of, 286; Tlingit, 300
Beads, 47, 135–36, 137, 158; blue, 38, **38**, 39, 47, **126**, 135, 138, 139, 165, 168, 178, 179, 184, 185, 232, **250**, 257, 258, **258**, 260, 263, 302; chief, 184; garnet, 234, **235**; glass, **10**, 38, **38**, **135**, **137**, 182, 184, 185, 187, 228; pony, 135, 178, 182, 185, 268; porcelain, 187; seed, 179, 185; shell, 38, 236; trading, 37, 45; wampum and, 270–73. *See also* Wampum
Beadwork, 138, 183, 315; Cheyenne, 180;

lane stitch, 178; Missouri River, 259; Plains, 178, 180, 321n11; Sioux, 180
Beak of the Bird of Prey, portrait of, **137**
Bear, Keith, 141, **141**; on flutes, 140, 142
Belitz, Larry, 160
Belts, 71, 108, 117–18, 120, 136; described, 267–68; Great Lakes, 121, 267; ornamented, 323n7; quilled, 120, **267**; sashes as, 267; Sioux, 267, 321n8; Woodlands, 267. *See also* Raven belt ornaments
Berne Historical Museum, 8, 157
Biddle, Nicholas, 20, 104, 115; pictographic robe and, 154
Big Belley/Big Bellies, 272, 320nn1. *See also* Hidatsas, Minnetares
Big White. *See* Sheheke
Billeck, Bill, 38, 259
Biographical war shirt, 60, 147–48, 154, 160–67, **161**, **167**, **169**, 170, 172–73
Bison calling, robes for, 191–92, 193, 195
Bison symbolism, 193
Black Cat (Posecopsahe), 152, 255, 257
Black Moccasin (Catlin), **36**
Blackfeet, 41, 42, 72, 132, 136, 165, 209; amulets from, 66–67; bison calling and, 192; calumets and, 216, 226; pipes and, 221, 249
Bladder bag, 231, **231**, **232**. *See also* Tobacco bags
Bodmer, Karl, 138, 168, 175, 213, 257; bow case/quiver and, 259; chokers and, 137; flutes and, 142; otter bag and, 90; pipes and, 215; portraits by, **137**, **180**, **259**; robes and, 266
Bonnets, 59, 61, 270
Boston Museum and Gallery of Fine Arts, 4, 5, 6, 9, 12, 69, 77, 161, 188, 191, 273; advertisement by, **79**; catalogues of, 79; curiosities at, 79, 80; described, 78–79; fire damage at, **80**; Lewis and Clark collection at, 11, 251, 252
Boston Society of Natural History, 79, 242
Bows, 72; collecting, 28; described, 255, 257, 259; elk antler, **256**, 257, **257**, **259**; Sioux, 252; wooden, 323n2
Bow cases, **258**, 259, 323n3
Brackenridge, Henry M.: on pipes, 210
Brass buttons, **10**, 46, **46**, **47**, 184–87, 255
Brasser, Ted, 164, 178, 193, 321n9; on Great Lakes painting, 145; on side-fold dresses, 177
Breads, 62, 72
British Museum, 8, 210, 257, 320n5
Broken Arm (Tunnachemootoolt), 59, 61, 162–63; Clark and, 204, 209–10, 239; pipes from, 204, 210, 236, 239
Brooklyn Museum, 8
Brules, 108, 115–16, 153; Corps of Discovery and, 115, 322n7; raven belt ornaments and, 106; raven skins and, 104
Bureau of American Ethnology, 218
Bureau of Indian Affairs, 283
Bustles. *See* Raven belt ornaments

Caddoans, 166
Calumets, 12, 55, 57, 75, 111, 201, **205**, **216**, **217**, **220**, 222, 228, 322n6, 322n9; ceremonialism, 217, 218, 219, 221, 242, 244; cylindrical, 218; described, 213–14, 221; diplomacy and, 215, 219, 221; exchanging, 202, 205, 218, 271; feathers on, **61**, **200**, 235; features of, 218, 221, 223, 226, 229; fragments of, **216**; friendship and, 215; Hako, 219, 221, 222–23; quillwork on, 234–36; stems, **219**, **222**, **223**, **224**, **227**. *See also* Pipes
Camas bulbs, 62, **62**, 72, 278
Cameahwait, 42, 43, 59, 75, 162–63; clothing of, 59, **74**, **75**; Lewis and, 162, 208; pipes and, 204; tippets and, 162, 163
Canadian Conservation Institute, 160
Canadian Museum of Civilization, 177
Canadian National Museum, 8
Canoes, 100, 285, 293
Caps, 71, 73; sailor's, 298–306, **300**; war, 60, 72, 270
Captain Meriwether Lewis in Shoshone Costume (Saint-Memin), **74**
Captains Lewis & Clark holding a Council with the Indians (Gass), **34**
Carver, Jonathan: on pipe of peace, 215
Cascade Indians, 308
Catlin, George, 8, 59, 105, 110, 115, 116, 121–22, 131, 138, 183, 205, 213, 221, 252, 295; Boston Society of Natural History and, 242; bows and, 257; calumets and, 215, 216; chokers and, 137; Clark and, 130, 292, 306; dresses and, 182; drawings by, **55**, **110**, **112**, **131**, **151**, **205**, **214**, 306; Indian gallery of, 292; pipes and, **214**, 215, 228, 230, 239, 244, 246; portraits by, **36**, **183**, **209**; projectile points and, 255; raven belt drawings of, **112**, 116; skirt from, 292; wampum and, 273
Catlinite, 242–44
Cedar bark, 93, 96, 99–100, 290, 291, 293–94, **293**, 306, 309; preparing, **293**
Cedar trees, 293–94; Western red cedar, 297; yellow cedar, 297
Celilo Falls, 276, 301, 302
Cerre, Pascal, 152–53
"Chain of friendship," 56
Chambers, Harmon A.: wampum and, 273
Chan-Chä-Uiá-Teüin (Woman of the Crow Nation), portrait of (Bodmer), **180**
Charbonneau, Baptiste, 136, 285
Charbonneau, Toussaint, 132, 151–52; arrows and, 253–54; on carved pipe bowls, 243; on warfare, 152–53
Chardon, Francis A., 183
Cherokees: calumets and, 226, 322n6
Cheyenne, 158, 178, 182; Arikaras and, 183; bison calling and, 192, 195; ceremonial objects of, 113; pipes and, 204; quillwork of, 160, 194, 195; robes and, 263
Chief, pictograph of, **54**
Chiefly gifts, 56, 270. *See also* Gifts
Chinooks, 96, 97, 276, 290, 292; Clark and, 46, 302; cradles of, 309; cranial modification and, 309; digging bags and, 278; European goods and, 303–4; language of, 323n12; Lewis and, 46
Chokers, 137–38, **137**
Choteau family, 153, 321n5; August, 272; Paul, 319n4; Pierre, 321n5
Churchill, Delores: basket weaving and, 300
Clark, George Rogers, 129
Clark, Harriet Risque, 128–29
Clark, Jonathan, 61, 172, 318nn3
Clark, Meriwether Lewis: Hutter and, 128
Clark, William, 58, 59, 61; baskets and, 279; biographical war shirt and, 160; on bustles, 124; Catlin and, 130, 291, 306; collecting by, 3, 4, 5, 11, 12, 26, 27–30, 70, 147; cradles and, 307; engraving of, **34**; on grizzly bear necklaces, 261; Hutter and, 6, 12, 127, 128, 129, 131; Indian people and, 27, 129,

131; museum of, 130–31; museum catalogue of, 130; natural history and, 130; observations by, 26; otter bag and, 90; on petticoats, 290; pipes and, 202, 204, 210, 236, 239; portrait of, **26**, 75; raven belt ornaments and, 104, 106, 114; robes and, 148, 154; on "Shahala Nation," 308; on shirts, 164; skirt from, 292; on smoking, 208, 209; on Wahkiacums, 288; Yanktons and, 105
Clark, William Joseph, 317nn3
Clatsops, 93, 96, 97, 290, 298, 300, 301; language of, 323n12; onion-dome hats and, 96; Lewis and Clark's troubles with, 42
Cloth, 31; pipe stems and, **31**, 228; trading, 37. *See also* Trade cloth
Clothing, 4, 10, 27, 28, 53, 56, 59, 65–66; Arikara, 173; autobiographical, 157–58; bartering and, 64–65; ceremonial, 290; Chinook, 306; Clatsop, 306; collecting, 28; dance, 107, 305; European, 302, 304, 305; Mandan, 165, 166; Northwest Coast, 303, 305; Shoshone, 75; Yankton, 204
Coats, 31, 219, 304
Coe, Ralph T.: on pictographic robe, 148
Coleman, Winfield: on bison calling, 191–92, 193, 195; on side-fold dresses, 182
Collars: Shoshone, 137
Collecting, 8, 20, 21, 24, 26, 28, 115, 134, 148–49, 306, 321n5; by Lewis and Clark, 27–30
Colonial eagle motif, 247, **248**, 249
Columbia Plateau, 164, 239, 260
Columbia River, 7, 23, 37, 48, 54, 63, 73, 93; bartering on, 45–46; basketry tradition on, 12, 287; exchanges on, 47; exploring on, 46, 302, 303; hats on, 298; pollution of, 286, 287
Commerce, 22, 35, 67
Commodities by destination, 55
Commodities by metamorphosis, 55
Confederated Tribes of the Warm Springs Reservation, 278, 282–83
Cook, James, 8, 23, 99, 319nn4
Corn: gift of, 206; Mandan, **42**; trade, 41–42, 45
Corps of (Northwest) Discovery, 4, 9, 12, 22, 44, 61, 62, 243; bicentennial of, 48, 313, 315; challenges for, 33; collecting by, 29–30, 288; commission of, 20, 40; diplomacy and, 40, 204–5; gifts and, 53; "List of Requirements" for, 161–62, 184; map of route of, 18–19; route of, 37; trade and, 53, 65, 302, 303
Coues, Elliot, 20
Cous roots, **62**, 72, 278
Covenant chain. *See* "Chain of friendship"
Cowrie shells, 183, 186–87, **187**
Cradles, 136, **307**, **308**, 310; Chinook, 306–9, 311
Cranial modification, 307–8, 309, **310**
Crees, 170, 175, 178; bison calling and, 192, 193; flutes and, 142
Cross, Amanda, **48**
Cross, Kara, **48**
Cross, Mike, **48**; essay by, 48–51
Cross, Thomas Martin Old Dog, 50
Crow, the, 105, 108, 116, 122, 124. *See also* Raven belt ornaments
Crows, 50, 135, 136, 318n2; bows and, 259; calumets and, 216; cradle and, 135; Hutter and, 133, 136; pipe bowls and, 249; robes and, 263; tobacco and, 206
Cultural identity, 9, 13, 98, 116–17, 157, 171
Curiosities, 26–27, 56, 63, 66, 70, 74, 75, 77, 78, 79, 80, 129, 262, 302, 319n2, 320n3, 321n8
Cuscalah, 46

Dakotas, 116, 322n7; calumets and, 217, 226; pipes and, 228, 231, 234, 242–43, 247; raven belts and, 117; robes and, 149, 195
"Dance of discovering the enemy," 122
Davidson, George: watercolor by, **301**
Dearborn, Henry, 184, 188, 254, 262
D'Eglise, Jacques, 64
De Laguna, Frederica, 305
De Montlezun, Baron, 149, 262
Densmore, Frances, 138, 296
Denver Art Museum: calumets at, 213, 215, 216, **216**, **217**, 231, 322n6, 322n9; otter bags at, 319n1; pipe bag at, **232**; robes at, 263, **264**
Design motifs: anthropomorphic figures, 166, 168, **169**; antlers, 297, **297**; chief figure, **54**; colonial eagle, 247–49, **248**; Coyote, Tsagaglalal and, 284; cross, 234; deer, 295; dogs, 281, 285; elk, whales and, 296, 297; eyes, 297, **297**; falcon-men, 108; "four winds," 244; harpooners, 98, 102, 296; hieroglyphics, 105; hourglass, 236; human effigy figures, 246–47, **247**; humpback whales, 100; quirt/man, 168; representational, 99, 156, 158; stars, 264; thunderbirds, **228**, 232, 236, 320n5; whales/whaling, 100–103; X-ray faces, **280**, 282; zoomorphic, 246–47, **247**
De Smet, Pierre Jean, 153, 231
Dhegiha Siouans, bison calling and, 192
Digging bags. *See* Sally bag
Diplomacy, 27, 31, 40, 44, 47, 211, 215, 218; cross-cultural, 11, 35, 57, 58; forest, 33–34, 37–38; Indian, 20, 34, 57, 130, 132, 205; gifts and, 11, 56, 64, 89, 115, 201, 203, 219; pipes and, 202, 204–10, 219, 221; protocols of, 35, 63; wampum and, 271
Dogbane, 277, 278–79, 285
Dog Society, 138, 139, 257
Dorsey, George: on pipes, 226
Douglas, Eric: on pipes, 215
Dove, Carla, 139, 220, **220**
Dresses, 73, 163, 180, 187; dance, 184; described, 171–72; evolution of, 173–74; interpretation of, 174; Lakota, 175; Mandan, 161; Nakota, 175; ornamental, 269; painted, **179**; shirts and, 172–73; Sioux, 104, 173, 175, 182, 321n8; social identity and, 175; special occasion, 184; strap, 173; yokes, 183. *See also* Side-fold dresses
Drouillard, George: Shoshones and, 66
Du Coigne, Jean Baptiste: gifts and, 203
Duff, Wilson: on sexual symbolism, 103
Duke Paul. *See* Friedrich Paul Wilhelm

Eagle bone whistle, 138–40, **139**
Earthlodges, 157, 275; drawing of, **55**; photo in, **141**
Elk, hunting/whaling and, 296–97
Elk, Mary: quillwork by, 315
Elk Dance, 297
Empire, 20–24
Enlightenment, 17, 21; America and, 24; Jefferson and, 22
Eppes, Francis, 129, 320n2

"Estimate of the Eastern Indians" (Lewis and Clark), 24, 28
Ethnography, 27, 81, 107
Ethnological Museum Berlin, 8, 117, 221, 267
Evans, John, 24, 44–45
Ewers, John C., 81, 131, 213, 264, 322n3; on chokers, 137; on pipes, 210; on robes, 148, 155–56, 321n2; side-fold dresses and, 173, 177; on Wilkinson, 211
Exchange, 10, 11, 33, 36, 37, 43–47, 55, 69, 97; intertribal, 57, 300; networks, 164; protocols of, 34, 43; success for, 10
Exploration, 20–24

Fabric. *See* Cloth; Trade cloth
Falcon-men, 108
Feathers, **61**, 72, 108, 113, 159, **200**, 208, 235; identifying, 220, **220**; roach, 268–70, **269**, 323n8; symbolism of, 270
Feder, Norman, 188, 197, 321n9; on biographical shirts, 163; on knife sheath, 145; on quillwork, 196; Schoch shirt and, 153–54; on side-fold dresses, 173, 174–75, 177, 180, 182
Feest, Christian, 140, 320n5
Fevret de Saint-Memin, C. B. J., 172
Field Museum of Natural History: belts at, 117
Findlay brothers: fur trade and, 153
Flageolet, 140, 142
Fletcher, Alice, 217, 218; Omaha culture and, 124–25; on pipe bowls, 243
Florence Museum of Anthropology and Ethnology, 8
Flutes, 141; wooden, 140, **140**, 142
Fogg Museum, 162
Food: Native, 61–62. *See also* Hospitality
Fort Berthold Indian Reservation, 50–51, 138, 141, 170, 178, 321n4
Fort Clatsop, 93, 97, 290; wintering at, 85
Fort Mandan, 24, 45, 64, 70, 97, 154, 163, 188; corn trade at, 41–42; letter/packing list from, 28–29, 54, 149, 171, 257, 262; Lewis and Clark at, 150, 171, 172, 192; pipes from, 203; wintering at, 39, 65–66, 254
Fort Pierre, 175, 205, 221
Fort Wood: wintering at, 88
Four Bears, 90, 273; feasting with (Catlin), **55**
Fox, Dennis, 12, **314**, 315
Foxes. *See* Mesquakies
Framing the West at Monticello: Thomas Jefferson and the Lewis and Clark Expedition, 313
Friedrich Paul Wilhelm, duke of Württemberg, 131, 196, 210; bow case of, 323n3; raven ornaments of, 114, 117–18, 120; chiefly gifts for, 59; collection of, 8; Hidatsas and, 257; knife sheath of, 144, 145; war whistle of, 140
Friendship, 60, 67, 215, 272, 304
From Generation to Generation (Gold basket), 287, **287**
Fulton, Scott, 166, 306
Fur trade, 22, 23, 150, 151, 182, 187, 273; clothing from, 302; Great Lakes/Woodlands, 268

Garments. *See* Clothing
Garreau family, 132
Garrison Dam, 50, 319n6
Garters, 184, 267–68, 321n8; beaded, 268, **268**
Gass, Sargeant Patrick, 45, engraving by, **34**
Gibson, James, 96, 304
Gifts, 53; ambassadorial, 85, 90; barter and, 39; chiefly, 56, 57–63, 270, 163; choosing, 39; diplomatic, 11, 56, 64, 89, 115, 201, 203, 207–8, 219; exchanging, 11, 46, 97; Indian-white relations and, 36; rituals of, 33–34; social relations and, 202–3; of state, 56, 57–63
Gifts of state, 56. *See also* Gifts
Globalization, 10, 182
Gilman, Carolyn, 318n3, 320n3, 321nn5, 6
Gold, Pat Courtney, **2**, 12, 99, **282**, **286**; baskets and, 12, 277, 280, **282**, **287**; essay by, 282–87; on dogbane, 279; on skirts, 291
Gourd rattle, 142–44, **143**
Grass Dance, 107, 110, **113**, 119, 270
Great Father, 36–37, 45, 207
Great Lakes peoples, 111, 116, 119–20, 121, 122, 133, 143, 235; flutes and, 140; migration of, 117
Grinnell, George Bird, 174
Grizzly Bear, portrait of (Catlin), **209**
Grizzly bear claw ornaments, 260–62, **260**, **261**
Gros Ventres, robes and, 193
Haidas, 300, 323n13
Hail, Barbara, 163, 174, 183
Hako, 217, 218, 219, 222–23
Hamilton, Alexander, 70, 263
Harpooners, **94**, **95**, 98, **101**, 102, 296, 319nn4
Harrington, M. R., 117; belt of, 118, 119, 121
Harrison, Joseph, Jr., 292, 306
Harrison, William Henry, 89, 211
Harrod, Howard L.: on hunting beliefs/rituals, 192
Ha-she-a (Cut Nose), 105
Hats, 4, 10, 53, 66, 304; black-rimmed, 295–98, **296**, **297**; Chinook, 96; Clatsop, 71, 299; collecting, 28; conical, 93; European-style, 298; Native, 306; rainproof, **65**, 66, 96, 99; top, 298–306, **299**; twined, 252; whaling, 12, **94**, **95**, 98, **99**, 100–102; winter/summer, 98; woven, 96. *See also* Caps; Knob-top conical hats; Onion-dome hats; Whaler's hats
Ha-wan-je-tah (One Horn), 59, **205**
Headdresses, 111, 113, 124; sashes as, **218**, **261**, 267
Heddles, 234, 236
Henry, Alexander, 173, 254
Hernández, Juan Josef Pérez, 99
Heye Foundation, 117
Hidatsas, 37, 42, 64, 134, 147, 150, 151, 152, 153; bison calling and, 192; bone whistles and, 138; bows and, 257, 259; corn trade and, 41; Corps of Discovery and, 254; history of, 48–51; Missouri River and, 50; robes and, 158, 193, 197; Sioux and, 149–51; smallpox and, 49; trade and, 150; tobacco and, 206; war against, 40; war whistles and, 140. *See also* Minnetarees; Three Affiliated Tribes
Hieroglyphics, 105, 162
Historical Society of Pennsylvania, 70
History of the Expedition under the Command of Captains Lewis and Clark (Biddle), 154
Hoadley, Bruce, 225, **225**
Hohâstillpelp, 261
Holdcraft, T. Rose, 236; on cloth, 31; on silk ribbons, 229; on wear patterns, 166
Holm, Bill, 295, 300, 305
Horse Capture, George: on tobacco bags, 232
Horse Capture, Joseph D.: on robes, 156–57

Hospitality, 36, 62, 63; food and, 207
Hourglass motif, 180
Howard, James H.: on bison calling, 195
Hudson's Bay Company, 186, 187
Hutter, Ann Carter, 320n1
Hutter, C. J., 135; ethnographic materials from, 150; Peale Museum and, 129
Hutter, Christian Sixtus, III, 320n1
Hutter, Edward Sixtus, 129
Hutter, Emma Cobbs, 129
Hutter, George Christian, 9, 150, 153, 157, 161, 163; arrows and, 253; Clark and, 6, 12, 127, 128, 129; collecting by, 12, 127, 134, 145, 147, 148, 160, 164, 251, 252; Corps of Discovery and, 147; diplomacy and, 132; dresses and, 175, 182; eagle bone whistle and, 138, 140; flutes and, 142; gourd rattles and, 143; knife sheath and, 144, 197; life of, 128–29; Peale and, 132–33; pipes and, 204, 228; portrait of, **128**; robes for, 148, 149, 188, 193; Sioux and, 154
Hutter collection, 38, 127, 131–34

Indian Department, 283
Indian Hall (Jefferson's), 12–13, 70, 262, 318–19n1; recreation of, **312**
Intertribal relations, 116–17, 205, 208, 242–43; wampum and, 271
Ioways, 7, 73, 88, 92, 105, 221; calumets and, 216; otter bags and, 90; pipes and, 211, 226, 228, 231, 234; study of, 116
Iroquois, wampum and, 270, 271, 272
Isch-nan-uanky, 59

Jackson, Donald: Sauks/Foxes and, 88–89
Jarvis, Nathan S., 8, 130, 180, 196; dress of, 175, 197
Jefferson, Thomas, 3, 28, 48, 54; the "Age of . . . ," 25; on arrows, 255; bear cubs for, 262; beliefs of, 25–26; buttons and, 46; collection of, 4, 5, 11, 55, 73, 313, 318nn2; Corps of Discovery and, 20, 89; Enlightenment and, 22; evolutionary theory and, 26; horticulture and, 62; Indian Hall of, 12–13, 262, 318–19n1; Indian peoples and, 27, 208, 211–12; instructions by, 24; Lewis and Clark collection and, 56, 252, 313; natural history and, 154, 196; Peace and Friendship Medal of, 45, **45**, **261**; Peale and, 70, 85, 149; pipes for, 203, 228; political/economic objectives of, 10–11, 35, 36; portrait of, **23**; raven belt ornaments and, 104, 114; robes for, 66, 148, 150, 188, 262; science and, 17, 25, 26; tobacco and, 206; wax model for, 75
Jefferson Peace and Friendship Medal, 45, **45**, **261**
Johnson, E. A., 153, 321n5

Kane, Paul: on cradles, 309
Kansas (Indians), 153, 211
Kapin (digging tool), 278, **278**
Kaskaskias, 7, 203, 273
Kennerly, James, 128–29
Kerlérec, Governor, 24
Keyser, James: on biographical war shirt, 165
Kickapoos, 153, 211, 273
Kimball, Moses, 77, 78, 79
King, Charles Bird, 115
King, J. C. H., 319nn4; 323n14; on knobbed form, 102; pipe bowls and, 244
Kinship, 37; fictive, 59, 202
Kipp, James, 130
Knife River, 50, 51, 136; Hidatsas and, 320–21n1; illustration of, **151**
Knife River Indian Villages National Historic Site, 320–21n1
Knife sheath, 144–45, **144**, 197, 320n5
Knob-top conical hats, 93, **94**, **95**, 97, **99**, **101**, 103, 319nn4
Köhler, Friedrich, 117
Kroeber, Alfred, 191

La Flesche, Francis, 218; Omaha culture and, 124, 125; on pipe bowls, 243
Lakotas, 104, 116, 150, 322n7; calumets and, 322n9; relations with, 40; robes and, 195, 196, 266. *See also* Teton Sioux
Larocque, François, 41, 45, 64; on Crow shirts, 165; wampum and, 273
Leggings, 31, 60, 71, 104, 163, 315, 321n8; Dakota, 196
Lessard, Dennis: on heddle work, 234
Lewis, Meriwether, 25, 59, 129, 130; Black Cat and, 255, 257; buttons and, 46, 186; Cameahwait and, 43, 54, 74, 75, 162–63, 208; Cameahwait's clothing and, 71, **74**, 75, 162; collecting by, 3, 4, 5, 12, 26, 27–30, 147; cradles and, 307; on cranial modification, 309; death of, 46, 203; diplomacy and, 35, 37–38; dresses and, 171–72; engraving of, **34**; on etiquette, 44; on exchange, 43; gifts and, 39; on hunting shirts, 162, 163; Indian peoples and, 27; instructions for, 23–24; on onion-dome hats, 96; packing list by, 184, 257; portrait of, **25**, **74**, 75, 172; science and, 25–26; on skirts, 290; tobacco and, 206; wampum and, 272; wax model of, 75
Lewis and Clark Bicentennial, 4–5; Indian Hall and, 13, **312**, 313
Lewis and Clark centennial, 80
Lewis and Clark journals, 5, 12, 30, **42**, 45, 55, **65**, 93, 96, 278, **278**, 307–9, **310**
Lewis and Clark National Historic Trail Interpretive Center, 158, 160
Lewis's woodpecker (*Melanerpes lewis*), **29**, 220
Like-a-Fishhook Village, 49, 50, 270
Lincoln, Levi, 22
Linden-Museum, 8
Long, Stephen H., 25, 105, 120; "dance of discovering the enemy" and, 122; Peale and, 74
Louis (Loise), Paul, 319n4. *See* Choteau, Paul
Louisiana Purchase, 5, 48, 88–89, 318n1
Love medicine: flutes and, 140, 142

Mackay, James, 24
Mackenzie, Alexander, 22–23
Madison, Dolley, 172, 321n6
Madison, James, 70, 172, 196, 321n6
Makahs, 96, 97, 99; cranial modification and, 309; elk/whales and, 296–97; hats of, 98, 295; paint for, 297, 298; thunderbirds and, 320n5; whaling and, 100, 101, 102, 297
Malaspina, Alejandro, 8, 23
Mandans, 37, 42, 44, 61, 132, 134, 136; adoption ceremonies among, 202; arrows and, 254; bison calling and, 192; bows and, 257, 259; calumets and, 216; corn of, **42**; Corps of Discovery and, 149, 152, 253, 254, 263; culture of, 7; flutes and, 142; gifts for, 41; history of, 48–51; Maximilian on, 168, 170; medals for, 45; Nuptadi, 152, 192, 255; pipes and, 215, 246; robes and, 148, 154,

158, 193, 197, 263, 321n2; shirts and, 8; Sioux and, 150–51, 152; smallpox and, 49; tobacco and, 206; trade with, 150–51; wampum and, 272; war against, 40, 152; whistles and, 140
Mandeh-Pahchu, portrait of (Bodmer), **137**
Mantles, 71, 191, 263, 322n13
Maquinnah (Webber), **99**
Markishtum, Ada, **293**
Marquette, Jacques, 202, 221
Maryhill Museum of Art, 282
Masukawadahi, 139
Mato Wamniomni (Whirlwind Bear), **113**
Maurer, Evan: on robes, 156–57
Maxidiwiac (Buffalo Bird Woman), 138, 140, 225
Maximilian, prince of Wied-Neuwied, 138, 152, 153, 213, 259; on arrows, 253–54; bear cubs and, 262; Bodmer and, 175, 180; bows and, 257; calumets and, 215, 216, 221, 226; Clark and, 130; dresses and, 175, 180, 182; expedition by, 8, 64; garters and, 268; knife sheath and, 144; on Mandans, 156, 160, 168, 170; otter bag and, 90; pipes and, 202, 246, 249; robes and, 158, 160, 193–94, 195; war whistles and, 140
McKenzie, Charles, 61
McLaughlin, Elaine, **91**
Medals, 44–45, **45**, 58, 73, 88, 130, 261, **261**
Medicine bags, 90
Medicine packets, 119
Mementos, 53–54, 56, 66, 131, 163
"Memorandum of Specimens and Artifacts." *See* Peale memorandum
Menominees, 89, 209; rattles and, 142, 143
Mesquakies (Foxes), 7, 9, 58, 73, 86, 117, 118–19, 121, 140; belts by, 118–19; Louisiana Purchase and, 88–89; necklaces by, 92; paintings/drawings of, 121–22; pipes and, 211, 228; sashes and, 268; Sauks and, 115–16
Metal: trading of, 45, 253, 254–55; drills, 271
Metal working, 39, 41, 254
Métis, sashes and, 268
Micmac: pipe bowl style, 249, **249**
Midewiwin, 89, 90; bags of, 89–90. *See also* Pinjigosauns
Milwaukee Public Museum, 117
Mineral specimens, 29, 73
Minnetarees, 150, 152, 154, 272, 320nn1; robes and, 263. *See also* Hidatsas
Mississippian culture, 107, 108
Mississippi River, 7, 88, 89, 143, 211, 217; traders on, 40, 151
Missouri Fur Company, 210
Missouri Historical Society, 130
Missouri River, 7, 23, 48, 49, 61; collecting along, 6, 29; damming, 159; drawing of (Catlin), **131**; ethnographic materials from, 148–49; exchange on, 40, 43, 134; exploring, 33, 34, 35, 37; glass beads on, 38; Indians along, 48; moonrise over, **316**
Missouris, 211, 212
Mitchill, Samuel Latham, 23, 196, 212, 322n14; Lewis and Clark and, 196; National Museum and, 322n14; on quillwork, 196
Moccasins, 53, 61, 71, 168, 202, 257, 315
Montagnais-Naskapi hunters: robes of, 192–93
Monticello, 3, 13, 54, 56, 70, 160, 188; Indian Hall at, 13, **312**, 313; Lewis and Clark collection at, 313; robes at, 149, 315, 321n2
Moses, Horace: skirts and, 288
Moulton, Gary, 20, 28–29
Musée de l'Homme, 8, 157; pipe stems at, 213, 322n10; side-fold dresses at, 177
Museo de América, 8
Museum für Völkerkunde, 117
Museum of Natural History (Harvard), 220

Nakotas, 116, 150, 195, 322n7
Naskapi, 177
Nationalism, 21, 22, 24
National Museum of Natural History (Smithsonian), 8, 38, 220, 259; cradles at, 306; skirts at, 291
National Museum of the American Indian, 117, 322n14
Natural history, 21, 27, 29, 67, 74, 132, 154, 196
Necklace: Mandan, 137
Neeshneparkkeook (Neeshneparkkeooh; Cut Nose), 62–63, 72, 318nn1
New England Museum, 78
New York Museum (Barnum's), fire at, 78
Nez Perce, 46, 57, 136, 163; chiefly gifts from, 61; Clark and, 261; Corps of Discovery and, 208–9; grizzly bears and, 261; pipes and, 204, 208–9, 236, 239; roots and, 62–63; trade with, 302; wampum and, 272
Nixluidix (Nik-loy-dik), 284, 302
No-che-ninga (No Heart of Fear), **261**
Nonesopretty, 31
Nootkas, 96, 98, 99, 103; Corps of Discovery and, 296; drawing of, **99**; elk/whales and, 296–97; hats for, 96, 98, 295; paint for, 297; thunderbirds and, 320n5; whaling and, 101, 102, 297
Nootka Sound, 98–99, 319nn4
North Dakota Council on the Arts, 92
North West Company, 41, 45, 318n2
Northwest Native American Basketweavers Association, 287
Northwest Passage, 22, 23
Nuu-chah-nulths. *See* Nootkas

Ojibwas, 89, 116, 196, 321–22n11; bison calling and, 192; calumets and, 216; flutes and, 142; gourd rattles and, 143; *pinjigosauns* and, 89; pipes and, 234, 239, 243, 246
Omaha Dance, 107, 119, 270
Omaha Hethu'shka (Heluska) Society, 107
Omahas, 37, 105, 116, 122, 124–25, 221; bow cases of, 323n3; Hutter and, 133; Poncas and, 124–25
Omaha Sacred Pipes of Fellowship, **218**
Oneota tradition, 244
Onion-dome hats, 65, **65**, 93, **94**, **95**, 96, 98–99, 295, 319nn4, 5; acquiring, 96; interpreting, 103
Ordway, John, 96, 207
Oregon Folklife Master-Apprentice Program, 283
Oregon Governor's Arts Award, 285
Ornaments, 72; bird-quill, **194**; grizzly bear claw, 260–62, **260**; hair, 304–5; prestige, 183; Sioux, 114. *See also* Raven belt ornaments
Osages, 116, 211, 212, 260
Otos, 65, 105, 116, 117; pipe bowls and, 247
Ottawas, 89, 246
Otter bags, *frontispiece*, 85, 89, 90, **87**, **90**, 92, 267. *See also* Tobacco bags
Ozette Village, 97

Painting, 122, 297, 298; Great Lakes, 145; hide, 159, 175; Plains, 178
Pallotepallers, 37, 62, 72, 73; leggings of, 60. *See also* Nez Perce
Parfleches, 144
Parshall, Jo Esther, 12, **91**; quillwork by, 91–92, **92**, 315
Pash-ee-pa-ho (Little Stabbing Chief), drawing of (Catlin), **112**
Passhequo-quaw-mash, 72
Pawnees, 116, 211, 212, 221; pipes and, 243, 247
Peabody-Monticello Native Arts project, 315
Peabody Museum of Archaeology and Ethnology, **2**, 54, 69, 70, 79–83, **340**; arrows at, **250**, **253**, **256**, **258**; baby carrier at, **38**, **126**, **135**; bark beater at, **294**; bark shredder at, **294**; basketry hats at, **12**, **94**, **95**, **101**, **296**, **297**, **299**, **300**; belt at, **267**; biographical war shirt at, **60**, **165**, **166**, **167**, **169**; bladder bag at, **267**; bows at, **256**, **257**; bowcase at, **258**; buttons at, **10**, **46**, **47**; calumets at, **52**, **61**, **200**, **219**, **222**, **223**, **224**, **227**, **229**; choker at, **137**; cradles at, **307**, **308**; dresses at, **171**, **176**, **179**, **181**, **185**, **187**; feathers at, 270; flute at, **140**; garter at, **268**; grizzly bear claw ornaments at, **260**; Hutter collection at, 127; knife sheath at, **144**; Lewis and Clark collection at, 3, 5, 6, 8–9, 85, 134, 147, 170, 220, 221, 252, 284, 313; Lewis and Clark exhibit at, 75; Northwest Coast cases at, **81**; otter bag at, *frontispiece*, **87**, **90**; painted and quilled robe at, **265**; painted box-and-border robe at, **266**; pipe bowls at, **241**, **242**, **243**, **244**, **245**, **246**, **247**, **248**; pictographic bison robe at, **32**, **41**, **54**, **146**, **148**, **155**, **157**, **314**; pipe stems at, **31**, **225**, **228**, **231**, **233**, **235**, **236**, **237**, **238**; quilled bison robe at, **189**, **190**, **194**, **198**, **324**; quiver at, **16**, **250**, **258**; rattle at, **143**; raven belt ornaments at, **84**, **109**, **110**, **111**; roach feather at, **269**; sally bag at, **274**, **280**; sash at, **267**; skirts at, **288**, **289**; staff of, **343**; tube pipe at, **240**; wampum at, **271**; war whistle at, **139**; whalers' hats at, **12**, **94**, **95**, **101**
Peace and Friendship Medal, 45, **45**, **261**. *See also* Medals; Jefferson, Thomas
Peale, Charles Willson, 4, 11, 53, 69, 105; Barnum and, 77; bear cubs for, 262; on belts/garters, 267; Corps of Discovery and, 73–74, 75, 85, 252; ethnographic materials and, 148–49; on grizzly bear claws, 262; Hutter and, 132–33, 134; Jefferson and, 149; medals and, 88; museum of, 70, 147, 319n1; pacifist beliefs of, 75; paintings by, **23**, **25**, **26**, **68**, 70, 75; pipes and, 203, 213, 215, 228, 322n2; raven belt ornaments and, 104, 114–15; robes and, 149–50; sally bags and, 276; silhouettes by, 212, **213**, 319n4; tribal identifications by, 319n2; wampum and, 273
Peale, Franklin, 161, 172, 260
Peale, Titian, 74, 105
Peale memorandum, 5, 58, 62, 88, 93, 149–50, 171, 172; belts/garters in, 267, 268; calumets in, 216–17; ornaments/necklaces in, 260; pipes in, 201, 204, 210, 211, 228; reproduced, 71–73; robes in, 149–50, 262, 263; sally bags in, 275; tobacco bags in, 232; Willoughby and, 270
Peale Museum, 4, 9, 11, 28, 56, 64, 67, 69, 70, 75, 79; acquisition ledger of, 54, 322n2; catalogue of, 5–6; Hutter and, 127, 128, 129, 134, 135, 137, 142, 144, 148, 173, 182, 191; labels from, **5**, 6, 80, **86**, **106**, 114, **139**, 144, 160–62, **161**, **162**, **172**, 173, **260**, 288; Lewis and Clark collection at, 37, 55, 56, 57, 60, 62, 65, 67, 134; Lewis and Clark exhibit at, 75; objects at, 58, 59–60, 145, 153, 160–61, 172, 201, 204, 211–12, 252, 260, 263, 269, 270, 273, 275, 288, 298, 299. *See also* Peale; Peale memorandum
Péhriska-Rúhpa (Two Ravens), 138, 158, 257; portrait of (Bodmer), **259**; whistle of, 139
Pendant hooves, **198**, 199
Penney, David W.: Mesquakie sashes and, 268
Philadelphia Museum. *See* Peale Museum
Pictographic bison robe, **41**, **146**, **155**, **157**, 170, 191, **314**; described, 148–58, 160
Pierre Choteau and Company, 321n5
Pigments, 159, **159**
Pinjigosauns, 89. *See also* Midewiwin
Pipe bowls, 213, **241–49**; basalt, 249, **249**; carving, 243, 246–47; catlinite, 242–43, 247, 249; described, 241–44, 246–47, 249; Great Lakes, 242, 246, 247, 249; human/animal effigy, 107, 246, 247, **247**; Micmac, 249, **249**; types of, 244, 246–47, 249; Woodlands, 242, 249. *See also* Pipes
Pipe containers, 90, 91
Pipes, 4, 10, 27, 28, 73, 202, 208–9, 212–13, 228, 230, 272; calumets, 213–28, **216–20**, **222–24**, **227**, **229**; ceremonies, 12, 201–3, 208, 218; clan, 226; classifying, 215; communication and, 202; diplomacy and, 201, 202, 203, 204–10; disk, 244, 246; effigy, 107; flat-stemmed, 228–39, **228**, **231**, **233**, **235**; forms of (Catlin), **214**; hair caps on, 232, 239; medicine, 170; miscellaneous, 228, 230–31; of peace, 207, 232; ovoid, 234–37, **236**, **237**; presentation, 240; quillwork on, 230, 234–36; returning, 203; sacred, 226; social relations and, 226; spiral, 238–39, **225**, **238**; trading on, 202; treaty, 58, 130, 230, 322n5; tube, 239, **240**. *See also* Calumets; Pipe bowls; Pipe stems
Pipe stems, 184, 213, **213**, 219, **225**, **228**, **231**, **233**, **235–38**, 239–42; fabrics and, 31, **31**; silk ribbons on, 229; wrapping/weaving, 234–36. *See also* Pipes
Pipestone, 242, 243, 246
Pipestone National Monument, 243
Pipe-tomahawks, 226, 249
Pish-quit-pahs (Pishquilpahs), 63, 72, 73, 275, 276, 323n9
Politics, 10–11, 20, 44; of encounters, 34–37, 39–43
Pompy's Tower/Pompey's Pillar, 154
Poncas, 116, 124
Poplar Forest, 129, 320n2
Potawatomis, 89, 133, 153, 211; calumets and, 216; pipes and, 228; study of, 116
Poulson's Daily Advertiser, 67, 104
Prince Maximilian. *See* Maximilian
Prucha, Francis Paul: on medals, 45
Pryor, Nathan, 206–7

Quahog clam, 271
Quillwork, 90, 92, 111–12, 113–14, 118, 156, 160, 163, 165, **232**, 266; crosshatch, 264; Parshall and, 91–92; on otter bag, *frontispiece*, **90**; on pipes, 228, 230, 232, 233–36, **233**, **235**, **236**, **237**, 239; on robes, 158, 150; techniques of, 90, 92;

Woodlands, 145, 230; woven, 236; zigzag, 168
Quin-na-chart Nation, 97
Quiver, **250**, 259, **259**

Radford, Harriet, 128
Radison, Pierre d'Esprit, Sieur de, 215
Raven belt ornaments, 12, 61, **84**, 85, 104–25, **109**, **110**, **111**, **112**, 203, 221–22, 231, 235; bustles, 104, 106, 112, **112**, 114, 115, 119, **123**, 208, 221–22, 231; bird-skin, 107; Dakota, 118; dance, 113, **113**, 120; feather, 108; Grass Dance, 110, **113**; Lakota, 113; Mesquakie, 117, 118–19; Omaha, 110, 120, 124; Oto, 114, 118, 120; Ponca, 124; Sauk/Omaha, 122; Sioux, 104; described, 104–8, 110–25; stylistic differences of, 119–21
Raven bustles. *See* Raven belt ornaments
Raven House, 305
Reciprocity, 35, 36, 58, 202, 203
Red Stick ceremony, 192
Ribbons. *See* Silk ribbons
Ritual objects, 188, 219
Roach feathers, 268–70, **269**, 323n8
Roach spreaders, 270, 323n8
Robes, 4, 10, 55, 59, 61, 66, 147–48, 149–50; battle, 152; bison (buffalo), 28–29, 72, 150, 188, 193; box-and-border, **182**, **264**, 265, 266, **266**; Cheyenne, 150, 191–92; crosshatching on, 264; described, 262–66; Hidatsa, 168, 188; Mandan, 28, 150, 158, 165, 168, 188; painted, 156, 158, 197, **265**; quilled, 156, 158, 188, **189**, **190**, 191–97, 199, **265**, 266; shaman's, 193; women's, 265. *See also* Pictographic bison robe
Rock art, 154, **283**
Ronda, James, 20, 22, 25, 41, 317nn1, 318n3; on sign language, 301–2; on trade, 302
Root digging, 72, 278, **278**; tool for, **278**
Root Feast, 278
Roots, 62–63, 72–73
Rosettes, 194

Sacagewea, 48. *See also* Sakakawea
Sacred bundles, 122
Sahaptin, 63, 276, 278, 323n9
Sailor's caps, 298–306, **300**
Saint-Memin, C. B. J. Fevret de: paintings by, **56**, **74**
Sakakawea, 42, 48, 132, 136, 151, 152, 154, 208, 243, 285, 318nn4
Sally bags, **274**, 282, **280**, **282**, 283, 284; described, 275–81; studying, 285–86
Sand Bar. *See* Tchon-su-mons-ka
Santees, 163, 195, 196
Sashes, 267, **267**, 268
Sauks, 58, 117, 120, 121, 122, 212, 257; Lewis and, 88; Louisiana Purchase and, 88–89; Mesquakies and, 115–16; necklaces by, 92; paintings/drawings of, 122; pipes and, 211, 228, 232
Scalp dances, 115
Schlick, Mary Dodds, 275, 282, 283; on baskets, 278; on dogbane, 279
Schoch, Lorenz Alphons, 8; biographical shirt of, 153–54, 157, 163
Science, 10, 17, 20, 22; early American, 21, 24–27
Sedge grass, 279, 280, 284, 285
Sellers, Charles Coleman, 4, 9, 221, 319n4, 321n2; biographical war shirt and, 160–61; on Hutter, 128, 129; sally bags and, 275
"Shahala Nation." *See* Cascade Indians
Shake Hand, 44, 63, 207, 208, 210
Shamans, **157**, 193, 195
Sheheke, 56, 61, 149, 152, 322n3; portrait of (Saint-Memin), **56**
Shields, 252, 257; Butch Thunder Hawk and, 159, **159**
Shields, John, 150, 254
Shirts, 160–69; Berne, 158–59; Crow, 165; deer leg, 164; dresses and, 161, 163, 172–73; exchanging, 162–66; extent/distribution of, 164; hunting, 161–62, 172; Kiowa, 163; Lakota, 164; linen, 161, 163; Mandan, 163, 166; Sioux, 163; war, 164. *See also* Biographical war shirt; War shirts
Shoshones, 37, 44, 46, 66, 136, 172, 209; Clark and, 272; Corps of Discovery and, 257; exchange with, 43; grizzly bears and, 261; Lemhi, 42, 162, 208; pipes and, 209–10; tippets of, 162; wampum and, 272
Shredders. *See* Bark shredders
Side-fold dresses, **10**, 147–48, 170–88, **171**, **176**, **179**, **181**, **183**, **187**; pattern of, **174**; interpretation of, 174; painted, 175, **176**, 177–80, **179**; portrait with, **180**; Sioux-type, 180–88, **180**, **181**, **183**; Yankton, 197
Sih-Sa, depiction of, 259
Silhouettes, 212, **213**, 319n4
Silk ribbons, 229, **229**
Sioux, 9, 40, 49, 50, 64, 65, 66, 73; Arikaras and, 149–51, 152, 153, 154–55; belts by, 117; calumets and, 216, 226; Corps of Discovery and, 115; garters and, 268; Hutter and, 154; pipes and, 204, 232; ritual practices of, 112; robes and, 193
Sixth Infantry, 128, 129, 131
Skinner, Alanson, 86, 90, 117, 118–19, 218; ethnography of, 116; Sauks and, 122
Skirts: cattail, 306–7; woman's fiber, 251, 288, **289**, 290–92, 295
Smallpox, 49, 50, 152
Smith, C. H., 195, 197, 322n14
Smithsonian Institution, 8, 91, 131, 139, 197, 220, 259, 292, 295; silhouettes at, 212
Smoking, 44, 54; diplomacy and, 202; ritual, 202, 206, 208, 315; symbolism of, 206
Snowy Owl ceremony, 192
Social identity: clothing and, 164, 175, 304
Social relations, 30, 35, 50, 131; gifts and, 202–3; lubricating, 205–6; pipes and, 226
Soldier societies. *See* Warrior societies
Spanish Missouri Company, 44
Spear points, 66, 73
Speyer, Arthur: collection of, 8
Standing Rock Sioux Reservation, 159
Stewart, Hilary, 290, 291, 294
Subarctic, 105
Suhtaios, 158, 182; bison calling and, 192, 195; quillwork of, 160, 194
Sully, Thomas: painting by, 79
"Summary Statement of Rivers, Creeks, and most Remarkable Places" (Clark), 288
Symbolism, 159, 272; bison, 193; feather, 270; sexual, 103; social, 10, 55, 271; tobacco, 206

Tahawarra, silhouette of, **213**
Tahirussawichi, 217, 219
Tar-ro-mo-nee, Shake Hand and, 44
Tchan-dee, **205**
Tchon-su-mons-ka, 183; portrait of (Catlin), **183**
Teton Sioux, 40, 132, 322n7. *See also* Lakotas

The Dalles, 279, 302
Thompson, David, 173, 254, 321n10
Three Affiliated Tribes, 48–51, 318n4
Thunderbirds, 118, **228**, 232, 236, 268
Thunder Hawk, Butch, 12, **159**, 314; on pipe caps, 239; painted bison robes and, 159; pigments of, **159**; story of, 159; on tobacco bags, 232
Tinklers, 184, **185**, **198**
Tippets, 162, 163
Tlingits, 300, 305, 323n13; "navy women," 305
Tobacco, 47, 48, 62, 207, 208, 257; gift of, 206; symbolism of, 206; trading, 37
Tobacco bags, 6, 54, 71, 86–92, **87**, **90**, 196, 232, 319n1, 323n7. *See also* Otter bags
Tomahawks, 226, 249, 254
Tools: metal, 254; root digging, **278**; trading, 37
Torrence, Gaylord, 6, 9, 12, 92, 140, 179, 264, 268, **342**; on beaded strips, 135; on belts, 267; on Great Lakes painting, 145; on raven belt ornaments, 106–25; on robes, 266; on side-fold dresses, 177–78
Trade, 46, 56, 97, 276, 279, 301; Chinese, 98; corn, 41–42; described, 302; fabric, 31; indigenous, 36, 37, 45, 64, 305; protocols of, 33–34
Trade cloth, 31, **31**, 37, 44, **126**, 135, **135**, 221, 223, 226, 228, 234, **250**, 257, **258**
Trade goods, 39, 53, 66, 170, 184
Traders, 40, 147, 151; Arikaras and, 152; feast for (Catlin), **205**; influence of, 304
Treaty of 1855, 282
Truteau, Jean Baptiste, 44–45, 64
Tsagaglalal (She Who Watches), 280, **283**, 284
Tule, harvesting, **286**
Tunnachemootoolt. *See* Broken Arm
Two Ravens. *See* Péhriska-Rúhpa

United Tribes Technical College, 91, 159, 314
University of Pennsylvania Museum, 317n2

Vancouver Island, 23, 96, 99, 103
Vermilion, 160, 163, 178
Victor-Howe, Anne-Marie, **2**, 12, 93, 282, 284; on black-rimmed hats, 295–98; on cedar, 293–94; on Chinook cradles, 306–9, 311; on sally bags, 275–81; on top hats/sailor's caps, 298–306; on woman's fiber skirt, 288, 290–92, 295

Wacochachi (Wa-Ko-Ba-Di-A), 121; autobiographical drawings by, 121–22, **123**
Wahkiacums, 288, 290
Wahkiacus, Sally, 276
Wahon'da (Power That Gives Life), crow/wolf and, 124
Wampum, 35, 39, 44, 58, 71, 119, 163, 209, 210, 236, **236**; described, 270–73; diplomacy and, 271; faux, 92; marine shell, 273; strings of, **271**; Woodlands, 273
Wands, 219, 221, 223, 228
War ax, **42**
War bundles, 117, 119, 120, 121, 122
War Dance, 270
War Department, 283
Warm Springs Reservation, 278, 282–83
Warriors: depiction of, 154, **155**, **157**
Warrior societies, 61, 119, 121, 122, 124, 125, 159, 166
War shirts, 59, **60**, 148, 164, 170; pictographs on, 153. *See also* Biographical war shirt
Wascos: culture of, 276, 278, 279, 302; baskets and, 280, 284, 285, 286, 287; belief system of, 281; Corps of Discovery and, 282–87; sally bags and, 12, 275, 283, 284
Washington, George, 70, 79; medals of, 45, 247; pipes and, 203; wampum and, 273
Watson, Rubie, 3
Wax figures, 75, 78
Webber, John, 99; portrait by, **99**
Whaler's hats, **12**, 93, **94**, **95**, 96–103, **101**
Whales, 100; elk and, 296, 297
Whaling, 97, 99, 100, 101, 102–3, 290, 298; hats for, 12, **94**, **95**, 98, **99**, 100–102
Whelk shell, 271
Whistles, 138–40, **139**
White Cloud: pipe from, 234
White Pigeon, 73, 211
White Skin, 73, 88
Wilkes expedition: Peale and, 74
Wilkinson, James, 211, 212, 315
Will, George, 42
Willoughby, Charles C., 4, 9, 79, 114, 115, 175; article by, 4, 80–81; classifications by, 171; Peale memorandum and, 191, 270; on pictographic bison robe, 148; on raven objects, 106; roach spreaders and, 323n8; on side-fold dresses, 177
Wilson, Gilbert, 138, 225, 257
Winnebagos, 89, 117, 118, 121, 133; pipes and, 211, 228
Winter counts, 157, 195
Winter Quarters (Davidson), **301**
Wishrams, 276, 279, 284, 302; basketry and, 280; belief system of, 281
Wolf Chief, 170, 257, 269
Woman's fiber skirt, 251, **288**, **289**; described, 288, 290–92, 295
Wood, identifying, 225, **225**
Woodlands culture, 7, 108, 111, 120, 168, 244; quillwork of, 90, 92, 235
Woodpeckers, **29**, 219, 220, 223, 233, 239. *See also* Lewis's woodpecker
Woody, Ken, 158, 160
Wright, Robin K.: on hats, 295

Yanktonais, 115, 157, 158; Clark and, 105; pipes and, 204; quillwork of, 160
Yanktons, 73, 132, 150, 196; calumets and, 216; Corps of Discovery and, 206, 207, 208; gifts for, 208; pipes and, 204, 205, 210, 231; robes and, 195
Yellowstone River, 48, 132, 154, 215, 315
York, 43, 46–47, 150

Zepp, Wolfgang: on roach spreader, 270

ARTS OF DIPLOMACY

Project direction by Joan K. O'Donnell

Copy editing by Jane Kepp

Design and composition by Kristina Kachele

Principal photography by Hillel S. Burger

Composed in FontFont Atma, FontFont Meta,

and Johann Sparkling display

Printed and bound in China by C & C Offset Printing Co., Ltd.

on 157gsm Japanese NPI matt art